Image Processing In Java

ISBN 0-13-974577-7

90000

9 780139 745775

Image Processing in Java

Computer Graphics in Java (forthcoming)

Image Processing In Java

Douglas A. Lyon

Prentice Hall PTR
Upper Saddle River, NJ 07458
www.phptr.com

Library of Congress Cataloging-in-Publication Data

Lyon, Douglas A.
Image processing in Java / by Douglas A. Lyon
p. cm.
Includes bibliographical references and index.
ISBN 0-13-97457707
1. Java (Computer program language) 2. Image processing--Digital techniques. I. Title.
QA76.73.J38L94 1999
006.6'6--dc21 99-10044
CIP

Editorial/production supervision: *Kathleen M. Caren*
Cover design director: *Jerry Votta*
Manufacturing manager: *Alan Fischer*
Marketing manager: *Lisa Konzelmann*
Acquisitions editor: *Bernard Goodwin*
Editorial assistants: *Diane Spina*

Prentice Hall books are widely used by corporations and government agencies
for training, marketing, and resale.

The publisher offers discounts on this book when ordered in bulk quantities.
For more information, contact: Corporate Sales Department, Phone: 800-382-3419;
Fax: 201-236-7141; E-mail: corpsales@prenhall.com; or write: Prentice Hall PTR,
Corp. Sales Dept., One Lake Street, Upper Saddle River, NJ 07458.

Printed in the United States of America
10 9 8 7 6 5 4 3

ISBN 0-13-974577-7

Prentice-Hall International (UK) Limited, *London*
Prentice-Hall of Australia Pty. Limited, *Sydney*
Prentice-Hall Canada Inc., *Toronto*
Prentice-Hall Hispanoamericana, S.A., *Mexico*
Prentice-Hall of India Private Limited, *New Delhi*
Prentice-Hall of Japan, Inc., *Tokyo*
Prentice-Hall (Singapore) Pte. Ltd., *Singapore*
Editora Prentice-Hall do Brasil, Ltda., *Rio de Janeiro*

This book is dedicated to Martin and Rheva Lyon,

for their love and support.

Contents

List of Figures and Tables

Acknowledgment

So many people help during the process of writing a book, it is impossible to keep track of them all. To all that I have forgotten to list, please don't hold it against me! First and foremost are the members of the Computer Science and Engineering Department at the University of Bridgeport. They have made UB an enjoyable place to work. Without the free time afforded me by the chairman, Steve Grodzinsky, I could never have finished yet another book project.

Thanks go to Victor Silva, for the first version of the Histogram class; Raul Mihali for his contribution of the FileDialog batch processing class, the test patterns and the Martelli algorithm to Dongyan Wang for his contribution of the prime factor FFT and IFFT algorithm; to Naoki Chigai, for his contribution of the HTMLGenerator applet; to Hong Shi for his contribution of the Graph class; to Hak Kywn Roh for his statistical color quantization "Wu" class and to my many other students, who put up with my beta code and manuscript efforts. I hope that this did not hurt my student evaluations too much!

Many researchers freely distribute their code on the Net. Such examples are a source of some of the ideas in this book. My thanks to Geert Uytterhoeven <http://www.cs.kuleuven.ac.be/~geert/> for supplying his excellent WALI (Wavlet LIfting) package for my perusal. Thanks too, go to Wayne Rasband for his contribution of the *MedianCut* class. Also for putting his *Image/J* program, with source code, into the public domain (available at <http://rsb.info.nih.gov/ij/>). Thanks go to Peter Walser for his contribution of the *idx3d* class. Also for putting his 3D program, with source code, on the Web (available at <http://www.vis.inf.ethz.ch/students/pwalser>). Thanks to Allan Greenier for helping out with the formatting of Java for the CD-ROM . Thanks to Don Gilbert for giving his permission to distribute the *dclap* package.

The excellent work of the library staff at the University of Bridgeport's Magnus Wahlstrom Library, particularly Shelly Roseman and Krystyna Kossarska, kept those books and journal articles flowing. Their tireless fulfillment of my inter-library loan requests played a central part in getting this book out the door!!

The proofreaders of the early drafts, particularly Mary Ellen Buschman, and Fran Grodzinsky, really helped me straighten out my writing.

Rob Gordon's comments were instrumental in my moving the book to the new AWT.

Sandy Hackenson, at Prentice Hall, first presented this book to the publisher. It was her encouragement that led me to approach the acquisitions editor, Bernard Goodwin. Bernard's guidance and encouraging words were instrumental in getting this book project out the door, ahead of schedule.

This work was made possible, in part, by a grant from the National Science Foundation DUE-9451520, the Larsen Foundation and by a grant from the Education Foundation of America.

This work was also made possible by support from the DeWitt Tool Brothers Company.

Foreword

I am pleased and flattered that my former Ph.D. student, Doug Lyon, has asked me to contribute a foreword for his new book. As I tell my students, you can't ever divorce your thesis advisor.

This foreword raises some questions and suggests some answers. The book tells you how to manipulate images, in the most ambitious computer environment ever devised, to find answers for yourself.

Where do digital images come from, and where do they go? What can be done with them? Where can one find the best tools for image processing and analysis? How does one know if one has done it right?

Clearly, the World Wide Web is the most plentiful source of images since cave drawings. There are millions of satellite photos, micrographs, maps, sketches, facsimiles of books and articles, pictures of objects for sale and of objects to be admired. One convenient taxonomy divides this bottomless database into

(1) graytone and color pictures of real objects and scenes (*photos*),
(2) contrived pictures (*graphics*, *virtual reality*, *scientific visualization*),
(3) scanned engineering drawings, maps, schematic diagrams (*line drawings*), and
(4) images of printed text (*text images*).

Each of these categories requires a somewhat different approach. This book concentrates on the first two types.

Not only are more images posted daily, but also the same images are massaged and recirculated. Digitized pictures are immortal! But the Web is not a closed ecosystem: it interfaces with all of science, art and society (not to mention business). They are the ultimate source and destination of this visual cornucopia. The book benefits from the author's having his feet firmly planted in all these worlds.

Image processing transforms one picture into another, while image analysis extracts "information" from a picture. The selection of topics is based on the consensus that has emerged over the last four decades about the most useful building blocks shared by image applications of both kinds.

To begin with, considerable effort may be necessary just to compensate for the effects of camera or scanner distortions. Calibration techniques for correcting sensor/transducer anomalies are based on digitized test charts with known properties. Image displays should also be calibrated. Unlike some of the more theory-oriented texts, this book shows examples of useful test charts.

The most elementary operations change the value of each pixel (picture element) depending on the distribution of pixel values in the entire picture or in some neighborhood. For example, filters attempt to suppress insignificant detail and preserve significant features. Edge detectors look for significant transitions, which may be combined by boundary following. Thresholding and segmentation seek to partition the image into relatively homogeneous regions. Texture analysis delineates deterministic (*brick*) or statistical (*grass*) regularities. Thinning, skeletonizing and distance transforms attempt to find economical representations of dominant shape features. Color coordinate transformations, a special strength of the author, link objective and subjective aspects of color. Geometric operations allow registering (overlaying) one picture of an object onto another picture of the same, or similar, object.

Following the success of frequency-domain analysis in communications and signal processing (also fertile grounds for Java, as lambently discussed in Doug's earlier book), two-dimensional integral transforms with sinusoidal, rectangular or wavelet kernels bring spatial frequency techniques into play. Mathematical morphology is, in some sense, the discrete analog of convolution in linear systems theory. Another important topic is that of compression. Over the years, specialized lossy and lossless methods have been developed for high-contrast (fax) documents, for gray-scale and halftones, for color, and for video.

In the sixties and seventies, image software was written mainly in assembly languages or in FORTRAN, and a typical image size was 256x256 pixels (except for satellite pics). In the 80's dawned the era of C, extended eventually by the advent of C++. As computer storage expanded, images grew to 1024x1024 pixels and then kept on growing. For a wallet-sized snapshot-quality picture, or an A-sized page of laser-printer quality print, 4096x4096 pixels are necessary. Digital video processing (as opposed to display) still requires a hefty installation.

Java, both language and environment, is likely to be the next universal paradigm. This book eloquently argues its merits, but is not blind to its shortcomings. The expansion of toolkits and libraries is gradually raising the level of abstraction at which the image programmer works, putting more complex projects within the reach of a single individual. At the same time, the portability of the Java language facilitates large-scale collaboration.

One difficult question that remains is how to know if you have done it right. Quantitative evaluations are hard to come by in this field, and assessments by panels of experts are expensive and confusing. It is always a good idea to compare the results of alternative methods on the same data: consistency is not *only* the virtue of small minds. The effort invested in collecting a large set of test images is well spent. But the separate components of an image-based system cannot be fully evaluated in isolation. At our current level of understanding, there is no real alternative to building a complete system and testing whether it fulfills its objectives.

Perhaps in another book (Pattern Recognition in Java?), the author will show us how to classify pixels, features, objects, and entire pictures. Applications include modeling and statistical characterization of images, indexing and content-based retrieval. Also just over the horizon is Computer Vision in Java for robots, telemedicine and industrial inspection.

The narrative that follows is written with verve and gusto. It draws deeply on the author's not-always-painless experiences in programming, picture processing, computer art, and teaching computer skills. The program snippets are elegant and translucent. The image processing techniques presented are copiously illustrated with examples based on the sensuous Mandrill photograph.

This may not be the very best book that will ever be written on image processing in Java, but it is the *first* book. It takes bold vision and fast footwork to be first. First books on a subject are often far more influential than more elaborate later works that follow well-traveled (and muddied) trails. We expect that this book will play a significant role in the current paradigm shift. Enjoy it!

George Nagy
Professor of Computer Engineering
Rensselaer Polytechnic Institute

October 1998, Uppsala, Sweden

Preface

This book presents various algorithms used for the implementation of image processing programs in Java. The difference between this book and other image processing books is that it is written in Java.

Why Should I Care About Java?

The *Java* difference is: the image processing programs in this book are as portable as the Java virtual machine. All the software has been tested on Suns, Macs and Windows 95/NT. The compile-once run-anywhere mantra means that we are no longer tied to expensive workstations (or a single development platform) for distributing our image processing algorithms. Research will be advanced as we emerge from the software Tower of Babel. It is now possible to perform the most elementary of laboratory behavior (i.e., repeat the experiment and duplicate the results) with a click of the mouse! People have tried to build portable GUI-based systems in the past [Watson]. But none have caught on like Java.

This book presents an object-oriented image processing program and, as such, may be the first book of its kind. It is certainly the first image processing book on Java that I know of.

Who Should Read This Book?

This book was written for the professional and the student. This book assumes some mathematical maturity in the reader. Elementary linear algebra and a first college calculus course should hold the reader in good stead. This book also assumes that the reader has some background in Java. The foundation upon which the book is built would require some detailed knowledge of the Java AWT. However, it is possible for the reader to ignore the low-level coding details and simply make use of the high-level classes. In this way, seasoned image-processing programmers can ease their introduction to Java. To further ease the transition, I have attempted to make the image processing code look as readable as possible. In the trade-off between clarity and speed, I have selected clarity. Speed, via optimization, will come as the code is transitioned to the new Java advanced imaging package (Early Release 1 has only just come out!).

Why Not Use the Advanced Imaging Package?

As of this writing, there are compelling reasons not to use the JAI (Java Advanced Imaging) package. One reason is that the package has a native implementation. As a result, there is no reference Java source code available (this is important for those who like to do research and or tinker). Use of the JAI (Java Advanced Imaging) package means that the Java will only run on Windows or Sun systems (i.e., no Macs, No SGI's, etc.). Take heart, however, as this situation may change soon. These are implementation quips; the algorithms for the image processing will continue to have currency. Also, teachable/understandable source code has value (particularly when compared with having no source code at all).

Full source code is disclosed for the entire book. It is possible to get the code and use it without reading the book. This is probably a big waste of your time! The book is reasonably priced and documents the code in detail. In any case, the publisher and I will thank you for buying a copy!!

Once the advanced imaging package is made available, the low-level code will probably change. I shall attempt to keep the API presented in this book as stable as possible, and this should help to protect the time we invest in the development of image processing software. My goal is to never deprecate my own API, if at all possible. Sun, on the other hand, deprecates API's regularly. I deprecate the practice of deprecation! ;)

This Is Not Just a Textbook

This is not just a textbook, yet it has many attributes of a textbook. For example, some chapters have suggested projects at the end. Some of these projects make for a fine homework assignment. Others could be used for a term project. The more advanced ones explore topics of current research. The projects cover topics that I would have addressed myself, if time and space permitted.

The reason that this is not just a textbook is that it contains a great deal of how-to material and serves as a practicum of image processing. In fact, all the algorithms described in this book are implemented and working (except for those listed in the Project sections at the back of the chapters).

What Chapters Do I Need to Read?

This book contains 15 chapters. Several of them are self-contained. Some are needed in order to provide a basic foundation in the Java software.

Chapter 1 is an elaboration of some of the points raised in this preface. We can consider this preface to be the executive summary of Chapter 1 and the rest of the book.

Chapter 2 provides information about event processing and it is central to the understanding of the event processing program used in the book code (called *Kahindu*). So, skim or skip Chapter 1, but Chapter 2 is important! Kahindu is a region in Kenya, known for its fine AA coffee.

Chapter 3 gives the basics for image display and processing within the Kahindu program. In many ways it is even more important that Chapter 2 in understanding the Kahindu model of image processing. Thus Chapters 2 and 3 should be read, and in that order.

Appendix B is an optional next read. Appendix B discusses the finer points of the interface design and icon construction. It is good for those who want to emphasize Java interface design and understand the inner workings of the Kahindu program.

Chapter 4 introduces histograms and histogram equalizations. This chapter may be skipped or delayed. It is designed to give a basic introduction to image processing by example.

Chapter 5 introduces many image processing concepts, including transform theory. It is important to the understanding of Chapters 8, 9, 10 and 15. It should probably not be skipped.

Chapter 6 is needed for an understanding of Chapters 7 and 15. It covers streams in Java and is very Java-oriented.

Chapter 7 speaks about the detail of writing various image file formats. It is important to cover some of Chapter 7 in order to understand the code in Chapter 15.

Chapter 8 covers convolution using direct methods. Coverage of Chapter 8 is central to an understanding of Chapter 9.

Chapter 9 covers applications of convolution to perform filtering. This could have been integrated with Chapter 10, but the applications of convolution grew too numerous to be contained in one chapter. Reading Chapter 9 will help with understanding Chapter 10.

Chapter 10 covers edge detection. This chapter could be skipped, particularly if there is little time or interest in the edge detection topic. However, it is better to read Chapter 10 after Chapter 9 and before Chapter 15.

Chapter 11 covers morphological filtering. This topic is not central for an understanding of the rest of the book and this chapter is optional.

Chapter 12 covers non-convolutional methods for edge detection. Sub-topics, such as heuristic search and Hough transforms, are not needed for an understanding of the rest of the book.

Chapter 13 covers color spaces. Some projects in Chapter 15 do refer back to the material contained in Chapter 13, but not in detail. This chapter's coverage is optional.

Chapter 14 covers digital image warping. The chapter is not required for an understanding of Chapter 15.

Chapter 15 covers unitary transforms (FFT and Wavelet). Most courses in image processing will want to cover Chapter 15.

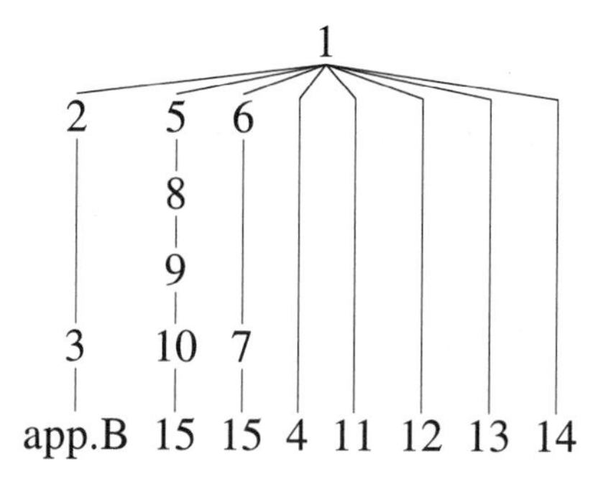

Fig. 0-1. Prerequisite Structure By Chapter

A one-term course could be taught *without covering Java* using Chapters 5, 4, 8, 9, 10, 11, 12, 13, 14, and 15. The professor could then ask the students to perform the projects using whatever language they prefer. Some students do not know Java and are more comfortable in another language.

A course that emphasizes the practical aspects of image processing in Java could cover the Chapters in order, skipping to the topics of interest.

How Can I Use the Web to Teach and Learn More Image Processing?

The Web has shown itself to be a flexible means for submitting such projects to the instructor. They also have the side benefit of increased visibility for the work.

There are Web resources that are used to support this book. These Web resources include mailing lists which readers are welcome to join. The *java-list* is used for announcing updates for the book and for general Java discussions. I shall try to answer questions as best I can. See Appendix A for the details.

Licensing of the Kahindu software is free for non-commercial use. For commercial use, please send me e-mail. Be sure to indicate if you want a site license or an individual license. The Kahindu program could be turned into the next "killer-app" with the right influx of capital.

If you have an idea that could make this a more useful book, or if you find a bug in an example or error in the text, please feel free to send me e-mail.

Praise my pocket protector and swill that Java!

– DL, lyon@DocJava.com

1. Introduction

The medium is the message means...a totally new environment has been created.

– Marshall McLuhan, 1964

The Web is the world's largest and fastest growing database. This database permits intelligence to be treated as a global resource. One goal of this book is to disclose techniques that enable the rapid deployment and development of algorithms that can help to speed delivery and processing of image data on the Web.

In addition, this book is designed to fill a need for an image processing book in Java that goes beyond a brief overview. The following sections detail the nature of the book, the type of reader the book intends to target, and the motivation for the subject.

1.1. What is This Book About?

This book covers the design and construction of applications that perform image processing. The topics in this book include restoration, compression, segmentation and representation. While the theoretical underpinnings are covered in some detail, the main purpose of the book is to show how to implement the described algorithms. Thus, the tone and nature of this book are those of a *how-to* book in image processing.

The following sections define the scope of the image processing treatment covered by this book. They also describe the limits to the discourse of the Java technology, and justify the use of Java in an image processing application.

1.1.1. What is Image Processing?

An image may be defined as any two-dimensional array of numbers. The numbers may represent a visible image, or they may represent a signal from any type of a transducer. For example, a sonar transducer (an underwater microphone) will convert a pressure wave in water into electrical energy. This type of signal forms a *sonar image*.

In fact, image processing data does not have to represent a *real energy signal*. As an example, consider that financial data may be used to describe a two-dimensional *signal space*, wherein the set of signals may consist of time-varying parameters such as inflation vs. average wage rate.

Further, some image processing techniques exploit the interframe coherence in an image sequence. This type of coherence exploitation is called image sequence processing and is a kind of three-dimensional image processing.

Most generally, image processing inputs data consisting of a discrete lattice over which some computation is performed. When the output of the image processing consists of a non-image representation (such as a boundary, a statistical summary or a motion estimation), then the image processing is performing an analysis.

1.1.2. What is Java?

Java is a computer programming language invented by Sun Microsystems. Java technology includes both the Java language and those systems that surround Java and enable its deployment and execution. Sometimes the Java language is confused with the Java technology. It is important to remember that the Java language is only one component of the Java technology. One aspect of Java technology is known as the Java Virtual Machine (JVM). The JVM is used to run the byte codes that are generated by Java compilers. Byte codes may be interpreted by a byte-code interpreter or may be executed directly in hardware. Such technology is described in more detail in [Lyon and Rao].

1.1.3. Why Do Image Processing in Java?

Image Processing is of central importance to the Web. Of all the data formats that appear to be in wide use—text, audio and images—the images take by far the greatest bandwidth. The web has an immediate need for image processing technology in Java.

New image formats appear all the time. Browsers are growing in size and complexity as a result of having to install plug-ins and helper applications that enable the decoding of the growing number of formats. Java obviates the need for deployment of bloated browsers or the gaggle of plug-ins and helper applications that clog our computing environments.

This book shows you how to use Java to restore, transform, warp, compress, segment and represent images. Some of the Java topics that are reviewed in this book include: use of the Abstract Window Toolkit (AWT), the design of custom image formats, and the implementation of CODECs (COders and DECoders). Thus, any programmer interested in using Java to distribute new image formats will need to implement at least some of the Java programs in this book.

1.2. Who Should Read This Book?

This book is designed for people who have some knowledge of Java programming and have a desire to experiment with and learn about image processing. The book is not designed for beginners in Java and is intended as a second book. There is really no point in cluttering up a book with a bunch of class descriptions that appear elsewhere. There are some excellent first books in Java, for example [Lemay and

Perkins] and [Winston]. There also are excellent books on image processing. Some have code in C [Meyler and Weeks], some have code in FORTRAN [Gonzalez and Wintz]. Some have no code at all but are no less excellent as a result [Pratt] and some touch upon image processing in Java [Lyon and Rao]. In fact, [Lyon and Rao] could be a good first book for advanced programmers who do not know Java or DSP.

1.3. What Will You Learn?

This book covers the entire spectrum of image processing, with emphasis on fundamentals and clear Java examples. Topics include enhancement, restoration, compression, in-line Java decoders, and Web deployment.

The topics have many image processing examples and the material has been tested extensively in the classroom. This book gives a unified, introductory treatment of two- and three-dimensional image processing concepts.

This book also shows some advanced topics in image processing, including how to perform multi-resolution compression. These techniques enable the viewing of a low-resolution overview of an image that gets progressively better as the download proceeds.

In addition, we introduce a mult-resolution image format that enables the streaming of variable compression ratio image sequences. The data is deployed with a Java decoder so that the distribution of new formats is automatic. This book is not a complete reference on Java. Only those API features needed to perform image processing are covered.

2. Event Processing and the AWT

... When, in the course of human events

– Thomas Jefferson. 1743–1826.

... Thus in the beginning the world
was so made that certain signs
come before certain events.

– Cicero. 106–43 B. C.

The Abstract Window Toolkit (AWT) is a collection of classes and methods designed by Sun. It has an Application Programmer Interface (API) that is only partly covered in this book, to the extent needed to enable the reader to construct a simple interface for the display and processing of images. There are two aspects to the AWT that are needed for image processing: image display and a simple Graphics User Interface (GUI).

When the Java environment is distributed, a set of classes is included that provide for the core API. The classes are grouped into packages. One core API package is known as the java.awt package. This chapter presents a subset of the java.awt package and some custom extensions.

The extensions to the java.awt package consist mainly of subclasses that provide an interface that is intended to simplify programming. The approach to software development is to write for code *reuse* as much as possible. For example, the code AWT has a *Frame* class. One subclass of this, the *ClosableFrame,* provides the service of interception and processing of a window close request. The *ShortCutFrame* is a sub-class of the ClosableFrame and provides the services of mapping keyboard events into strings that are embedded into menu items. Such an architecture provides layers, each one adding additional services. Eventually we evolve the services into a level high enough to make image processing in Java a much simpler task.

The reader is apt to ask two basic questions about this approach. Why supply so many layers? Why not just use the Sun API directly? The answer is software reuse, the ability to write short, readable programs and, just as important, ***insurance!*** The Sun API is evolving and will probably cause code to break in several places as it changes. We have seen this before. The Sun JDK 1.0.2, for example, contained several methods that became *deprecated*, i.e., they were given disapproval indicate that future support might be withdrawn. When this occurs, code that uses the deprecated methods will no longer work. Thus we protect our code from the changes in the API by providing an extra layer around those parts of the AWT that appear to be changing. As the changes take effect, the number of lines of code that require alteration should be minimized by the extra layer. For example, JDK 1.1 broke some code in [Lyon and Rao]. When this occurred, the code in [Lyon and Rao] became deprecated and broken. The fixes were made to a few centralized classes and the majority of the old code was salvaged. Such changes are particularly discouraging to programmers who want to see their code compile (without change) in the future. The rapid obsolescence of source code is a relatively new phenomena in the computer science community. In fact, legacy systems include source code that is decades old.

Just as important, Sun may release another version of the JDK at any time. When a new JDK is released, it can have a new API. The new API may not be part of the core Java classes and so not all browser vendors are required to support it. We suggest that Java code be made available that can be used when non-core APIs are relied upon. Further, when the non-core APIs are implemented in a platform, they may be used to accelerate the code. Using a few kernel classes as a layer to interface to the Sun API is sound software engineering that will pay off in reduced maintenance cost.

Additionally, the custom classes presented in this book are examples of procedural abstraction. Such examples encourage standard software engineering discipline, such as information hiding and modular decomposition. In Java, it is not difficult to write bad programs. In fact, with large class libraries changing rapidly, sound software engineering practice becomes essential to the continued survival of the code (plan on Sun's deprecating on your programs!).

Finally, if an applet uses a new API, an older browser that does not support the new API will not run the applet. Indeed, this type of incompatibility has already occurred in previous releases of the JDK, making non (feature *dujour*)-compliant browsers unable to run a modern applet.

2.1. The Frame Class

In the java.awt package there is a class called the *Frame* class. The Frame class is used to provide event handling and display services. Classes that extend the Frame class typically add Graphics User Interface (GUI) elements called *widgets*. Widgets include elements such as *check boxes*, *scroll bars* and *menuItems*. When widgets are added to a frame, they alter both the appearance and functionality of the frame. Typically, a drawn widget is an instance of a class that extends the *Component* class. The reader should know that some writers will call a collection of instances of classes that sub-class the *Component* class *components*. Generally a widget is the appearance of an instance of a class, but this does not have to be the case. For example, in [Lyon and Rao] widgets are presented that bypass the (then-current) event handler in favor of a (now-current) subject-observer model.

User interaction with widgets causes an *event* to be instanced. In the old JDK 1.0.2 event model, The handleEvent method is passed a single argument. This argument is an instance of the *Event* class. In the new JDK 1.1 event model, instances interested in an event must *listen* for the event. Typically, a class will implement a *Listener* interface. The *Listener* interface requires an implementation of *action* methods that are invoked when an event occurs. Instances of classes that sub-class the *Component* class will keep a list of interested class instances.

The relationship between event generators and event listeners is like the *observer* design pattern in object oriented programming. The observer pattern is a behavioral pattern that defines a one-to-many notification list between instances. For example, suppose that the user clicks on a button. An instance of the *Button* class will update all instances on its notification list. Thus, the button has a subscription list for any events that it generates. The subscribers must be qualified before they can obtain a subscription. To be qualified, they must implement a listener interface. A qualified class is subscribed to an event broadcast by use of an special *add* method. Communication is typically one-way. The observer design pattern is sometimes called the *observer-observable* relationship. It is also called the *subject-observer* relationship.

The publish-subscribe design pattern is available in some word processors. For example, in Microsoft Word, several documents can subscribe to the same file (i.e., a chart in a spread-sheet). This is different from the observer design pattern (as implemented in the AWT) because the document must pull in the changes when it is activated. Under JDK 1.1's event model, the changes are pushed into the interested objects, using a method invocation. Thus, the publish-subscribe dependency in Word has a non-automatic update mechanism. In comparison, the JDK 1.1 event model has an *automatic* update by the subject to the subscribing observers.

In order to create an abstract coupling between the subject and the observer, the observer must implement an *interface*. The interface requires implementations of update methods that will be invoked by the subject. The subject has a list of observers that conform to the requirement that they implement the interface.

The JDK 1.1 event model implements the observer design pattern using a series of interfaces. Each interface is designed around a special event. For example, there is an interface for observers of the *ActionEvent* class. This interface is called the *ActionListener*. Thus, a subscriber to the subject that publishes *ActionEvent* instances must implement the *ActionListener* interface before a subscription can be processed. The *ActionListener* interface requires that the *actionPerformed* method be implemented.

Thus, it is the programmers' responsibility to implement the *actionPerformed* method to process an *ActionEvent*. The following example shows a listing of the *ClosableFrame* class:

```
1.    package gui;
2.    import java.awt.*;
3.    import java.awt.event.*;
4.
5.
6.    public class ClosableFrame extends Frame
7.      implements WindowListener {
8.
9.    public static void main(String args[]) {
10.     ClosableFrame cf = new ClosableFrame("Closable Frame");
11.     cf.setSize(200,200);
12.     cf.show();
13.    }
14.
15.
16.    // constructor needed to pass
17.    // window title to class Frame
18.    public  ClosableFrame(String name) {
19.     // call java.awt.Frame(String) constructor
20.     super(name);
21.     setBackground(Color.white);
22.     addWindowListener(this);
23.    }
24.
25.    public void windowClosing(WindowEvent e) {
26.     dispose();
27.    }
28.    public void windowClosed(WindowEvent e) {};
29.    public void windowDeiconified(WindowEvent e) {};
30.    public void windowIconified(WindowEvent e) {};
31.    public void windowActivated(WindowEvent e) {};
32.    public void windowDeactivated(WindowEvent e) {};
```

```
33.     public void windowOpened(WindowEvent e) {};
34.
35.
36.
37.
38.     }   // end class ClosableFrame
39.
```

Recall that an observer must be qualified to subscribe to a subject's events. The qualification for a *WindowEvent* subscription is the implementation of the *WindowListener* interface:

```
6.     public class ClosableFrame extends Frame
7.       implements WindowListener {
```

In order to keep the class from being abstract, the following implementations are required (even if they do nothing!):

```
25.     public void windowClosing(WindowEvent e) {
26.      dispose();
27.     }
28.     public void windowClosed(WindowEvent e) {};
29.     public void windowDeiconified(WindowEvent e) {};
30.     public void windowIconified(WindowEvent e) {};
31.     public void windowActivated(WindowEvent e) {};
32.     public void windowDeactivated(WindowEvent e) {};
33.     public void windowOpened(WindowEvent e) {};
```

The *ClosableFrame* instance is able to subscribe itself to its own window events, using the constructor (see line 22):

```
18.     public  ClosableFrame(String name) {
19.      // call java.awt.Frame(String) constructor
20.      super(name);
21.      setBackground(Color.white);
22.       addWindowListener(this);
23.     }
```

Thus, the *ClosableFrame* is both an observer and a subject. This is actually rather typical of the design approach taken in the *gui* package.

Other frame classes in the *gui* package inherit the ability to be closable by extending the *ClosableFrame* class. For more details on design pattern usage, see [Gamma].

2.2. Interaction

This section addresses the problem of handling events generated by a user. The topic of computer-human interaction is a deep one and is addressed only to the point of permitting the construction of elementary graphic user interfaces. For more details on this subject, the reader may consult [Geary].

There are only two input devices needed in this book: the mouse and the keyboard. This section shows how to build a simple pop-down menu interface with mouse input and keyboard shortcuts.

2.2.1. The EventTester

A *Frame* instance in Java typically becomes the center of activity for the Java programmer. As features are added to a program, more components are added to a frame and more event handling dispatches are added to the code. As a result, many Java books grow large classes that extend Frames and become depositories for a huge event dispatch. Typically, there is one dispatch for every MenuItem instance in the Frame and a parallel dispatch for each keyboard short-cut. As demonstrated in [Lyon and Rao], such software engineering design is bad practice. We propose to create a new API based on a method that can match keyboard and mouse events.

There are going to be several different kinds of events that can trigger a single response. For example, a keyboard event (called a short-cut) can be used to open a file. Also, a menu selection made by the mouse can open a file. The AWT typically requires that the same procedure be invoked from two different places in the source code. This is called a *parallel source-code dispatch.*

To prevent parallel source-code dispatch, we constrain the MenuItem instances so that they contain a string that is isomorphic with respect to a keyboard event. This way, either the MenuItem instance selection or a keyboard event can be used to trigger a response. The idea is that we will construct an instance of a class that provides the services of a *ClosableFrame*, draws a string in its center, has a main menu bar and several hierarchic menu items, and is able to process keyboard short cuts. These services are the bases for all the programs in this book.

An example of the *Event Tester* output shown in Fig. 2-1.

Fig. 2-1. The *EventTester* Frame

To continue with the broad overview of the program, consider the pop-down hierarchical menu shown in Fig. 2-2.

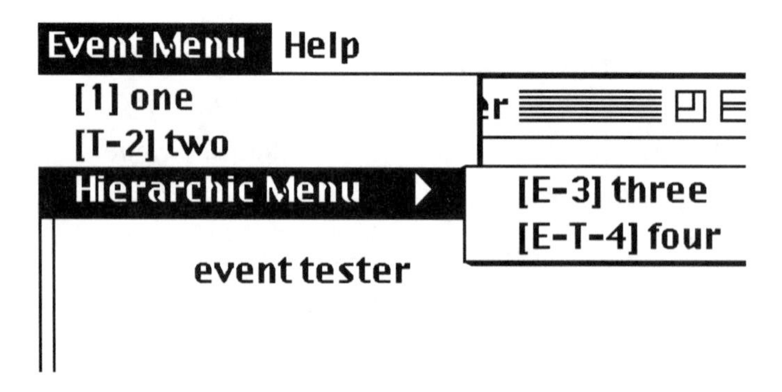

Fig. 2-2. The Hierarchical Menu of the *EventTester*

Fig. 2-2 shows a non-standard prefix keyboard short-cut system which is used in the Kahindu program. This is discussed in more detail later. The following code snippet illustrates the process of creating the menu:

```
...
7.      Menu m1 = new Menu("Event Menu");
8.      MenuItem item1_mi = addMenuItem(m1,"[1] one");
9.      MenuItem item2_mi = addMenuItem(m1,"[T-2] two");
10.
11.     Menu hierarchicMenu = new Menu("Hierarchic Menu");
12.     MenuItem itemH1_mi = addMenuItem(hierarchicMenu,"[E-3]
        three");
13.     MenuItem itemH2_mi = addMenuItem(hierarchicMenu,"[E-T-4]
        four");
14.
```

Lines 7-14 show the creation of the menu items and two different menus. The first menu is called *m1* and the second menu is called *hierarchicMenu*. The idea is that the *hierarchicMenu* is added to the *m1* menu. The *addMenuItem* method takes care of some house-keeping, including the subscription of interest in the new *MenuItem* instance returned by the *addMenuItem*. The *addMenuItem* method also inspects and stores the strings in the brackets (i.e., “[E-T-4]”). These represent keyboard shortcuts. These are discussed in Section 2.2.1.2. To initialize the menu bar, we use

```
25.     public void initMenuBar() {
26.         MenuBar mb = new MenuBar();
27.         m1.add(hierarchicMenu);
28.         mb.add(m1);
29.         setMenuBar(mb);
30.     }
```

Note that the menu bar is built from the bottom up. That is, the lowest level sub-menu, *hierarchicMenu*, is added to the parent menu *m1*. Finally, the *m1* menu is added to the menu bar instance *mb*. After this is complete, the menu bar is set using line 29.

As a program increases in complexity, the number of menu items and menus on the menu bar will grow and so will the amount of code needed for processing events. Subclassing enables the addition of features to the event processing while keeping the size of the event handling method small. This is enabled by the invocation of *super.actionPerformed* at the end of the event handling.

As the number of menu items increases, the screen space gets used up. Menus with more than 10 items make scanning the list tedious. Also, many of the items may be irrelevant to the users' current need [Laurel]. This is the primary motivation for the use of a hierarchic menu. The hierarchic menu allows a grouping of items into relevant sub-menus. The danger is that the hierarchical menu becomes so deep that the user gets lost. According to Tognazzini, the hierarchical menu offers hundreds of extra options, but it takes much longer for the user to make a selection. Additionally, Tognazzini says that menu selections are generally faster than keyboard short cuts [Tog]. As far as we know, there has been no study of the trade-off between keyboard short-cut selection speed and deep hierarchical menu selection speed.

As a result, keyboard short cuts are embedded into every menu item. This enables the user to decide how to invoke a process. Also, to increase the number of available characters, keyboard modifiers are used. These are discussed in Section 2.2.1.2.

2.2.1.1. Intercepting Menu Events

An event processing program must assign semantic meaning to an event. The implementation of our event interpreter is complicated by the goal of creating an automatic linkage between keyboard shortcuts and menu item event processing. The advantage of such a custom linkage is that we have control over the number and type of keyboard events that can be used to control our system. Another advantage is that keyboard shortcuts are embedded directly into the string representations of a menu-item label. This automates the consistency between a label and its short-cut.

As a matter of taste, we have placed the shortcuts at the head of the menu-item label. This can easily be changed, according to preference. We use the prefix system to differentiate from the AWT's post-fix system (which currently has several problems, discussed in Section 2.2.1.2).

An instance of an Event class has a class variable that contains a reference to an instance of the selected menu item. So, to handle and process the menu item selections (using the mouse or keyboard as the input device), we need to add only a few lines to the code:

```
35.     public void actionPerformed(ActionEvent e) {
36.         if (match(e, item1_mi))
37.             System.out.println("Item 1!");
38.         else if (match(e, item2_mi))
39.             System.out.println("Item 2!");
40.         else if (match(e, itemH1_mi))
41.             System.out.println("Item h1!");
42.         else if (match(e, itemH2_mi))
43.             System.out.println("Item h2!");
44.         super.actionPerformed(e);
45.
46.     }
```

The *actionPerformed* method has been invoked by a set of custom classes that will be discussed shortly. To decode the keyboard events we use the procedure described in the following section.

2.2.1.2. The *ShortCutFrame* - Make Me a Match!

It has long been known that users say that keyboard shortcuts accelerate menu item selection. It has also been established that keyboard shortcuts are generally not faster [Tog]. In order to create a visual interface that is consistent with the keyboard shortcuts, we take an approach that embeds the shortcuts into an instance of a MenuItem's label.

There are several modifier keys available on a keyboard: *control, alt* and *meta*. None of them behaves in a consistent cross-platform manner! Any keyboard short-cut used by the operating system or the JVM will be intercepted before being passed to the event handler. Such shortcuts must be off-limits on all platforms to create truly portable Java code. For example, moth-{n, k, o, r and w} are all used by the MetroWerks JVM on the Mac.

The official JDK 1.1.5 keyboard handler will only permit the use of the control key [Geary]. Worse, under some VMs, we have found that the key values passed to the event handler are altered by their keyboard modifiers. Alt and control keys cause the key values to change by an offset that is a function of which key is pressed.

To add insult to injury, key offsets are not platform independent. In fact, the problem is so bad that there is currently no way to tell the difference between *cntl-j* and *cntl-c*, on the Mac. This could well be a bug in the JVM.

As a result of these problem, we have given up trying to use the standard modifiers. Instead, we have devised a scheme to employ multiple keystrokes using the *escape* and *tab* keys. Thus, in this book, when the term *modifier keys* is used, it refers to the escape and tab keys.

We have selected a labeling scheme for representing the modifier keys in a MenuItem instance. The labeling scheme results in a mapping that is shown in Table 2-1.

Modifier	Code
esc	E -
tab	T -
esc-tab	E - T -

Table 2-1. Keyboard Shortcuts

The scheme that we have selected is based on the observations of Mullet and Sano who state that the modifier keys should not visually compete with the alphabetic characters in a menu item [Mullet].

Examples of some keyboard shortcuts are shown in Fig. 2-3 and Fig. 2-4.

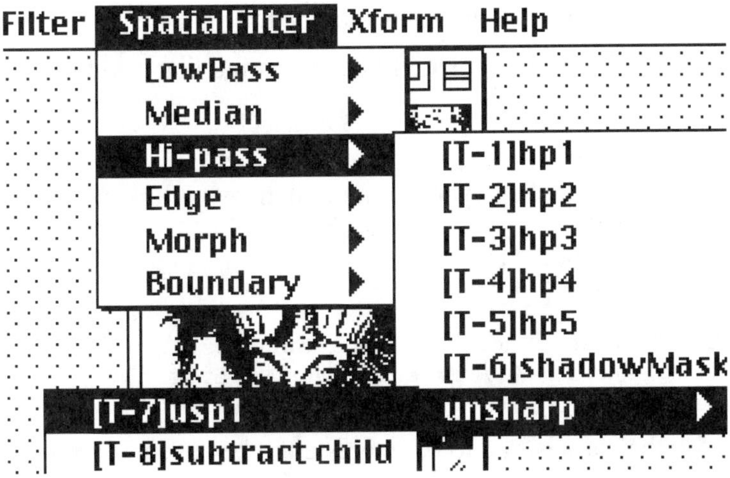

Fig. 2-3. Keyboard Shortcuts Nested with the Hierarchic Menus

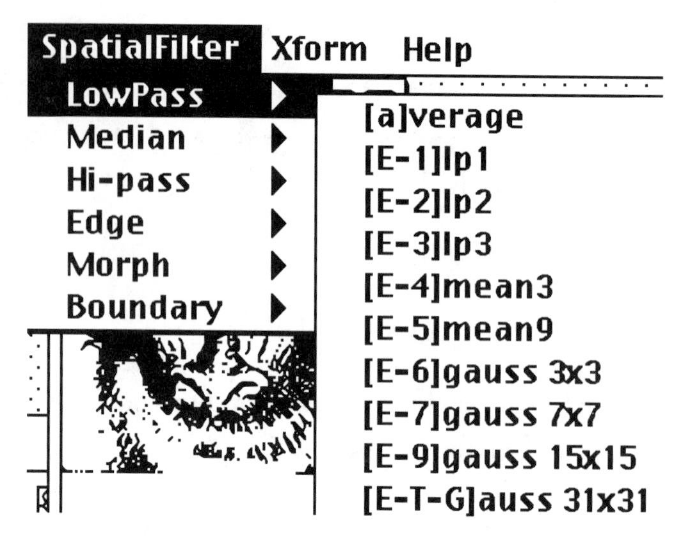

Fig. 2-4. Three-Key Keyboard Shortcuts.

The last example shown in Fig. 2-4 indicates that the user must press 3 keys in order (*escape, tab* and *G*). The programmer should keep in mind that asking the user to press 3 keys is going to slow down the user.

The code snippet for providing the keyboard-menu item event linkage follows:

```
1.    package gui;
2.    import java.awt.*;
3.    import java.awt.event.*;
4.
5.     public class  ShortCutFrame extends ClosableFrame
6.         implements KeyListener, ActionListener {
```

Note that in lines 5-6 the class is said to be interested in key events and in action events. The *ShortCutFrame* class becomes a qualified observer of the key and action events by implementing the *KeyListener* and *ActionListener* interfaces. Line 5 shows that the *ShortCutFrame* supports the close event by inheritance from the *ClosableFrame*.

The idea is that the *ShortCutFrame* will keep track of keyboard events and, after processing, will invoke the *actionPerformed* method. The *actionPerformed* method is thus invoked for menu item selection and is invoked by the *ShortCutFrame* when qualified keyboard events appear. The keyboard events are tracked and stored in a kind of data structure called a *Petri* net. This topic is left to Section 2.2.1.3.

2.2.1.3. Petri Nets Eat Events-Yum!

The *ShortCutFrame* code tracks keyboard events using a *Petri net.* A Petri net is *bipartite digraph.* The term bipartite means that it has two types of nodes, places and transitions. The term *digraph* is short for *directed graph.* This means is that all the arcs in the graph have direction.

A Petri net diagram is a graphical representation of a Petri net. The *place*, *transition* and *arc* are shown in Fig. 2-5.

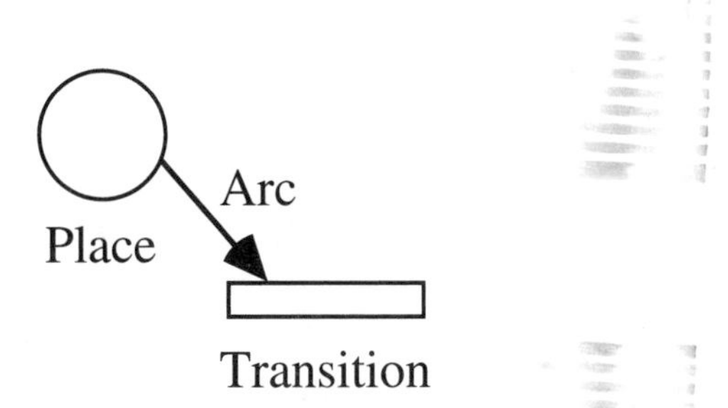

Fig. 2-5. The Petri Net Components

Places can have multiple inputs and multiple outputs. Transitions have only a single output.

For any problem, it is important to have an English language specification. Thus, the basic problem may be stated as follows:

Let *esc* denote the detection of an escape key. Similarly, *tab* will denote the detection of a tab key. Let a non-modifier key be denoted as *!mod*. Thus we may say that *if !mod* means "if the user types a non-modifier character" (i.e., not *tab* and not *esc*). Let *c* = char typed. We may state the problem as a series of rules:

```
1. if !mod then return "[c]"
2. if esc and !mod return "[E-c]"
3. if tab and !mod return "[T-c]"
```

```
4. if esc and tab and !mod return "[E-T-c]"
```

Given the problem specification, we may now formulate the Petri diagram, shown in Fig. 2-6.

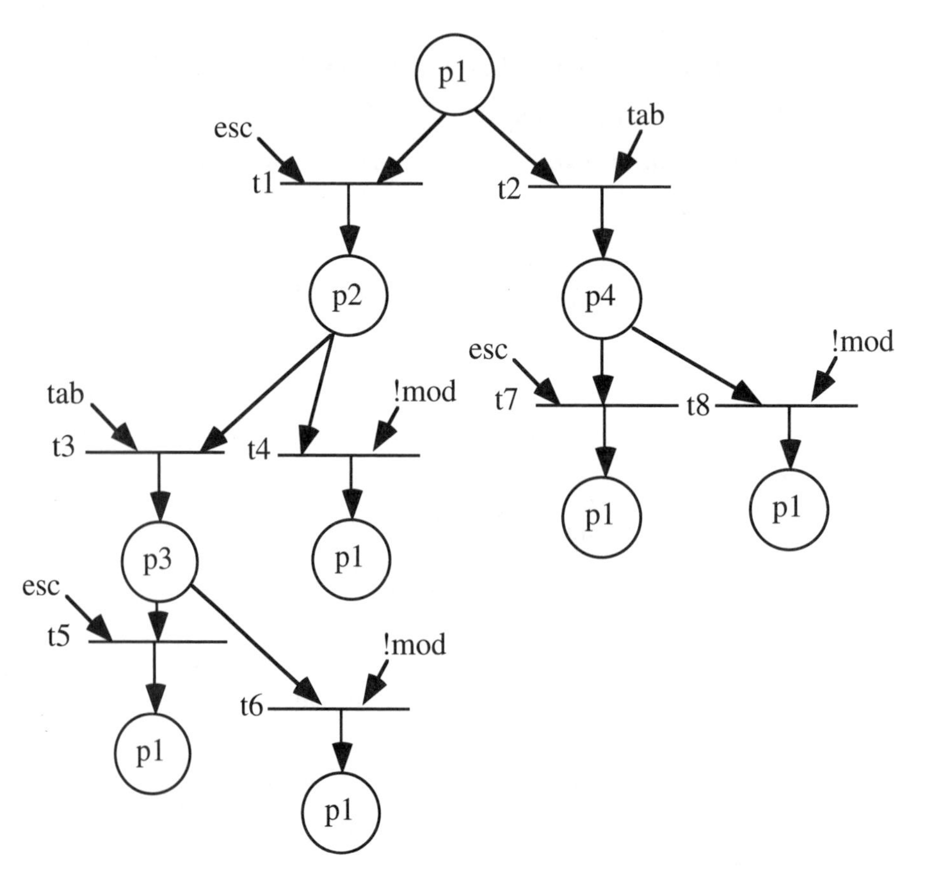

Fig. 2-6. Petri Diagram for Key Events

The transitions and places in Fig. 2-6 are numbered. A transition may have side effects. A Petri net may be represented graphically, using a diagram, or it may be represented textually using a Petri net table. The Petri net table is a more compact form which is more suitable for computer implementation (particularly for more complex examples). The Petri table for the diagram shown in Fig. 2-6 appears in Fig. 2-7.

Name	Place	Enabling Token	Transition	Actions	Next Place
start	p1	esc	t1	null	p2
	p1	tab	t2	null	p4
got esc	p2	tab	t3	null	p3
	p2	!mod	t4	"[E-c]"	p1
got esc tab	p3	esc	t5	null	p1
	p3	!mod	t6	"[E-T-c]"	p1
got tab	p4	esc	t7	null	p1
	p4	!mod	t8	"[T-c]"	p1

Fig. 2-7. The Petri Table

There are several implementation methods available for a Petri table. The choice of implementation methods depends upon the application. One approach is to use a switch statement to control the dispatch as a function of the current place. This is shown in the code below:

```
56.     private String petriMap(int c) {
57.     String s ="[" + (char) c + "]";
58.     switch (place) {
59.         case 1:
60.             if (isEsc(c)) {
61.                 // t1
62.                 place=2;
63.                  return null;
64.             }
65.             if (isTab(c)) {
66.                 // t2
67.                 place=4;
68.                  return null;
69.             }
70.              return s;
71.         case 2:
72.             if (isTab(c)) {
73.                 // t3
74.                 place = 3;
75.                  return null;
76.             }
77.             if (isEsc(c))
78.                  return null;
79.             // t4
80.             place = 1;
81.              ps = "[E-"+(char)c+"]";
82.              return (ps);
83.         case 3:
```

```
84.                 if (isTab(c))
85.                     return null;
86.                 if (isEsc(c)) {
87.                     // t5
88.                     place = 1;
89.                      return null;
90.                 }
91.                 // t6
92.                 place = 1;
93.                 ps = "[E-T-"+(char)c+"]";
94.                 return (ps);
95.             case 4:
96.                 if (isEsc(c)) {
97.                     // t7
98.                     place = 1;
99.                      return null;
100.                        }
101.                         if (isTab(c)) return null;
102.                        // t8
103.                        place = 1;
104.                         ps = "[T-"+(char)c+"]";
105.                         return (ps);
106.                    }
107.                    ps = s;
108.                     return ps;
109.                }
```

To implement the *KeyListener* and *ActionListener* interfaces, the *ShortCutFrame* must implement several methods:

```
21.     public void keyPressed(KeyEvent e) {}
22.     public void keyReleased (KeyEvent e) {}
23.     public void actionPerformed(ActionEvent e){};
24.
25.     public void keyTyped (KeyEvent e) {
26.         int key = e.getKeyChar();
27.         ps = petriMap(key);
28.         actionPerformed(new ActionEvent(e,key,key+""));
29.     }
```

Note that line 25 shows the only method that actually does something, the *keyTyped* method. Line 26 uses the *getKeyChar* method to obtain the integer representation of the key character. Line 27 uses the *petriMap* invocation to map keystrokes into a string. Finally, line 28 invokes the *actionPerformed* method with a new *ActionEvent*. The arguments are the event instance. Recall that the classes which extend the *ShortCutFrame* implement the *actionPerformed* method. For example, in the *EventTester* class, we saw:

```
35.     public void actionPerformed(ActionEvent e) {
36.         if (match(e, item1_mi))
37.             System.out.println("Item 1!");
38.         else if (match(e, item2_mi))
39.             System.out.println("Item 2!");
40.         else if (match(e, itemH1_mi))
41.             System.out.println("Item h1!");
42.         else if (match(e, itemH2_mi))
43.             System.out.println("Item h2!");
44.         super.actionPerformed(e);
45.
46.     }
```

Thus, in order to implement the *match* method, we use

```
16.     public boolean match(AWTEvent e, MenuItem mi) {
17.      if (e.getSource() == mi) return true;
18.      if (ps == null) return false;
19.     return mi.getLabel().startsWith(ps);
20.     }
```

Line 17 assumes that if the source of the event is the same as the *MenuItem* instance, *mi*, then we have a good match. Line 18 asserts that if the *ps* string (a representation computed from the *petriMap*) is null, then the match must fail. Line 19 obtains the string from the menu item label and compares the prefix with the string stored by the Petri net. The result of that comparison determines if the match will succeed or not.

A full listing of the *ShortCutFrame* follows:

```
1.     package gui;
2.     import java.awt.*;
3.     import java.awt.event.*;
4.
5.      public class  ShortCutFrame extends ClosableFrame
6.          implements KeyListener, ActionListener {
7.      private int place = 1;
8.      private String s = null;
9.      private String ps = null;
10.
11.     public ShortCutFrame(String title) {
12.         super(title);
13.          addKeyListener(this);
14.     }
```

```
15.
16.     public boolean match(AWTEvent e, MenuItem mi) {
17.     if (e.getSource() == mi) return true;
18.      if (ps == null) return false;
19.     return (mi.getLabel().startsWith(ps));
20.     }
21.
22.     public void keyPressed(KeyEvent e) {}
23.     public void keyReleased (KeyEvent e) {}
24.     public void actionPerformed(ActionEvent e){}
25.
26.     public void keyTyped (KeyEvent e) {
27.         int key = e.getKeyChar();
28.         ps = petriMap(key);
29.         actionPerformed(new ActionEvent(e,
30.             ActionEvent.ACTION_PERFORMED,key+""));
31.    }
32.    public MenuItem addMenuItem(Menu aMenu, String itemName)
{
33.        MenuItem mi = new MenuItem(itemName);
34.        aMenu.add(mi);
35.        mi.addActionListener(this);
36.        return(mi);
37.    }
38.
39.     private boolean isEsc(int c) {
40.        return ( c == 27);
41.     }
42.     private boolean isTab(int c) {
43.        return (c == '\t');
44.     }
45.
46.     private boolean isMod(int c) {
47.        return (isEsc(c) || isTab(c));
48.     }
49.
50.    private String petriMap(int c) {
51.      String s ="[" + (char) c + "]";
52.      switch (place) {
53.          case 1:
54.              if (isEsc(c)) {
55.                  // t1
56.                  place=2;
57.                   return null;
58.              }
59.              if (isTab(c)) {
60.                  // t2
61.                  place=4;
62.                   return null;
```

```
63.                 }
64.                  return s;
65.         case 2:
66.                 if (isTab(c)) {
67.                     // t3
68.                     place = 3;
69.                      return null;
70.                 }
71.                 if (isEsc(c))
72.                      return null;
73.                 // t4
74.                 place = 1;
75.                 ps = "[E-"+(char)c+"]";
76.                 return (ps);
77.         case 3:
78.                 if (isTab(c))
79.                      return null;
80.                 if (isEsc(c)) {
81.                     // t5
82.                     place = 1;
83.                      return null;
84.                 }
85.                 // t6
86.                 place = 1;
87.                 ps = "[E-T-"+(char)c+"]";
88.                 return (ps);
89.         case 4:
90.                 if (isEsc(c)) {
91.                     // t7
92.                     place = 1;
93.                      return null;
94.                 }
95.                  if (isTab(c)) return null;
96.                 // t8
97.                 place = 1;
98.                 ps = "[T-"+(char)c+"]";
99.                 return (ps);
100.                    }
101.                    ps = s;
102.                     return ps;
103.                }
104.    }
```

Note that the constructor:

```
11.     public ShortCutFrame(String title) {
12.          super(title);
13.           addKeyListener(this);
```

```
14.     }
```

declares the short-cut frame as an observer of its own keyboard events. Also, every time a menu item is added:

```
32.     public MenuItem addMenuItem(Menu aMenu, String itemName)
{
33.      MenuItem mi = new MenuItem(itemName);
34.      aMenu.add(mi);
35.       mi.addActionListener(this);
36.       return(mi);
37.     }
```

Line 35 declares that whatever frame has extended the *ShortCutFrame* is going to subscribe to the action events generated by selecting the menu item. Hence, extending the *ShortCutFrame* is sufficient for processing keyboard shortcuts and implementing a null version of the *actionPerformed* method. The programmer must supply the *actionPerformed* implementation and invoke *super.actionPerformed* at the end of the processing.

To recap, we have experimented with various techniques to obtain a consistent cross-platform behaviour for modifier key usage. For whatever reason (bug in the virtual machines, bug in the AWT, bug in the brain, etc.), this did not work properly.

This concludes the discussion on processing keyboard events. Using the Petri net model, we can create long event sequences. Further, probabilities may be assigned in the Petri table, making it a stochastic Petri table. This enables the generation of Nth order Markov event streams (a topic that is beyond the scope of this book) [Lyon 1997]. The Petri table processing of events is not a new idea, although this is the first appearance of the idea as implemented in Java, as far as we know [Lyon 1995].

Petri nets are able to model concurrency and interrupts, two features that are notably lacking in the finite state machines. This section describes the implementation of an elementary Petri net for processing key-stroke events. Key-stroke events are not concurrent and, as the example in this section is elementary, a finite state machine could be used as well as the Petri nets. Java has built-in multi-threading and this gives currency to the modeling of concurrency.

2.2.2. BooLog, the Boolean Dialog

BooLog (Boolean diaLog) provides a means to obtain a boolean response from a user to a yes-no question. BooLog provides a class abstraction that isolates the programmer from needing to program the AWT directly. Further, BooLog provides a template to follow when implementing more complex dialogs.

In summary:

```
public class BooLog extends Dialog
   implements ActionListener {
  public Button yesButton;
  public Button noButton;
  public void actionPerformed(ActionEvent e)
  public BooLog(
       Frame frame,
      String title,
      String prompt,
      String yes, String no)
  public static void main(String args[])
 }
```

The action method is typically modified so that some meaningful action is performed. A class interested in getting notification would have to implement the *ActionListener* and register with the components that are interesting. For example:

```
BooLog bl = new BooLog(this, "boo", "scared?", "yes", "no");
bl.noButton.addActionListener(this);
bl.yesButton.addActionListener(this);
```

An *actionPerformed* method is then required. For example:

```
public void actionPerformed(ActionEvent e) {
   Object o = e.getSource();
   if (o == bl.noButton) {
         System.out.println("too bad");
         return;
   }
   if (o == bl.yesButton) {
         System.out.println("Good!");
         return;
```

```
    }
}
```

The constructor:

```
public BooLog(
    Frame frame,
    String title,
    String prompt,
    String yes, String no)
```

takes an instance of a Frame class. It is an error to pass null in for the Frame instance. The *title* is used as the title of the dialog. *Prompt* contains a *String* instance that is to contain the yes-no type question. The *yes* string is used for the label of the yes button. The *no* string is used for the label of the no button. Thus, the Dialog instance may use any labels for the button, such as *ok-cancel* or *yes-no* .

The

```
public static void main(String args[])
```

method provides a way to test the class. For a deeper understanding of the AWT, there are books that discuss the classes in detail [Geary] [Chan and Lee][Lyon and Rao].

2.2.3. ExpandoLog, the Expandable Dialog

ExpandoLog is the name of a class that enables the user to input a sequence of strings, with prompts. ExpandoLog provides a method that blocks the execution of the invoking program until the user enters the required input or selects the *cancel* button.

ExpandoLog is like BooLog in that it is used as a means to isolate the programmer from the API of the AWT. The best way to show how to use the *ExpandoLog* is by an example:

```
public static void main(String args[]) {
```

```
        String title = "Rotation Dialog";
        int fieldSize = 6;

        String prompts[] = {
             "X (degs):",
             "Y (degs):",
             "Z (degs):"
        };

        String defaults[] = {
             "1.0",
             "2.0",
             "3.0"
        };

        ExpandoLog xpol = new
                ExpandoLog(
                   new Frame(),
                   title,
                   prompts,
                   defaults,
                   fieldSize);
    }
```

The output of the example is shown in Fig. 2-8.

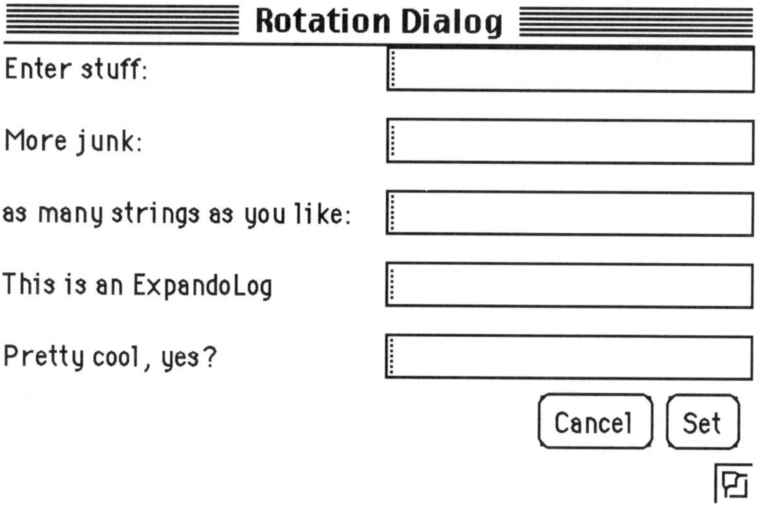

Fig. 2-8. The ExpandoLog Dialog

The main method of the ExpandoLog tester controls the appearance of the ExpandoLog dialog. Line 6 establishes the title for the dialog:

```
5.          public static void main(String args[]) {
6.              String title = "ExpandoLog";
```

Line 7 is used to specify the number of spaces to be used for the text fields. The text fields are all of uniform size and the user proceeds from one text field to the next by depressing the tab key:

```
7.              int fieldSize = 9;
```

Lines 8-14 establish the labels for each text field. Also, the number of labels is always equal to the number of prompts. Every text field has a label:

```
8.              String prompts[] = {
9.                "Enter stuff:",
10.               "More junk:",
11.               "as many string as you like:",
12.               "This is an ExpandoLog",
13.               "Pretty cool, yes?"
14.             };
```

The ExpandoLog class instance, *xpol*, is constructed in lines 16-18. Once this constructor is invoked, a *modal* dialog box appears. A modal dialog box prevents interaction with any other window, forcing the user to enter input:

```
16.             ExpandoLog xpol = new
17.                     ExpandoLog(title,
18.                      prompts, fieldSize);
```

The thread of execution is blocked with the invocation of the *getUserInput* method shown in line 20:

```
20.             String userInput[] = xpol.getUserInput();
```

This simplifies the implementation of the image processing code. To see if the user canceled the operation, we check the boolean member variable, *processData*:

```
21.             if (xpol.processData)
22.                 for (int i=0; i<userInput.length; i++)
23.                      System.out.println(userInput[i]);
24.             else System.out.println("User canceled");
25.         }
```

In order to provide defaults for the text fields, the constructor for the ExpandoLog class is overloaded. If a second array of string is passed into the constructor, it is taken as the default value for the field.

2.2.4. Class Summary

```
package gui;
public class ExpandoLog extends
  Dialog implements ActionListener {
  public  Button cancelButton
  public  Button setButton
  public ExpandoLog(
      Frame frame,
     String title,
     String prompts[],
     String defaults[],
     int _fieldSize)
 public void printUserInput()
 public String [] getUserInput()
 public static void main(String args[])
 public void actionPerformed(ActionEvent e)
```

2.2.5. Class Usage

The following code generates the output shown in Fig. 2-9:

```
1.     import gui.*;
2.
3.     public class ExpandoLogTester {
4.
5.         public static void main(String args[]) {
6.             String title = "Rotation Dialog";
7.             int fieldSize = 6;
8.             String prompts[] = {
9.               "X (degs):",
10.              "Y (degs):",
11.              "Z (degs):"
12.             };
13.
14.             String defaults[] = {
15.              "1.0",
16.              "2.0",
17.              "3.0"
18.             };
19.
20.             ExpandoLog xpol = new
```

```
21.                     ExpandoLog(
22.                      title,
23.                      prompts,
24.                      defaults,
25.                      fieldSize);
26.
27.             String userInput[] = xpol.getUserInput();
28.             if (xpol.processData)
29.                 for (int i=0; i < userInput.length; i++)
30.                     System.out.println(userInput[i]);
31.             else System.out.println("User canceled");
32.         }
33.     }
```

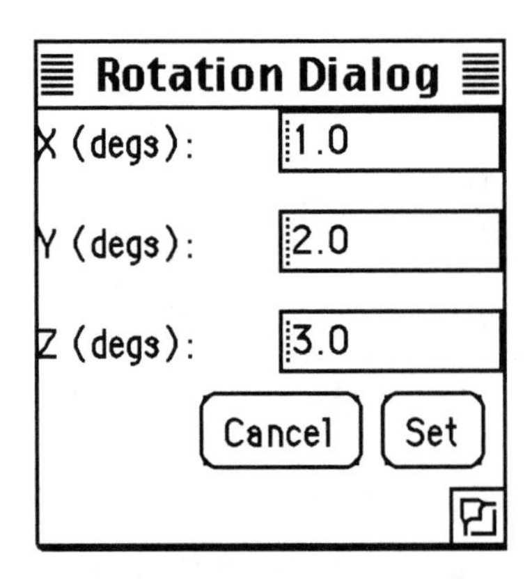

Fig. 2-9. ExpandoLog with Default Values

Lines 8-18 are used to specify the labels and defaults for the ExpandoLog instance:

```
8.              String prompts[] = {
9.                "X (degs):",
10.               "Y (degs):",
11.               "Z (degs):"
12.             };
13.
14.             String defaults[] = {
15.              "1.0",
16.              "2.0",
17.              "3.0"
18.             };
```

If the *defaults* array is shorter than the *prompts* array, an exception will be thrown. The *ExpandoLog* is not a particularly good interface, but it is simple to use.

ExpandoLog performs no checking of user input, that is up to the programmer. The *ExpandoLog* dialog is a type-in user interface with a simple layout and a specific purpose.

Typically the ExpandoLog class is subclassed. The main method has an example of code that shows the dialog. The *actionPerformed* method only processes the *cancelButton* by setting the dialog visibility to *false*.

It is beyond the scope of this book to discuss issues such as layout managers, GUI design details, or even to list the AWT components. These topics are covered in [Geary] [Horstmann and Cornell] and [Lyon and Rao].

2.3. Summary

The ExpandoLog class introduces the idea of using a simple abstract and modular interface to obtain a kind of procedural abstraction. The idea is to help programmers develop the skills needed to hide complexity. This leads to an ease of maintenance in the face of a quickly changing API. This can be of direct assistance to the new Java programmer who is often overwhelmed by the large number of Java class libraries.

The idea of procedural abstraction for teaching a new language is not a new one [Roberts].

For an excellent book on building interfaces using Java, see [Geary and McClellan].

For an update to the JDK 1.1 AWT model, see [Geary and McClellan 2].

For more general coverage of the class libraries, see [Cornell and Horstmann] and

[Roberts]. For a very readable treatment on interface design (though not a Java book, *per se*) see [Tog].

2.4. Suggested Projects

1. Add the modifier keys Meta, Alt and Control, to the Kahindu menu. One technique for mapping modifier keys, appears below:

```
private boolean contains(String str1, String str2){
  return str1.regionMatches(0,str2,0,str2.length());
}
public String mapModifiers(KeyEvent e) {
  String modString =
        KeyEvent.getKeyModifiersText(
              e.getModifiers());
  String newMods = "";
  if (contains(modString,"Ctrl")) newMods += "^-";
  if (contains(modString,"Alt")) newMods += "A-";
  if (contains(modString,"Meta")) newMods += "M-";
  if (contains(modString,"Shift")) newMods += "S-";
  return newMods;
}
 public void keyPressed(KeyEvent e) {
  String mods = mapModifiers(e);
  String key = e.getKeyText(e.getKeyCode());
  if (mods.equals("")) return;
  ps = "["+mods+key+"]";
  System.out.println(ps);
  actionPerformed(new ActionEvent(e,
        ActionEvent.ACTION_PERFORMED,key+""));
 }
```

Keep in mind that the above is non-working code. Try the above and see if you can get consistent results across several platforms. A menu item such as

```
MenuItem medianOctagon5x5_mi =
  addMenuItem(medianMenu,"[^-o]catgon 5x5");
```

should enable Cntrl-o to be used as a keyboard shortcut. It does not work now. Try the code and find out why.

2. There is a JDK 1.1 AWT method for adding keyboard shortcuts. The technique is described in [Chan and Lee2]. Modify the Kahindu program to get this working. What happens on different platforms?

3. The command pattern, described in [Gamma], says that a MenuItem should be able to invoke the command associated with itself. One technique for doing this involves *introspection* (a topic beyond the scope of this book). Research the introspection topic and get keyboard shortcuts working for the following code example (which is on CD-ROM in BeanTester.java):

```
1.    package gui;
2.    import java.awt.*;
3.    import java.awt.event.*;
4.    import java.beans.*;
5.    import java.lang.reflect.*;
6.    import java.util.*;
7.
8.
9.    public class BeanTester extends BeanFrame {
10.
11.     public static void main(String args[]) {
12.         BeanTester bt = new BeanTester("BeanTester");
13.     }
14.     public BeanTester(String title) {
15.         super(title);
16.         initMenuBar();
17.         setSize(300,300);
18.         show();
19.     }
20.     public void item1() {
21.         System.out.println("item1");
22.     }
23.     public void item2() {
24.         System.out.println("item2");
25.     }
26.     public void itemH1() {
27.         System.out.println("itemH1");
28.     }
29.     public void itemH2() {
30.         System.out.println("itemH2");
31.     }
32.
33.     public void initMenuBar() {
34.         Menu m1 = new Menu("Event Menu");
35.
36.         addMenuItem(m1,"[E-T-1]item1","item1");
37.         addMenuItem(m1,"[T-1]item2","item2");
38.
39.         Menu hierarchicMenu = new Menu("Hierarchic Menu");
40.         addMenuItem(hierarchicMenu,"[E-h]itemH1","itemH1");
41.         addMenuItem(hierarchicMenu,"[h]itemH2","itemH2");
42.
43.         MenuBar mb = new MenuBar();
44.         m1.add(hierarchicMenu);
45.         mb.add(m1);
46.         setMenuBar(mb);
```

```
47.      }
48.      public void paint(Graphics g) {
49.          g.drawString("bean tester",150,150);
50.      }
51.
52.    }
53.
```

3. Displaying and Filtering Images in Java

If you would not be forgotten,
As soon as you are dead and rotten,
Either write things worthy reading,
Or do things worth the writing.

– Benjamin Franklin

This chapter shows how to create a class called the *ImageFrame*. The ImageFrame creates a simple two-choice menu upon construction. The choices allow the user to open a GIF format image or to revert to the original image. The ImageFrame is shown in Fig. 3-1.

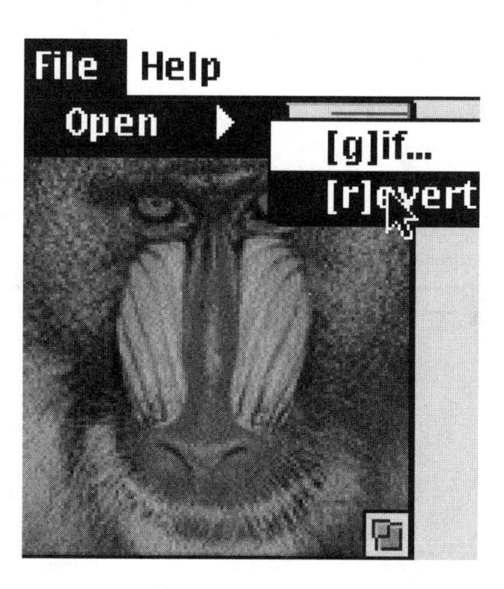

Fig. 3-1. The ImageFrame

In order to address the problem of image display, we must first discuss the issue of image representation. We will then take a bit of time for an exposition on arrays in Java. Finally we will bring together both topics in the *ImageFrame* class.

3.1. Image Representation and the Space-Time Balance

The choices that are made for the representation of an image must be lived with for the life of an image processing program. Such choices represent the classic trade-off between optimizing for space and optimizing for time. If we choose to optimize for time (i.e., minimize the time it takes to process an image) then we cannot optimize for space (i.e., minimize the space it takes to store an image). The converse is also true; we can pack an image very tightly into a small space, but then we cannot process the image at top speed.

The Java AWT model for storing an image minimizes the storage required. The implementation of this policy causes code to become both slow *and* unclear.

In Java, a pixel is stored as a 32 bit *int*. The *int* consists of 4 packed bytes. These bytes represent the alpha, red, green and blue planes, (ARGB) as shown in Fig. 3-2.

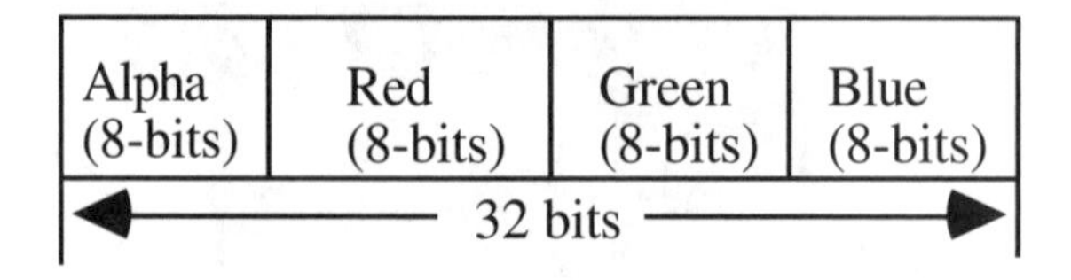

Fig. 3-2. A Packed Pixel

The packed pixel shown in Fig. 3-2 is the most tightly packed storage technique available for a 32 bit pixel. However, there are problems with this approach. For example, encoding and decoding both require bit shift and masking operations, increasing computation time.

Suppose you are given a packed color pixel, as shown in Fig. 3-2, and you want to convert it to a monochrome pixel by averaging the 3 colors. A method that accomplishes this task follows:

```
public int filterRGB (int rgb) {
  int red   = (rgb & 0xff0000) >>16;
  int green = (rgb &   0xff00) >>  8;
  int blue  =  rgb &     0xff;
  int gray  = (red + green + blue) / 3;
  return (
       0xff000000 |
       (gray << 16) |
       (gray <<  8) |  gray);
}
```

The *filterRGB* method unpacks the pixel, performs the filtering operation, then repacks the pixel. Three AND operations, three OR operations and four SHIFT operations are needed to access the three colors. Consider that images typically range in size from a few hundred bytes to a few megabytes. In fact, million pixel images are not that uncommon, particularly with modern scanner and camera technologies. This unpack-process-repack approach places overhead on pixel access.

Worse still, the code has become less clear. To encapsulate these operations, the AWT provides a series of methods that reside in an instance of a *ColorModel* class. These accessor methods add even more overhead to pixel access.

The astute reader will probably conclude that we have missed something. Java has a scalar data type called a *byte*. Why not unpack all the packed ints into bytes and then just use bytes? Been there, done that! Java does not have an *unsigned* byte! Thus, when two bytes are combined, they must both be checked for their sign. This adds even more overhead than unpacking and repacking.

In addition to the problem of access of color pixels, in Java, we find that Java image processing programs typically use a one-dimensional array to store pixel values. This is done, in part, because the Java language specification does not mention if two-dimensional arrays are stored in row-major or column-major order. This is a deep problem in Java. To understand why this is such a problem, we must understand how arrays are stored in memory.

3.2. Scalar Numeric Data Types in Java

Java has a number of scalar numeric data types. A scalar data type, in Java, is stored in a variable that is passed by value. This means that a copy of the data type is made during the invocation of a method. The scalar numeric data types of Java are *byte, char, short, int, long, float* or *double*.

A *byte* is an 8-bit signed quantity. *Char* is a 16 bit, unsigned quantity. *Short* is a 16-bit signed quantity. *Int* is a 32-bit signed quantity. *Long* is a 64-bit signed quantity.

The floating point scalar data types of Java are *float* and *double*. The float is a signed 32-bit IEEE-1985 floating point quantity. The double is a signed 64-bit IEEE-1985 floating point quantity [IEEE]. Java lacks an unsigned 8-bit quantity. This is a design choice that was made without any discernible justification. An 8-bit unsigned quantity is important to have when storing unpacked image data, as we shall see.

In summary, when scalar data types are passed as arguments to a method, they are passed by *value*. Scalar data types are discussed more fully in [Lyon and Rao].

Experience is a hard teacher
because she gives the test first,
the lesson afterward.

– Anon.

3.3. Arrays in Java

In Java, an array instance is of reference type. This means that a *reference* (32-bit address) is passed when an array name is used as an argument to a method or as a method return. Compare this with scalar data types in Java, such as *byte, char, short, int, long, float* or *double*.

An instance of an array has a member variable called *length*. The *length* member is read-only and yields the number of elements in the array. Arrays are numbered starting at zero and end at element number *length* - 1. Access beyond the end of the array throws an exception.

Every element in an array must hold the same data type. Array length is fixed and, memory permitting, there is no limit to the number of dimensions or size of an array. For example:

```
int a[] = new [100];
for (int i = 0; i < a.length; i++)
   a[i] = i;
```

An array may be filled with constants at compile time. A one-dimensional array looks like this:

```
int b[] = {1, 2, 3, 4};
```

A two-dimensional array may be initialized like this:

```
double mask[][] = {
                      {0.25, 0.25},
                      {0.25, 0.25}};
```

To pass the *mask* array, shown above, we invoke a method called *convolution* using the reference to the *mask* array:

```
convolve(mask);
```

An array is a reference data type that is indexed by *short, byte, char* or *int* type values. When values such as *short, byte,* and *char* are used, they are subjected to unary numeric promotion to the *int* type. It is a compile-time error to use *long* type variables to index an array. You may make an array of any type you like, even an array of *char*. However, an array of *char* is not a *String*.

Multi-dimensional arrays can have dimensions of different size. To make a 2x3x4 of type *int,* use

```
int C[][][] = new int[2][3][4];
System.out.println(C.length+", "
 + C[0].length+", "+C[0][0].length);
```

Which outputs:

```
2, 3, 4
```

To print all the elements in an array;

```
public static void print(double  a[][]) {
    for (int i = 0; i < a.length; i++) {
          for (int j = 0; j < a[0].length; j++)
            System.out.print(a[i][j]+" ");
          System.out.println();
    }
}
```

The number of elements in each dimension of the array may be found with the *length* member. The number of dimensions, however, can not be found so easily. Typically, a method is overloaded to take arrays with different numbers of dimensions.

Consider the following code:

```
public static Mat3 shear(Point2d sh) {
 double m[][] = new double[3][3];
 m[0][0] = 1;
 m[1][1] = 1;
 m[2][2] = 1;
 m[0][1] = sh.x;
```

```
    m[1][0] = sh.y;
    return new Mat3(m);
   }
```

The *m* array is a two-dimensional array that has three rows and three values in each row. The initial value for each of the elements in an array is defined as zero.

Unlike C, there is no way to know if Java stores arrays in *row major order* or *column major order*. This is a major problem for image processing programmers. For a two-dimensional array, this means that we cannot know which index to vary more quickly. If arrays are column major then increment the column more quickly:

```
    public void print() {
            for (int i = 0; i < 3; i++) {
                 for (int j = 0; j < 3; j++)
                   System.out.print(a[i][j]+" ");
                 System.out.println();
            }
    }
```

If arrays are row major, then increment the row more quickly:

```
    public void print() {
            for (int j = 0; j < 3; j++) {
                 for (int i = 0; i < 3; i++)
                   System.out.print(a[i][j]+" ");
                 System.out.println();
            }
    }
```

Memory in the computer is arranged in a linear, one-dimensional array. If we access the memory using a two-dimensional array and are not efficient, we will jump around in memory, as shown in Fig. 3-3.

0,0	0,1	0,2
1,0	1,1	1,2
2,0	2,1	2,2

0,0	0,1	0,2	1,0	1,1	1,2	2,0	2,1	2,2	Row Major
0,0	1,0	2,0	0,1	1,1	2,1	0,2	1,2	2,2	Column Major

Fig. 3-3. Row Major vs. Column Major Order

If an array is stored in row major order but accessed in column major order the memory is being accessed out of order. If the array is stored completely in RAM, this will not generally degrade speed. However, a large data structure, such as an image, may fit in main memory. When a data structure does not fit in main memory, it must be stored in a mixture of main memory and *virtual* memory.

Virtual memory maps a single processor's address space into two parts; main memory and auxiliary memory. If data is stored in auxiliary memory, then it must be swapped into main memory before it is accessed by the processor. When an array is stored in virtual memory, it is possible that indexing from pixel to pixel will cause the auxiliary memory to be accessed.

Modern computers typically use RAM for main memory and disk for auxiliary memory. If virtual memory is accessed using non-sequential addresses, then different parts of RAM can be continually swapped into and out of the disk. This is called *thrashing* and leads to a performance degradation. RAM is roughly a million times faster than disk storage and so the degradation can cripple a program's performance. Thus optimal virtual memory operations occur when algorithms access data using closely clustered addresses [Ralston].

As a result, it is typical, in Java, to *simulate* a two-dimensional array with a one-dimensional array. Suppose that *pels* is an *int* array of dimension *width x height*. Then a method to access a pixel by its location might look like:

```
public int getPixel(int x, int y) {
   return pels[y*width + x] ;
}
```

The *getPixel* method seems to solve the problem. We now know that the x-coordinate should be indexed more rapidly than the y-coordinate. But upon closer examination, we see that the cost is that of an extra multiplication and an addition every time a pixel is accessed.

Recall that an image can have millions of pixels. To access their colors will now require millions of unpacks, packs, multiplies and adds.

We address both the packing and array problems with a change in assumptions. We have decided to optimize for speed rather than for space. We have also assumed that an image is stored in RAM, not virtual memory. Finally, our design goal is to improve the readability of our code. The assumption that virtual memory is not required to store an image comes from the advent of inexpensive RAM chips that are making their way into computers at a rapid pace. A class that is implemented around these assumptions is called the *ImageFrame* and is shown in the next section.

3.4. Drawing in Java

In this section we provide a brief summary of the Java AWT and some of its key services. For a complete tutorial on this topic, see [Campione and Walwrath]. The AWT in Java provides several services:

- Drawing in 2-D
- Paint, repaint, and update
- Image display

The *paint*, *repaint* and *update* methods are known as *Component* class methods. The Component class resides in the *java.awt* package and is the base class of all classes in the AWT. The *repaint* method has 3 common forms:

```
public void repaint();
public void repaint(long t);
public void repaint(int x, int y, int w, int h);
public void repaint(long t, int x, int y, int w, int h);
```

An invocation of *repaint* with an argument for time, *t*, will schedule an invocation of the *update* method within *t* milliseconds. Update is a method that repaints a component by invoking the *paint* method. The *paint* method is called asynchronously upon initialization of a class or if the screen display is damaged. A sketch depicting the methods and their relationship is shown in Fig. 3-4.

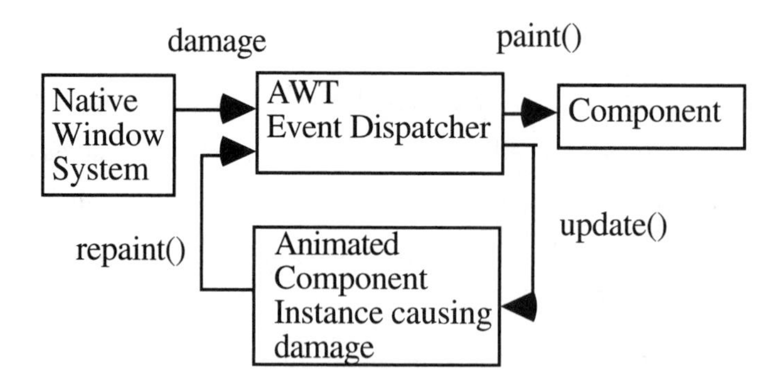

Fig. 3-4. Relationship between Drawing Methods.

It is typically up to the window system to invoke the *paint* method. An exception to this rule will be seen later, in the *GrabFrame* class. The GrabFrame class invokes the *paint method* in order to draw on an off-screen buffer. In general, the programmer will not invoke the *update* method either. The *update* method is invoked by the AWT event dispatcher. Typically the programmer schedules *update* and *paint* invocations with an invocation of the *repaint* method.

3.5. The ImageFrame Class

The *ImageFrame* is a class that resides in the *gui* package. The ImageFrame class is used as a base class that provides image open and display services. These services isolate the programmer from some of the AWT calls. The isolation makes image processing code look more like textbook image processing programs. The service includes access to images stored in 2-D arrays and optimizes for speed of execution, trading off for space.

The ImageFrame base class is the base class for most of the image processing operations in this book. It uses 6 bytes for each pixel, rather than the packed 4-byte pixels of the AWT. This has the net effect of providing a 150% increase in required

RAM. In trade, the pixels may be accessed without the multiply and add of a simulated array and without the 4 shifts and 6 masking operations needed to pack and unpack each pixel. It has also been shown, as a result of the space-time trade-offs, that virtual memory can slow the resulting code considerably. On the other hand, unpacking of a pixel occurs only once, when an image is read in. Repacking occurs only when an image is to be displayed. Finally, arrays are indexed without using the multiply-add combinations that have appeared so frequently in Java literature [Lyon and Rao] [Espeset] [Deitel and Deitel].

3.5.1. Class Summary

```
public class ImageFrame extends ShortCutFrame {
  public  short r[][];
  public  short g[][];
  public  short b[][];

  public int width = 128;
  public int height = 128;

  public ImageFrame(String title)
  public void setImage(Image i)
  public Image getImage()
  public void short2Image()
  public void image2Short()
  public void openGif()
  public void openGif(String fn)
  public static void main(String args[])
}
```

3.5.2. Class Usage

The ImageFrame class has a main method that may be invoked for testing:

```
public static void main(String args[]) {
        ImageFrame imgFrm =
            new ImageFrame("Image Frame");
}
```

To invoke this method, set *gui.ImageFrame* as the class of execution. Also note that the constructor of the *ImageFrame* requires a title string. Recall that the *ImageFrame* extends the *ShortCutFrame* as shown below:

```
public class ImageFrame extends ShortCutFrame {
```

This has the effect of permitting embedded keyboard shortcuts in the ImageFrame and its subclasses. The policy of the ImageFrame is to allow designers of subclasses access to the internal data structures that are used to represent a pixel, without having to pack and unpack the pixel for each access. Towards this end, there are 3 arrays provided (for red, green and blue color planes). Each array is 2-D:

```
public  short r[][];
public  short g[][];
public  short b[][];
```

One service provided by the ImageFrame class is that it tracks an image and waits for it to be loaded. Further, the ImageFrame keeps parameters, such as the width and height instance variables, consistent:

```
public int width = 128;
public int height = 128;
```

The programmer is advised to alter these values consistently with respect to the *r, g and b* arrays. The width and height are set automatically upon reading a new image file.

The ImageFrame takes a constructor that requires a String argument as a title:

```
public ImageFrame(String title)
```

If the title string should be null, then the ImageFrame instance will have an untitled frame.

The ImageFrame contains a private copy of an instance of a class known as the *java.awt.Image*. Accessor methods are provided for flexibility, but, generally speaking, they should not be needed. Programmers are advised to use these methods to alter the internal instances. The first accessor method has the following prototype:

```
public void setImage(Image i)
```

This has the effect of setting the height and width, resizing the frame, displaying the image, and maintaining consistency with the *r*, *g* and *b* arrays.

The second accessor method has the following prototype:

```
public Image getImage()
```

It is used to return the internal representation of the *Image* class instance.

The primary methods to be invoked by programmers of subclasses of the *ImageFrame* are *short2Image* and *image2Short*. The method *short2Image* takes the 3 arrays of *short* (*r*, *b* and *b*) and creates an instance of a *java.awt.Image* class. In addition, the method performs the service of updating the display. It should be invoked only after the image processing operation is done. There is a long packing and memory allocation process involved with *short2Image*. The prototype follows:

```
public void short2Image()
```

The method *image2Short* takes the private instance of the *java.awt.Image* class and creates 3 arrays of *short*, *r*, *g*, and *b*. In addition, it performs memory allocation process and updates the width and height data structures. The prototype follows:

```
public void image2Short()
```

The method *openGif* creates an file-open dialog box that is used to select a file. The only files supported at this level are of type JPEG and GIF. These image file types, and others, will be discussed in a later chapter. The prototype follows:

```
public void openGif()
```

In the case that the programmer already has a string representation of a full path name, the *openGif* method has been overloaded to take a String instance. It does not create an open-file dialog, but is, in all other respects, identical to the previous form of the *openGif* method. The prototype follows:

```
public void openGif(String fn)
```

One final point about the ImageFrame class, useful to those speeding development, is the default file name. A private instance variable, *fileName*, may be set to a hard coded path name. When this occurs, the image will be loaded and displayed by default. This instance variable is normally set to *null*, causing an open-file dialog to appear upon class instantiation.

```
21.     // A default file name..set to null
22.     // to start with file open dialog.
23.     // Set to string to start with an image.
24.     // Use a fully qualified path name, in quotes.
25.     private String fileName = null;
```

A full source listing of the *ImageFrame* class appears on the CD-ROM and in the following section. There are several classes in the ImageFrame class that are beyond the scope of this book. See [Lyon and Rao] or [Chan and Lee] for more details on their use.

As shown in Fig. 3-1, the ImageFrame class has a hierarchic menu called the *File* menu. A sub menu of the File menu is the Open menu. Two items on the open menu are the *getGif_mi* and the *revert_mi.* In addition, the ImageFrame establishes a menu bar. A snippet of code follows:

```
...
27.     private MenuBar menuBar = new MenuBar();
28.     Menu fileMenu = new Menu("File");
29.
30.     Menu openMenu = new Menu("Open");
31.     MenuItem openGif_mi = addMenuItem(openMenu,"[g]if...");
32.     MenuItem revert_mi =  addMenuItem(openMenu,"[r]evert");
```

These interface items are added during the instantiation phase, and a snippet for performing this task appears below:

```
50.      fileMenu.add(openMenu);
51.      menuBar.add(fileMenu);
52.      setMenuBar(menuBar);
```

We note that the order is important and that the *fileMenu* shown on line 28 is left with default visibility so that subclasses may add items to it in the future. Full source for the *ImageFrame* class is on the CD-ROM .

3.6. The FilterFrame Class

The *FilterFrame* class extends the ImageFrame class and resides in the *gui* package. It inherits the abilities to open, process and display images. The FilterFrame class provides the service of converting a color image into a gray scale image.

The main purpose of the FilterFrame is to provide an example of how to use the ImageFrame subclass.

3.6.1. Class Summary

```
class FilterFrame extends ImageFrame {
    public void gray()
    public FilterFrame(String title)
    public static void main(String args[])
}
```

3.6.2. Class Usage

As does the *ImageFrame*, the *FilterFrame* has a main method that permits testing and invocation:

```
public static void main(String args[]) {
  FilterFrame imgFrm =
      new FilterFrame("FilterFrame");
}
```

The *FilterFrame* constructor takes a string argument. This argument is optional and will be used to title the frame, if present. The prototype for the constructor follows:

```
public FilterFrame(String title)
```

The *FilterFrame* also has a method called *gray*. The purpose of this method is to set each pixel to the average of the color planes.

```
public void gray()
```

3.6.3. Class Implementation

In this section we describe the details of the implementation of the *gray* method. This is of interest to the reader who intends to subclass the FilterFrame and write their own image processing methods.

To write an extension of the ImageFrame, it is typical to have the class reside in the same package as the ImageFrame (gui):

```
5.     package gui;
6.     import java.awt.*;
7.     import java.awt.image.*;
```

The filter frame establishes a hierarchic menu structure. This is accessible from sub-classes (as we shall see in later sections). It is typical practice, as mentioned in Chapter 2, to incorporate keyboard shortcuts directly into the menu items:

```
11.      Menu filterMenu = new Menu("Filter");
12.
13.      Menu rgbMenu = new Menu("RGBto");
14.      MenuItem gray_mi =
15.      addMenuItem(rgbMenu,"[E-g]ray");
16.
```

The event handler inherits the *match* method from the *keyBoardShortCut* frame. This makes the event matching and keyboard short-cut detection trivial:

```
public void actionPerformed(ActionEvent e) {
   if (match(e,gray_mi)) {
         gray();
         return;
   }
   super.actionPerformed(e);
}
```

The heart of the FilterFrame, and the main point of this section, is the implementation ease of the *gray* method. Since the ImageFrame has provided 2-D data structures for representing the pixels, we take the following approach:

1. Write a double-nested for-loop to index into all the pixels
2. Modify the pixels, as needed.
3. Display the pixels.

We note that step 3 is required only after all image processing is complete. The act of display requires a packing and copying of data-structures. It is a process that requires some time. Here is an implementation of the *gray* method:

```
26.      public void gray() {
27.      for (int x=0; x < width; x++)
28.       for (int y=0; y < height; y++) {
29.            r[x][y] = (short)
30.               ((r[x][y] + g[x][y]  + b[x][y]) / 3);
31.                 g[x][y] = r[x][y];
32.                 b[x][y] = r[x][y];
33.       }
34.       short2Image();
35.      }
```

The display step is performed in line 34. Also note that on line 29 the result of the division on line 30 must be cast to *short*. This is because Java promotes arithmetic integer type expressions to *int* and floating point type expressions to *double.* Full source for *FilterFrame.java* is on the CD-ROM .

3.7. The GrabFrame Class

The *GrabFrame* class resides in the *gui* package and extends the FilterFrame. The *GrabFrame* class provides the service of taking an instance of a Frame (or any provides Container class that implements its *paint* method) and copying all the pixels that appear on the screen into the internal arrays of the *ImageFrame.*

Existing programs that are able to draw to the screen are typically started from within the GrabFrame, or its subclass. The method, *grab,* is used to extract the pixel data from the draw program (typically an *Applet* or *Frame* subclass). As we shall show in the following sections, existing code may be launched and the image resized and "grabbed" under user control.

3.7.1. Class Summary

```
public class GrabFrame extends FilterFrame {
   public void grab(Container f)
   public void testPattern()
```

```
    public static void main(String args[])
    GrabFrame(String title)
}
```

3.7.2. Class Usage

The GrabFrame class has a main method that illustrates an example of usage:

```
46.     public static void main(String args[]) {
47.             GrabFrame gf = new GrabFrame("Grab Frame");
48.             gf.testPattern();
49.     }
```

The GrabFrame constructor takes a string argument that is used for the title of the frame. The testPattern method starts a class called *SnellWlx*. The SnellWlx test pattern class creates its own frame and menu system. The user may resize the frame and interact with the menu until the display is satisfactory.

```
14.     public void testPattern() {
15.      sw = new SnellWlx();
16.      sw.init();
17.      sw.start();
18.      sw.resize(256,256);
19.     }
```

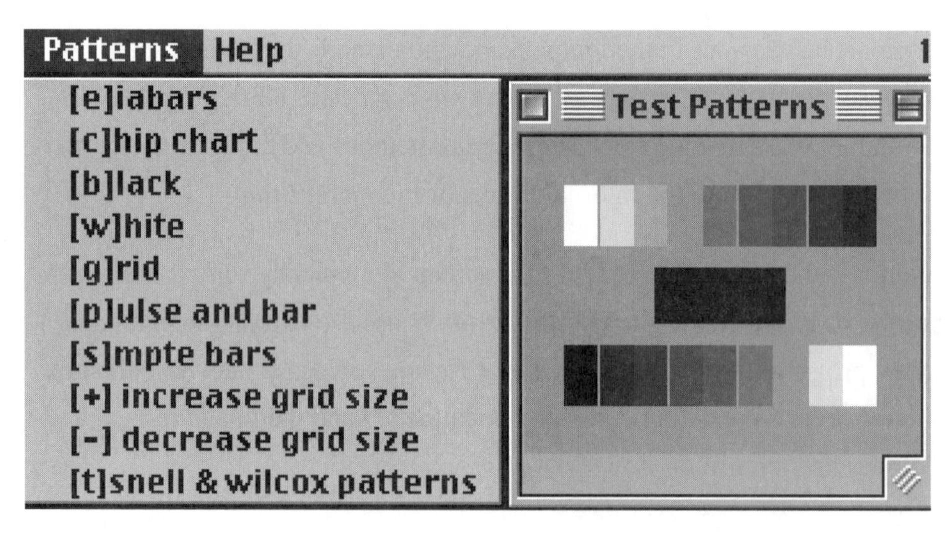

Fig. 3-5. The SnellWlx Frame

We note that the menu items use single key-stroke short-cut. This is due, in part, to the few choices presented to the user. When a new frame is created in this way, the menus and event handling are totally independent of the invoking frame.

3.7.3. Class Implementation

The menus added by the GrabFrame reflect the services provided in this substrate. There are only two menu items added to the main menu bar. Both menu items are added to the *rgbMenu* of the FilterFrame. The code for this follows:

```
8.     private  MenuItem testPattern_mi =
9.               addMenuItem(rgbMenu, "[T]est Patterns");
10.    private  MenuItem grab_mi =
11.               addMenuItem(rgbMenu, "[E-G]rab Patterns");
```

The effect of lines 8-11 is shown in Fig. 3-6.

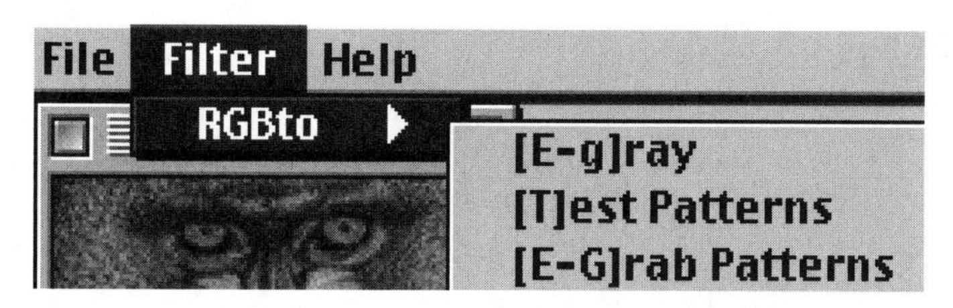

Fig. 3-6. The GrabFrame MenuBar

A *parent* frame is a frame that makes an instance of a new frame. A *child* frame is the new frame instance made by the parent. Note that the relationship between parent and child is different from that of base class and sub-class. The child does not extend the parent, it is only *created* by the parent. Perhaps the most interesting implementation note on this GrabFrame class is how it is able to grab the image data from a child frame. This is a multi-step process:

1. Keep a reference to the child frame.
2. Provide a way for the user to create the child frame.
3. Provide a way for the user to grab the child frame.

To follow step 1, we keep a private instance variable that acts as the reference to the child frame:

```
6.      private SnellWlx sw;
```

For steps 2 and 3, we created an interface, as shown in Fig. 3-6. To handle the events that this interface generates, we use

```
public void actionPerformed(ActionEvent e) {
    if (match(e, testPattern_mi)) {
            testPattern();
            return;
    }
    if (match(e, netImageSelector_mi)) {
            netImageSelector();
            return;
    }
    if (match(e, grabTestPattern_mi)) {
            grabTestPattern();
            return;
    }
    super.actionPerformed(e);
}
```

The full source for the *GrabFrame* appears on the CD-ROM .

3.8. SnellWlx Class and Test Patterns

The *SnellWlx* class resides in the *gui* package and extends the *Applet* class. Some books call an instance of a class that extends the Applet class an "Applet". This abuse of nomenclature has become common. The previous section disclosed a technique that enables the *GrabFrame* to grab the display of an instance of a *Container*. A diagram depicting the position of the *SnellWlx* class in the class hierarchy, so far, is shown in Fig. 3-7.

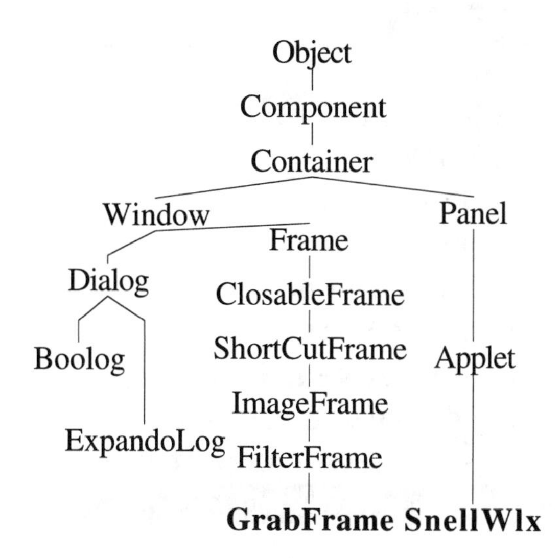

Fig. 3-7. Class Hierarchy for SnellWlx

The following section shows the class summary. The *start* method, as used in the previous section, is inherited from the Applet class. Perhaps the most remarkable thing about the SnellWlx class is that we need know nothing at all about its implementation details. In fact, we can incorporate Container subclasses into the GrabFrame sub-class instances, secure in the knowledge that we need only access the *paint* method.

3.8.1. Class Summary

```
public class SnellWlx extends Applet {
 public void init()
 public void paint(Graphics g)
}
```

3.8.2. Class Implementation

The series of paint and draw methods needed to create test patterns is of no particular interest; however, it is interesting to list and show the test patterns available from the SnellWlx class. The patterns list is given as a part of the menu in Fig. 3-8.

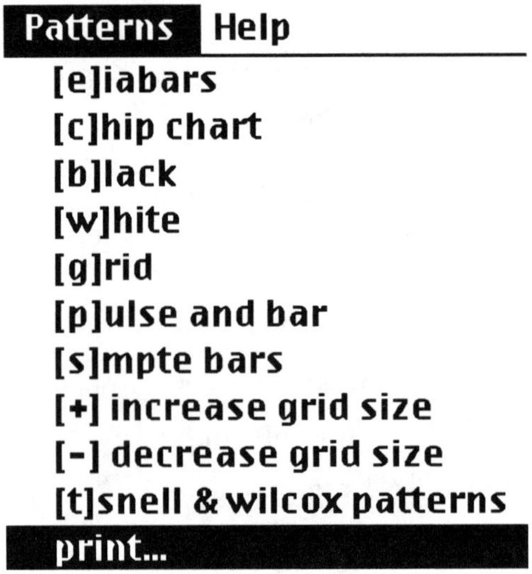

Patterns Help

[e]iabars
[c]hip chart
[b]lack
[w]hite
[g]rid
[p]ulse and bar
[s]mpte bars
[+] increase grid size
[-] decrease grid size
[t]snell & wilcox patterns
print...

Fig. 3-8. Test Pattern List

The test patterns are called the EIA Bars (Electronic Industrial Associates color bars), chip chart, black, white, grid, pulse and bar, SMPTE Bars (Society of Motion Picture and Television Engineers) and a Snell and Wilcox pattern.

4. Homogeneous Point Processing

Technology is dominated by two types of people:
those who understand what
they do not manage,
and those who manage what they do not understand.

– Unknown

Point processing is a kind of image processing that maps an input image into an output image without making any geometric changes. Point processing is the vehicle by which we wrap up our discussion of Java, using examples that perform low-level image processing operations.

We shall introduce several qualitative terms to describe the quality of an image. Nomenclature such as *washed out*, *brighter* and *contrast* are employed without a formal definition. The terms relate to the human visual system, a topic that is presented in Chapter 5. We shall see that the direction of the image processing field is guided, in part, by the human visual system. This is partly due to programmers who influence the development of their algorithms in accordance with a subjective image quality criteria [Stockham].

After an image is digitized, it is sometimes enhanced using a low-level preprocessing technique. The goals of preprocessing typically include noise suppression and enhancement [Hussain]. The class of algorithms known as *histogram transforms* fall's into the category of low-level image processing.

Histograms have also been used as a feature in color image matching and retrieval. This typically works with a metric for computing the distance between two histograms. Several metrics have been proposed for comparing histograms [Mehtre et al.].

Another application of the histogram is in the reduction of the number of values needed to represent an image. This is a topic that has been discussed at length in the literature [Kuo et al.].

The histogram of an image is a statistical measure that (when normalized to vary from zero to one) is also know as the Probability Mass Function (PMF) of an image. The PMF is computed by adding the total number of pixels of a particular value and then dividing by the total number of pixels in the image.

$$PMF = p(V = i) = p_V(i) = \frac{1}{WH} \sum_{x=0}^{W-1} \sum_{y=0}^{H-1} \begin{cases} 1 & if\ v_{xy} = i \\ 0 & if\ v_{xy} \neq i \end{cases} \tag{4.1}$$

Where

$$v_{xy} = \text{value of pixel at location } x, y \tag{4.2}$$

with

$$\begin{aligned} W &= \text{ width of the image, in pixels} \\ H &= \text{ height of the image, in pixels and} \\ p(V = i) &= \text{ probability that a particular value occurs} \end{aligned} \tag{4.3}$$

For example, suppose that the sample space of possible pixel values consists of an integer that ranges from 0 to 255, inclusive. Suppose that, of the 64 pixel values, 32 are 0, 16 are 128, and 16 are 255. Then the histogram for the image is given by Fig. 4-1.

Value	PMF
0	32/64
128	16/64
255	16/64

Fig. 4-1. Histogram of a Simple Image

One attribute of the PMF is that the probabilities will always sum to one.

$$1 = \sum_{i=0}^{K-1} p_V(i) \tag{4.4}$$

Where

$$K = \text{total number of values, typically } K = 256 \tag{4.5}$$

Another attribute is that the range on the PMF is constrained to fall within the interval from zero to one.

The Cumulative Mass Function (CMF) is a statistical measure that provides the probability that the pixel color is of a given value or less:

$$P_V(a) = \sum_{i=0}^{a} p_V(i) = p(V \le a) \tag{4.6}$$

Where

$$a = \text{a particular value such that } a \in [0 \ldots K-1] \tag{4.7}$$

For example, the CMF for the histogram of Fig. 4-1 is shown in Fig. 4-2.

Value	PMF	CMF
0	32/64	0.5
128	16/64	0.75
255	16/64	1

Fig. 4-2. The PMF and the CMF

A histogram modification is a pixel-level transform technique that alters the PMF of the image by making use of the CMF as a look-up table. As we shall see in the following sections, the histogram modification technique describes a broad class of algorithms able to perform a variety of image enhancements.

Histograms also have a role in image sequence analysis. For example, to detect a scene change, intensity histograms have been used to measure the difference between frames [Ahn et al.].

The PMF is also useful to the user, who, with training, can become good at finding relationships between histogram features and image features. The service of drawing the PMF of an image is performed by the *Histogram* class.

4.1. The Histogram Class

The Histogram class resides in the *gui* package. The Histogram class provides the service of computing and displaying the Probability Mass Function (PMF) of an image's intensity. Computation of the histogram involves adding up all pixels of a particular value. This is generally a trivial task for a 256 color image. However, modern color image processing typically uses images with at least 24 bits per pixel. Such pixels have $2^{24} = 16{,}777{,}216 \approx 16$ million colors. Histograms based on the PMF of 16 million colors are typically sparse. The reason is that an image typically has many fewer pixels than the number of colors. For example, a square image that is 4096x4096 will have 2^{24} pixels. Such an image will probably have many fewer than 2^{24} colors. Thus, even for a larger image, the PMF will be sparse.

There are some compact ways to store a sparse PMF, but they are beyond the scope of this discussion [Lyon 95]. The typical approach in implementing a histogram computation is to divide the image into its color components. Also, it is typical to take the color components as having only 256 possible values each. Since this does ease the computational burden and is in keeping with the kind of images that we presently support, we shall abide by this assumption.

Thus, the histogram is partitioned into *K* equal bins (where *K*=256):

```
histArray = new double[256];
for (int i=0; i<width; i++)
   for (int j=0; j < height; j++)
     histArray[plane[i][j] & 0xFF]++;
```

After all the values are stored in the *histArray*, we compute the maximum value and then normalize the histogram so that all values in the *histArray* will sum to one. Thus the probability that a pixel will have a value is stored in the *histArray* is:

```
double max = 0.0;
for(int i=0; i<256; i++)
   if (histArray[i] > max)
      max = histArray[i];
// Normalize
for(int i=0; i<256; i++)
   histArray[i] = (histArray[i] / max);
```

The mandrill, along with its histogram (normalized to vary from zero to one), is shown in Fig. 4-3.

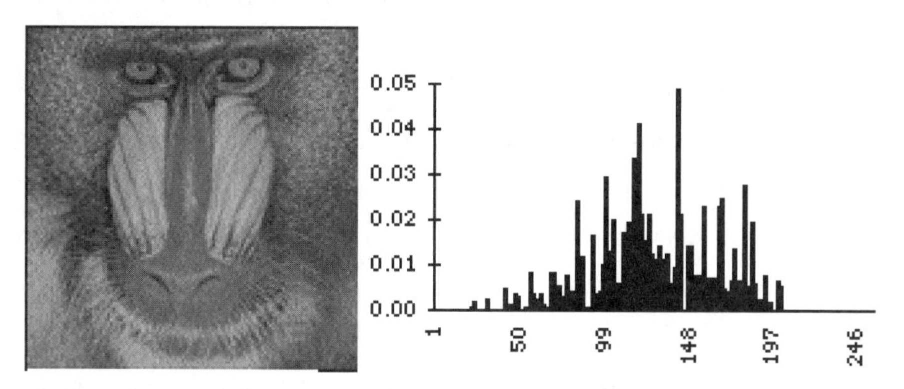

Fig. 4-3. The Mandrill and the Histogram

4.1.1. Class Summary

```
package gui;
public class Histogram extends ClosableFrame {
    public double[] getPMF()
    public double[] getCMF()
    public void printPMF()
    public void show()
    public Histogram(short plane[][], String title)
    public void paint(Graphics g)
 }
```

4.1.2. Class Usage

An example of the use of the *Histogram* class may be found in a class called the *NegateFrame*. The *NegateFrame* extends the *GrabFrame* (see Sect. 3.7) and so inherits the two-dimensional arrays for the red, green and blue color planes. These may be passed into the *Histogram* constructor directly. The following code snippet shows the set-up and invocation of the *Histogram* class:

```
private Histogram rh,gh,bh;
public void histogram() {
 rh = new Histogram(r,"Red");
 gh = new Histogram(g,"Green");
 bh = new Histogram(b,"Blue");
 rh.show();
 gh.show();
 bh.show();
}
```

Since a *Histogram* instance represents a *Frame* it is important to keep the reference to the *Histogram* instance non-local. If, for example, the *Histogram* were declared locally, then the life of the reference variable would be over after the invoking method returned. Once the variable's life is over, the data that the variable refers to could be reclaimed by the garbage collector at any moment. When this occurs, the frame will disappear from view. Thus, a non-deterministic visual behavior may result from local frame reference variables. The method

```
public void printPMF()
```

is used to output the PMF to the console for the purpose of debugging the data. It is also useful when exporting the data to an external plotting program (e.g., Excel).

The methods

```
public double[] getPMF()
public double[] getCMF()
```

return the PMF and CMF. The PMF and CMF are useful statistical measures of the image data, as we will see in the histogram equalization of Section 4.2.2.

To show the Histogram frame, the method:

```
public void show()
```

is used. It is the case that the programmer may not want to show the histogram frame. This is particularly true if all that is wanted is the CMF or PMF. The CMF and PMF are created upon Histogram class construction. The constructor for the Histogram class follows:

```
public Histogram(short plane[][], String title)
```

Note that the Histogram requires a two-dimensional array of *short*. The title string is used to title the window.

4.2. Homogeneous Point Processing Functions

A *homogeneous* point operation is one that alters the pixel's color as a function of its color. Such operations are independent of pixel location and are expressed by

$$v'_{ij} = f(v_{ij}) \qquad (4.9).$$

A point operation is one that alters a pixel's color as a function of both its location and color:

$$v'_{ij} = f_{ij}(v_{ij}) \qquad (4.10)$$

Even when point operation's are inhomogeneous, and (4.10) is in effect, they are always independent of the neighboring pixels.

When (4.9) is used for a color pixel, we assume all three colors (r, g, b) are normalized so that they will range from [0...255]:

$$v_{ij} \in [0...255] \qquad (4.11)$$

Point operations are also useful for filtering, compensating for sensor non-linearities, small intensity amplification, segmentation and coding [Hussain] [Jähne]. Histogram based processing (such as histogram equalization) can improve the brightness and contrast of an image [Myler].

A linear map will result in no change in the image. Such a relation is given by

$$f(v_{ij}) = v_{ij} \qquad (4.12).$$

We will apply an equation of the form of (4.12) to all three colors as they are stored in the 3 two-dimensional arrays of *shorts*, r, g and b.

The original image, with linear map, is shown in Fig. 4-4.

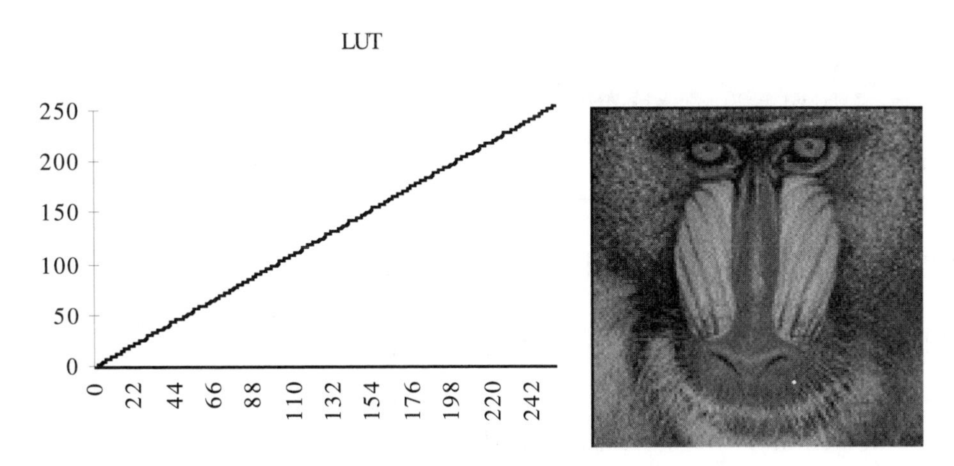

Fig. 4-4. Original Image with Linear Map

The negative image is formed from the input value using

$$f(v_{ij}) = 255 - v_{ij} \tag{4.13}$$

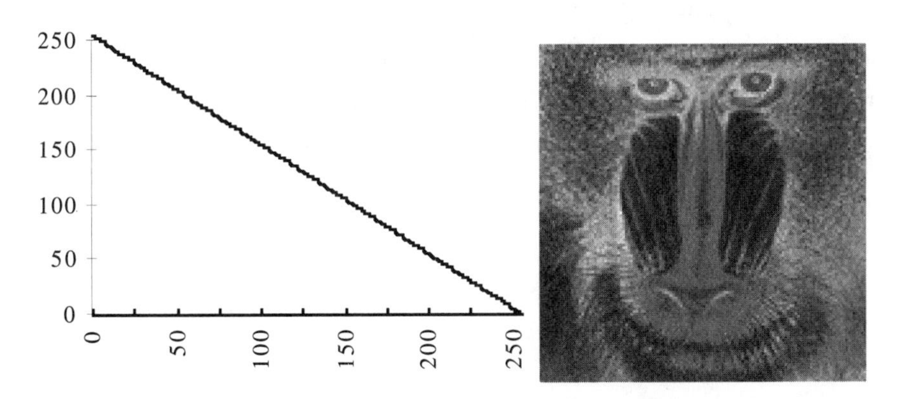

Fig. 4-5. The Negated Image

The code snippet for implementing (4.13) follows:

```
public void negate() {
 for (int x=0; x < width; x++)
```

```
        for (int y=0; y < height; y++) {
                r[x][y] = (short) (255 - r[x][y]);
                g[x][y] = (short) (255 - g[x][y]);
                b[x][y] = (short) (255 - b[x][y]);
        }
      short2Image();
    }
```

4.2.1. Using the Pow Function to Brighten or Darken

One of the techniques used in the negate method is to *cast* the result of the computation to *short* integer. The default integer data type for Java is *int*. The *int* is a signed, 32-bit quantity. The *short* is a signed 16-bit quantity. Thus, explicit casting from the *int* to the *short* is required, or a compilation error results.

A power function may be described in terms of a non-normalized coordinate system, assuming that the function f ranges from 0...255:

$$f(v_{ij}) = 255\left(\frac{v_{ij}}{255}\right)^{pow} \tag{4.14}$$

For $pow < 1$, a brightening results. In the case where $pow = 0.9$, for example, the curve and resulting brightening are subtle. The result of applying (4.14) with $pow=0.9$ is to brighten the image slightly. Repeated application makes the image appear brighter, but it also results in a loss of contrast. The plot of (4.14) and the result of its repeated application are shown in Fig. 4-6.

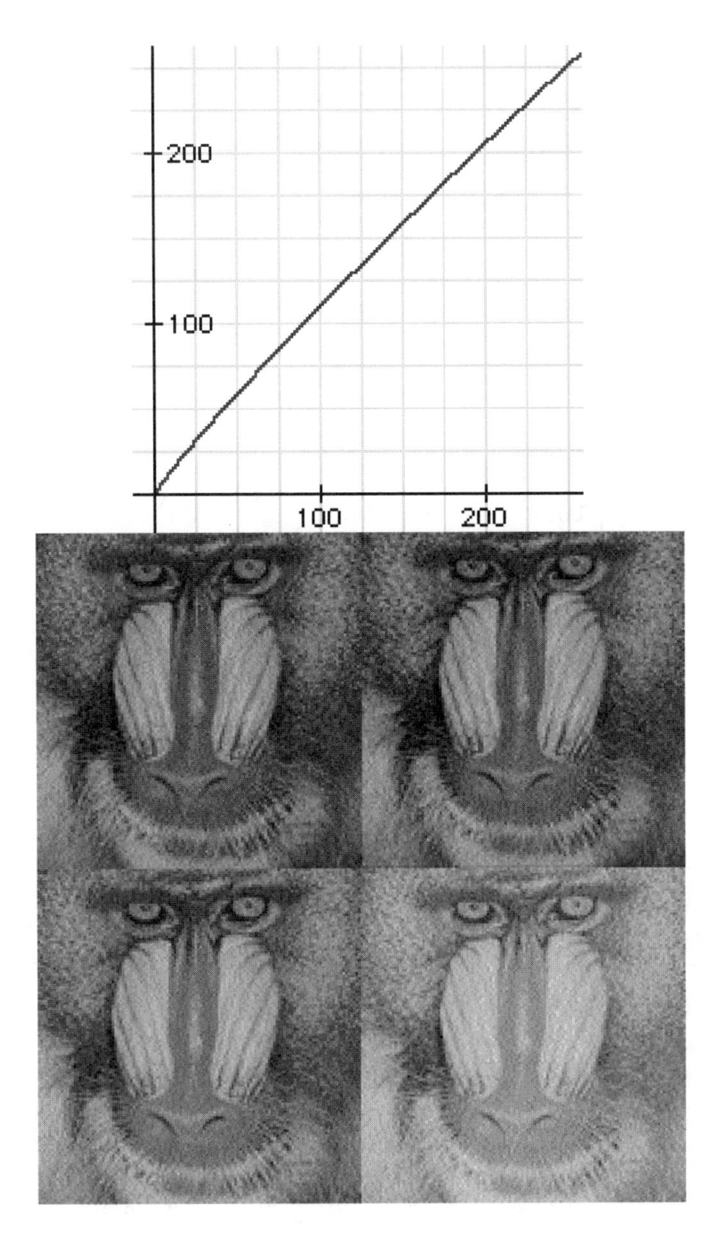

Fig. 4-6. Repeated Application of (4.14)

We reapply equation (4.14) to the original image in the upper left corner of Fig. 4-6 for seven times. The intermediate and final results are shown, with the most washed out (but brightest image) shown in the lower right.

When *pow* > 1, a darkening of the image occurs. Fig. 4-7 shows a graph depicting (4.14) with *pow*=1.5.

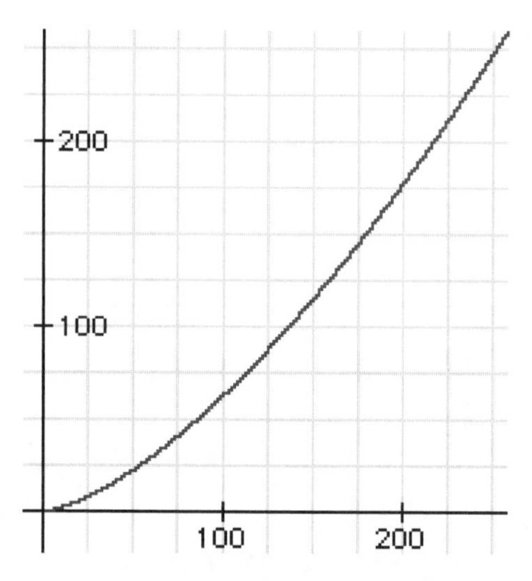

Fig. 4-7. Pow = 1.5, a Darkening Curve

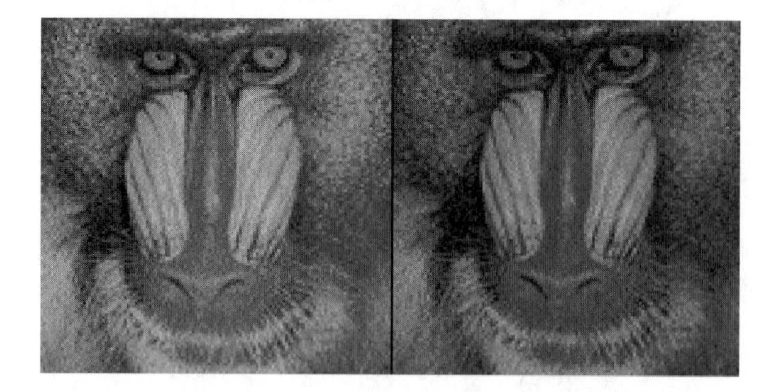

Fig. 4-8. Image after Application of Darkening Curve

The effect of the darkening curve shown in Fig. 4-7 appears in Fig. 4-8.

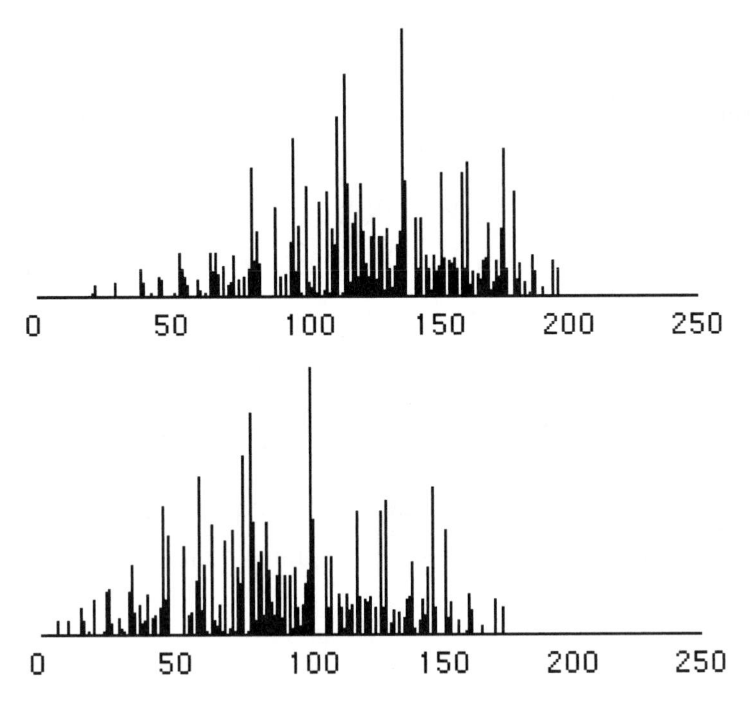

Fig. 4-9. Histogram Before and After Darkening

One of the attributes of the darkened image is that its histogram has shifted to the left and the distance between the histogram peaks has decreased. This is aesthetically pleasing because of the logarithmic response of the human visual system, a topic discussed in Chapter 5. The following code snippet, taken from the NegateFrame, illustrates how an image may be raised to a power, in Java:

```
public void powImage(double p) {
  for (int x=0; x < width; x++)
   for (int y=0; y < height; y++) {
      r[x][y] = (short)
                (255 * Math.pow((r[x][y]/255.0),p));
      g[x][y] = (short)
                (255 * Math.pow((g[x][y]/255.0),p));
      b[x][y] = (short)
                (255 * Math.pow((b[x][y]/255.0),p));
        }
        short2Image();
}
```

4.2.2. Using Linear Transforms to Alter Brightness and Contrast

In this section we present the techniques used to perform parametric linear brightness and contrast adjustments. Typically, interactive programs permit a user to enter a piece wise-linear transformation that specifies

$$v_{ij}' := f_{ij}(v_{ij}) \quad (4.12).$$

The formula is called a linear gray level scaling and is typically of the form

$$v_{ij}' = cv_{ij} + b \quad (4.15)$$

Where

$c =$ contrast
$b =$ brightness

Typically, the higher the value of the slope, the higher the contrast. Also, when the range on (4.15) falls out of range, the f function will clip against the minimum and maximum values. For example, (4.15) is plotted with $c = 2, b = -90$ in Fig. 4-10.

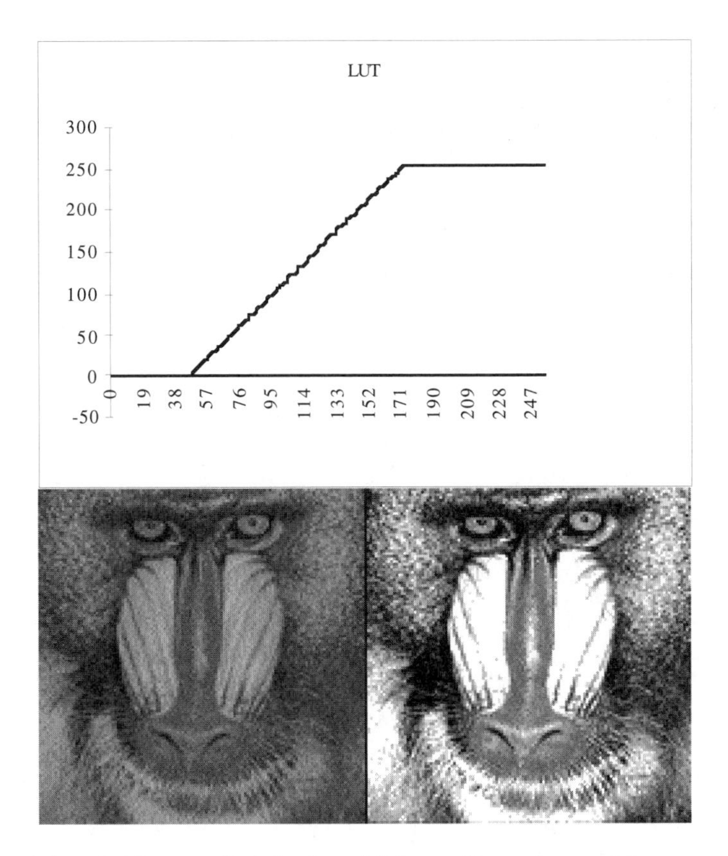

Fig. 4-10. Original and Linear Transformed Image

The values for c and b can be computed automatically for an image with a dynamic range that is exceeded by the dynamic range of the display. Suppose the dynamic range for the display is

$$\Delta D = D_{\max} - D_{\min} \tag{4.16}$$

where

$D_{\min}$ = minumum value that can be displayed and

$D_{\max}$ = maximum value that can be displayed

Let the dynamic range for the image be given by

$$\Delta V = V_{\max} - V_{\min} \tag{4.17}$$

where

V_{min} = minumum value in the image and

V_{max} = maximum value in the image

Then, to scale an image whose dynamic range is positive and smaller than that of the dynamic range of the display, i.e., $0 < \Delta V < \Delta D$, we use

$$v_{ij}' = cv_{ij} + b \tag{4.17a}$$

Where

$$c = \frac{\Delta D}{\Delta V} \tag{4.17b}$$

and

$$b = \frac{D_{min}V_{max} - D_{max}V_{min}}{\Delta V} \tag{4.17c}$$

Substituting (4.17c) and (4.17b) into (4.17a) yields

$$v_{ij}' = \frac{\Delta D}{\Delta V}v_{ij} + \frac{D_{min}V_{max} - D_{max}V_{min}}{\Delta V} \quad (4.18).$$

For example, suppose that $D \in [0...255]$ so that $D_{min} = 0$ and $D_{max} = 255$. Then (4.18) becomes:

$$v_{ij}' = \frac{255}{\Delta V}v_{ij} - \frac{255V_{min}}{\Delta V} \quad (4.19).$$

Also, if $v_{ij} \in [0...255]$ so that $V_{min} = 10$ and $V_{max} = 90$, then (4.19) becomes:

$$v_{ij}' = \frac{255}{80}v_{ij} - \frac{2550}{80} \quad (4.20).$$

A plot of (4.20) is shown in Fig. 4-11.

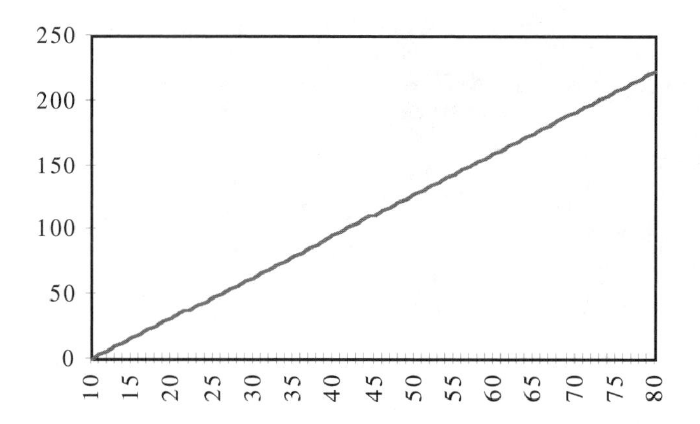

Fig. 4-11. Plot of Rescaled Range.

Fig. 4-12 shows a dialog for altering the contrast and brightness. The default values shown are computed using (4.18) and automatically appear in the text fields.

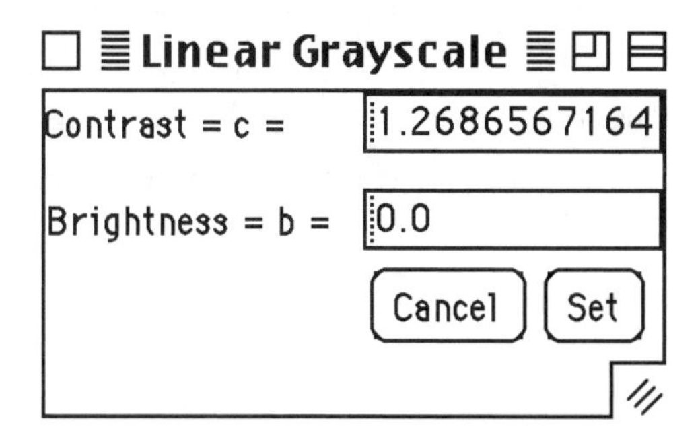

Fig. 4-12. Linear Grayscale Dialog Box with Defaults

The effect of applying the linear transform to the mandrill image is shown in Fig. 4-13. Also shown are the effects on the histogram of the image. Note how the histogram has been spread out to the maximum value displayable. The scaling operation appears to improve the general appearance of the image by making better use of the dynamic range of the display.

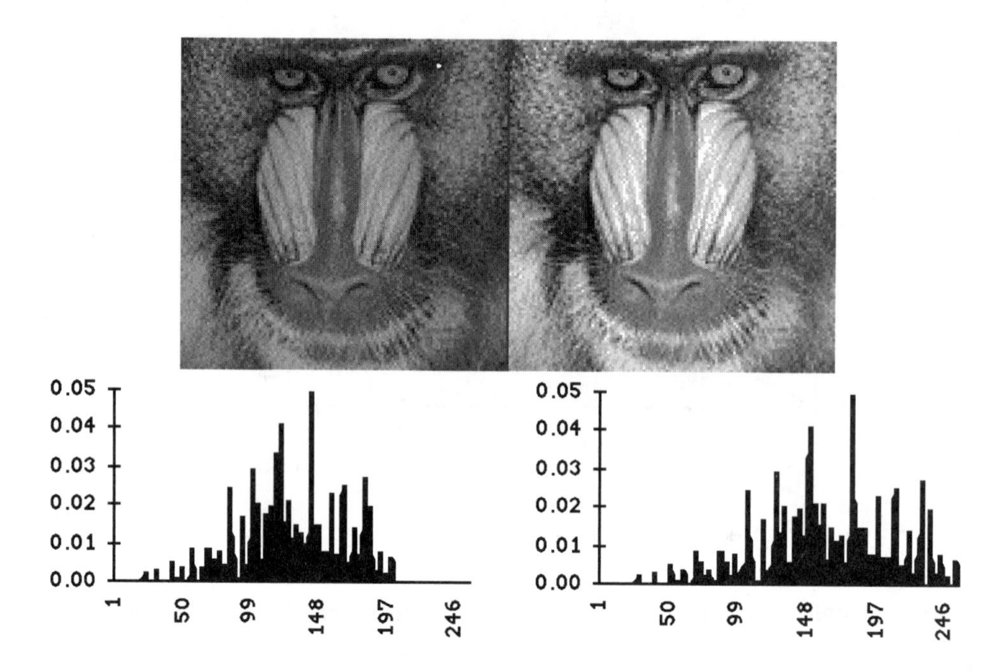

Fig. 4-13. Automatic Linear Contrast Enhancement

The linear transform is implemented in the code given below, note that a look-up table is computed and that this look-up table (called *lut*) is used to speed the application of the transform.

```
public short linearMap(short v,
        double c, double b) {
    // scale gray value to 0..1 range
    double f = c * v + b;
    // scale f into 0..255 range
    // clip f into range
    if (f > 255) f = 255;
    if (f < 0) f = 0;
    return (short) f;
}

public void linearTransform(double c, double br) {

  short lut[] = tt.getLut();
  for (short i = 0; i < 256; i++)
      lut[i] = linearMap(i,c,br);
  for (int x = 0; x < width; x++)
       for (int y = 0; y < height; y++) {
         r[x][y] = lut[r[x][y]];
         g[x][y] = lut[g[x][y]];
         b[x][y] = lut[b[x][y]];
       }
       short2Image();
}
```

The *tt* reference is an instance of a class known as the *TransformTable.* The TransformTable is described in Section 4.3.

4.2.3. The Uniform Non-Adaptive Histogram Equalization

The last section showed that low-contrast images result when pixel values are not far enough apart. Thus, the clustering of pixel values can hide detail from the user. Histogram modification is another technique for improving the contrast of an image [Hall 1974].

In this section we disclose one of the most commonly use techniques for the histogram modification of an image, the uniform non-adaptive histogram equalization (UNAHE). Some literature will describe the UNAHE as a "histogram equalization" but, as we will shortly see, histogram equalization is a term that can be used to describe a class of algorithms. The idea behind the UNAHE is to create a uniform

PMF from an image that has a non-uniform PMF. This is done using a scaled version of the CMF as a look-up table for the original image.

Recall that the CMF is given by

$$P_V(a) = \sum_{i=0}^{a} p_V(i) = p(V \le a) \tag{4.21}$$

Where

$$a = \text{a particular value such that } a \in [0 \ldots K-1] \tag{4.22}$$

The goal is to find a function,

$$v_{ij}^{'} = f(v_{ij}) \quad (4.23).$$

that has a uniform PMF. By setting the transfer function in (4.23) to the scaled CMF we obtain

$$v_{ij}^{'} = f(v_{ij}) = \Delta V P_V(v_{ij}) \tag{4.24}$$

We note that in the interval from $\left[V_{\min}, V_{\max}\right]$ that $f(v_{ij})$ is both single values and strictly monotonic. Further that $V_{\min}^{'} \le f(v_{ij}) \le V_{\max}^{'}$ for $V_{\min} \le v_{ij} \le V_{\max}$. For a proof that (4.24) yields a uniform distribution, we follow [Woods and Gonzalez].

Proof of (4.24)

In the following proof, we assume that the input and output are continuous values whose ranges are conditioned so that

$$V_{\min} \le v \le V_{\max}, V_{\min}^{'} \le v' \le V_{\max}^{'}$$

holds true. Let *v* be the continuously valued input and *v'* be the continuously valued output. Just as in the discrete case, we let

$$\Delta V = V_{\max} - V_{\min} \tag{4.25}$$

In the continuous domain, the function that corresponds to the PMF is called the PDF (probability density function). The PDF's of the input and output are related by

$$p_{V'}(v') = \left\{ p_V(v) \frac{dv}{dv'} \right\}_{v=f^{-1}(v')} \tag{4.26}$$

We assume that the inverse of the transfer function, $v = f^{-1}(v')$, is available, for the purpose of this proof. A numerical technique using look-up tables is used when we proceed to the discrete domain. Since we want the output PDF to be uniform, we would like

$$p_{V'}(v') = \begin{cases} \dfrac{1}{\Delta V} & \text{for } V_{\min} \le v' \le V_{\max} \\ 0 & \text{otherwise} \end{cases} \tag{4.27}$$

Starting with (4.27), we intend to derive an appropriate transfer function. Cross multiplying (4.26) by dv' we obtain

$$p_V(v)dv = p_{V'}(v')dv' \qquad (4.28).$$

Integrating both sides of (4.28) over the valid input and output ranges yields

$$\int_{V_{\min}}^{v} p_V(h)dh = \int_{V'_{\min}}^{v'} p_{V'}(w)dw \tag{4.29}$$

Where $h$and w are dummy variables of integration.

If the right-hand side of (4.29) is of uniform PDF, then it should have the shape of a rectangle. Every value in the PDF should be a constant equal to $1/\Delta V$. We may therefore solve the right-hand side of (4.29) to obtain

$$\int_{V_{\min}}^{v} p_V(h)dh = \frac{v' - V'_{\min}}{\Delta V} \qquad (4.30).$$

For the purpose of display, set $V'_{\min} = 0$ and solve for the output value:

$$v' = \Delta V \int_{V_{\min}}^{v} p_V(h)dh \tag{4.31}$$

If we have many intensities, then we can approximate the cumulative density function (CDF) of (4.31) with a cumulative mass function, (CMF):

$$v'_{ij} = f(v_{ij}) = \Delta V P_V(v_{ij}) \quad (4.32).$$

The more values we have, the more uniform the resulting PMF. Otherwise, we can expect the result to be somewhat less than uniform [Hall 1974].

Q.E.D.

The UNAHE is implemented using the *unahe* method in the *NegateFrame*:

```
1. // Uniform Non Adaptive Histogram
2. // Equalization
3.       public void unahe() {
4.   short lut[] = tt.getLut();
5.   double h[] = getAverageCMF();
6.   for (short i = 0; i < 256; i++)
7.       lut[i] = (short)(255*h[i]);
8.   applyLut(lut);
9. }
```

The above code snippet makes reference to several methods, including *getAverageCMF*, *getLUT* and *applyLut*. These are explained in more detail at the end of the chapter. Suffice it to say, for now, that the getAverageCMF returns the CMF for the red, green and blue color channels, and that the *applyLut* method transforms the red, green and blue color channels with the *lut* array. When rewriting the histogram technique, only line 7, an implementation of (4.32), is of direct interest. We shall show an automatic technique for deriving line 7 in Section 4.2.4.

The effect of UNAHE on the mandrill image is shown in Fig. 4-14. The original image is on the left, the UNAHE image is on the right.

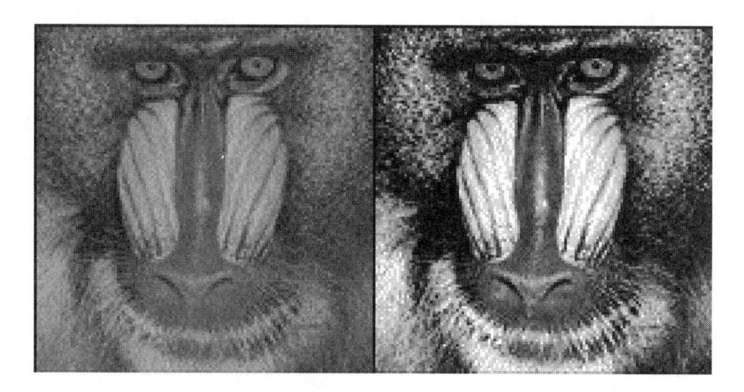

Fig. 4-14. Effect of the UNAHE on the Mandrill

Perhaps of direct interest is the effect which UNAHE has on the histogram of the image. Fig. 4-15 shows that the histogram is spread out and more uniform.

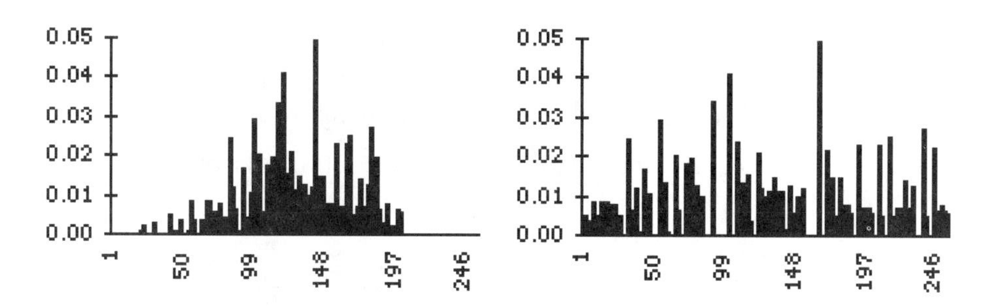

Fig. 4-15. Effect of UNAHE on the Histogram

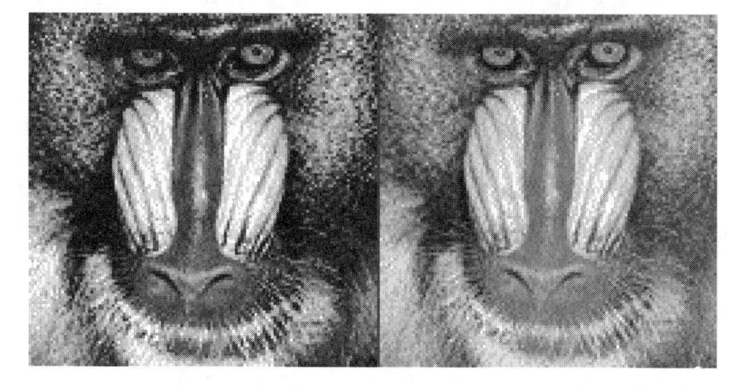

Fig. 4-16. UNAHE vs. Automatic Linear Transform

Fig. 4-16 shows the UNAHE transformed image on the left with the automatic linear transformed image on the right. The UNAHE has a much higher contrast, but it is not clear that it is of higher quality. Such an assessment requires a well-posed quality criteria.

The transform table, based on a scaled version of the CMF, is shown in Fig. 4-17.

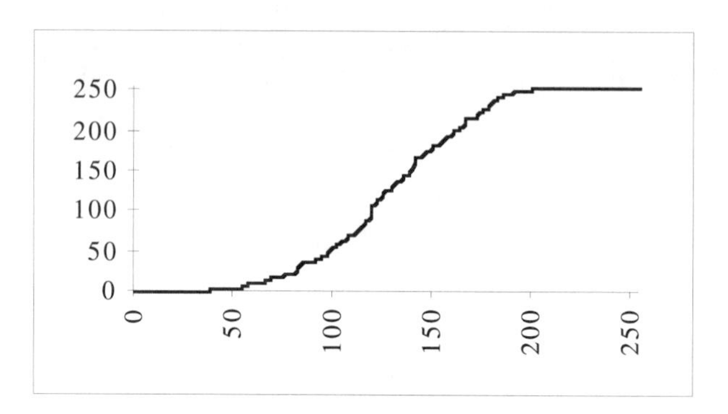

Fig. 4-17. Transform Table of the UNAHE CMF

One strength of the UNAHE technique is that it can enhance low-contrast detail. The weakness of the UNAHE is that it may also increase the contrast of noise [Hall 1974].

4.2.4. Maple and Exponential Non Adaptive Histogram Equalization

Section 4.2.3 used a transfer function that was able to yield a uniform, non-adaptive histogram equalization (UNAHE). This section shows how to extend the technique to yield an arbitrary PDF from an output image by deriving a transform function.

Recall (4.29)

$$\int_{V_{\min}}^{v} p_V(h)dh = \int_{V'_{\min}}^{v'} p_{V'}(w)dw \qquad (4.29).$$

On the left, we noted that we have the CDF of the input image and so we rewrite (4.29) as

$$P_{V'}(v') = \int_{v_{\min}}^{v'} p_{V'}(w)dw \qquad (4.32).$$

Suppose that we would like to specify the PDF of the output image. Once $p_{V'}(v')$ is known, we need only solve the integral in (4.32) to obtain the transfer function. Assuming that the output and input are well-conditioned so that

$$V_{\min} \le v \le V_{\max}, V'_{\min} \le v' \le V'_{\max}$$

Let's specify the PDF of the output so that it has an exponential distribution:

$$p_{V'}(v') = \alpha e^{-\alpha(v'-V_{\min})} \qquad (4.33)$$

By (4.32) we obtain

$$P_{V'}(v') = \int_{v_{\min}}^{v'} \alpha e^{-\alpha(w-V_{\min})}dw \qquad (4.34)$$

So now all we have to do is integrate (4.34) and solve for *v'* to obtain the new transfer function. Actually, (4.34) is a pretty easy integral, but it is really much more fun if we can automate the transfer function derivation process. One technique is to make use of a symbolic manipulation program. There are several excellent programs available for such tasks, but the examples in this book are based in Maple™ [Char et al.].

To revisit the problem of uniform histogram equalization, we present the following Maple output:

```
1. > restart;
2. > readlib(C):
3. > readlib(optimize):
4. > p:=1/(Vmax-Vmin):
5. > lut:=collect(solve(h[i]=int(p,w=Vmin..v),v),h[i]);
6. > lut := (Vmax - Vmin) h[i] + Vmin
7. > C(lut,optimized);
8.      t3 = (Vmax-Vmin)*h[i]+Vmin;
```

Line 1 restarts Maple and clears out local variable storage. Lines 2 and 3 read in libraries for transforming Maple results into optimized C programs. Maple optimization makes use of auxiliary variables to speed computation. These variables are typically declared to be of type double. Line 4 establishes a uniform PDF for which the variables *Vmin* and *Vmax* are used in place of $V'_{\min}$ and $V'_{\max}$. Mathematical notation often suffers at the hands of programmers! The term p is used for the PDF. Line 5 computes the CDF, setting it equal to the computed histogram of the input image (*h[i]*). Line 5 also collects the terms that contains the *h[i]* variable and factors them out. Finally, line 5 sets the result to the *lut* variable. The *lut* represents the look-up table. The symbolic output appears on line 6. Line 7 directs Maple to create optimized C code, the output of which appears on line 8.

Now we apply this technique to the exponential distribution function. Replacing line 4 of the Maple source with the Maple equivalent of (4.33) results in

```
p:=alpha*exp(-alpha*(w-Vmin)):
```

which, when integrated and reformatted for humans, results in

$$lut_i = V'_{\min} - \frac{1}{\alpha}\ln(1 - h_i) \tag{4.35}$$

Based on a minimum value for $V'_{\min}$ of zero, we formulate the following code to implement the *enahe* (Exponential Non Adaptive Histogram Equalization) method:

```
1. public void enahe(double alpha) {
2.    short lut[] = tt.getLut();
3.    double h[] = getAverageCMF();
4.    for (short i = 0; i < 256; i++)
5.          lut[i] = (short)
6.           (255*(-Math.log(1.0-h[i])/alpha));
7.    tt.clip();
8.    applyLut(lut);
9. }
```

Note that on line 6 we had to scale the result so that the look-up table would range to 255. Also, line 7 imposes a clipping method that sets any negative values to 0 and any values higher than 255 to 255. The user is able to enter an arbitrary value for alpha, so clipping is a necessary precaution. Fig. 4-18 shows two ENAHE look-up tables. The one at the top results from first performing a UNAHE, then a ENAHE. The one on the bottom results from a ENAHE, with no UNAHE. Both images use $\alpha = 4$.

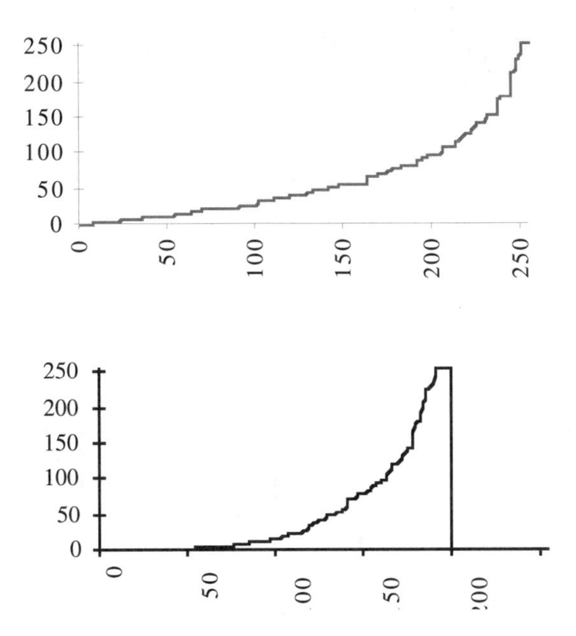

Fig. 4-18. Look-up Tables for a ENAHE Image

Fig. 4-19 shows an ENAHE image with associated histogram. The ENAHE has accentuated the darker values in the image. To counteract this effect, we first negate the image, then apply the ENAHE. The result of this manipulation is shown in Fig. 4-20.

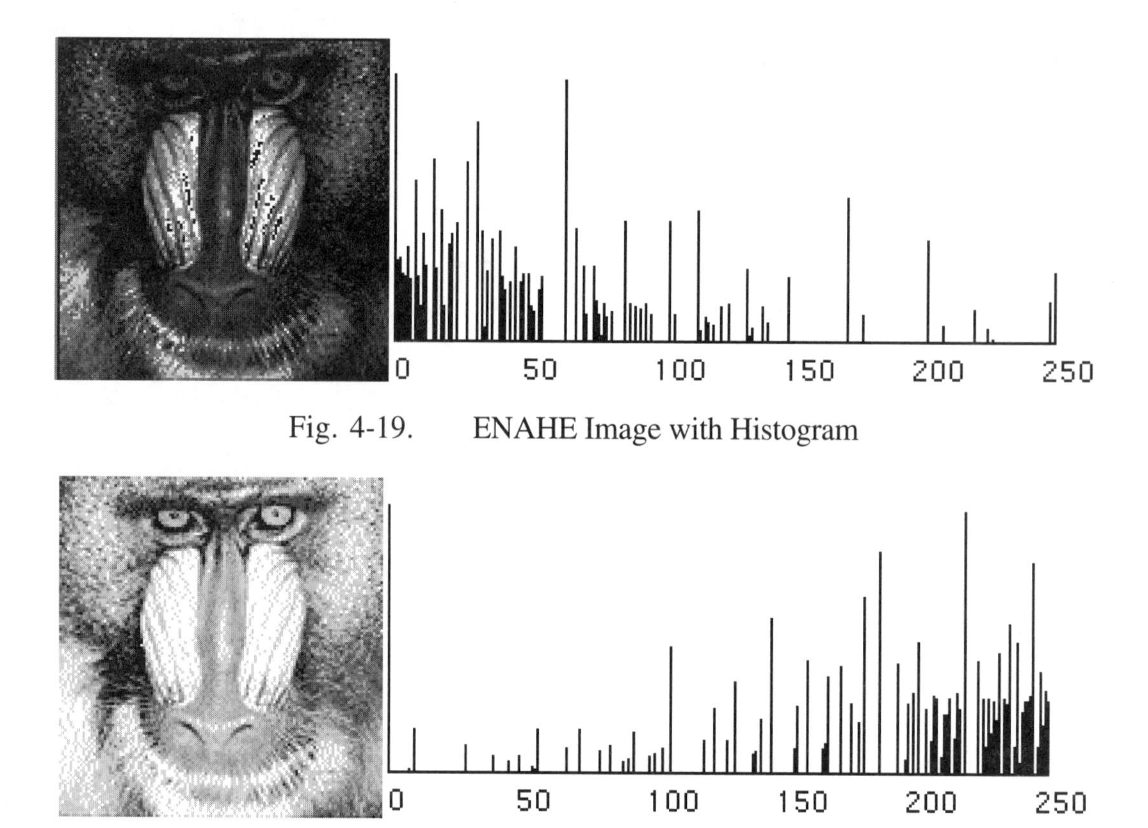

Fig. 4-19. ENAHE Image with Histogram

Fig. 4-20. Negate the Image Before the ENAHE, Then Negate Again.

As a final example, Fig. 4-21 shows a non-adaptive histogram equalization. This is created using the Rayleigh probability density function as the input into the Maple procedure.

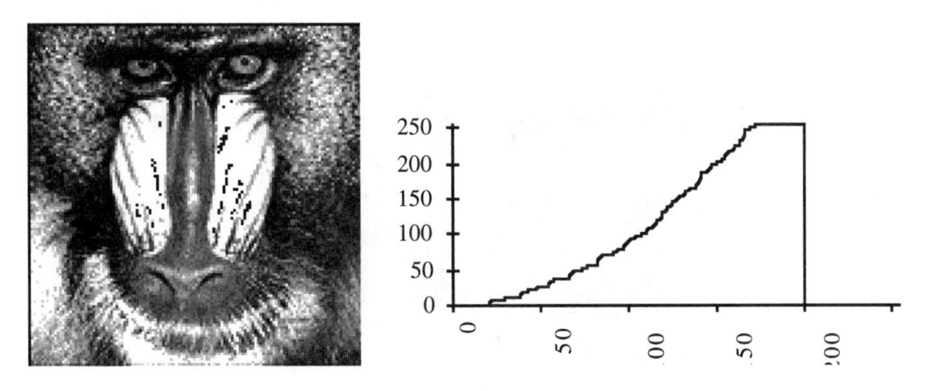

Fig. 4-21. Rayleigh Non-adaptive Histogram Equalization

The PDF of the Rayleigh equation is given by

$$p_{V'}(v') = \frac{v' - V'_{\min}}{\alpha^2} e^{\left[\frac{-\left(v' - V'_{\min}\right)^2}{2\alpha^2}\right]} \tag{4.36}$$

The transfer function is

$$lut_i = V'_{\min} + \sqrt{2\alpha^2 \ln\left(\frac{1}{1-h_i}\right)} \tag{4.37}$$

4.2.5. Adaptive Histogram Equalization

Adaptive histogram equalization is based upon dividing an images into smaller images, then performing a non-adaptive histogram equalization on the smaller images. Thus the image is processed on a block-at-a-time basis. Some authors show adaptive histogram equalization for rather fine-grained subdivisions. For example, [Umbaugh] uses a 7x7 matrix and [Pratt] shows a 2x2 example. The coarser-grained subdivisions are also of interest.

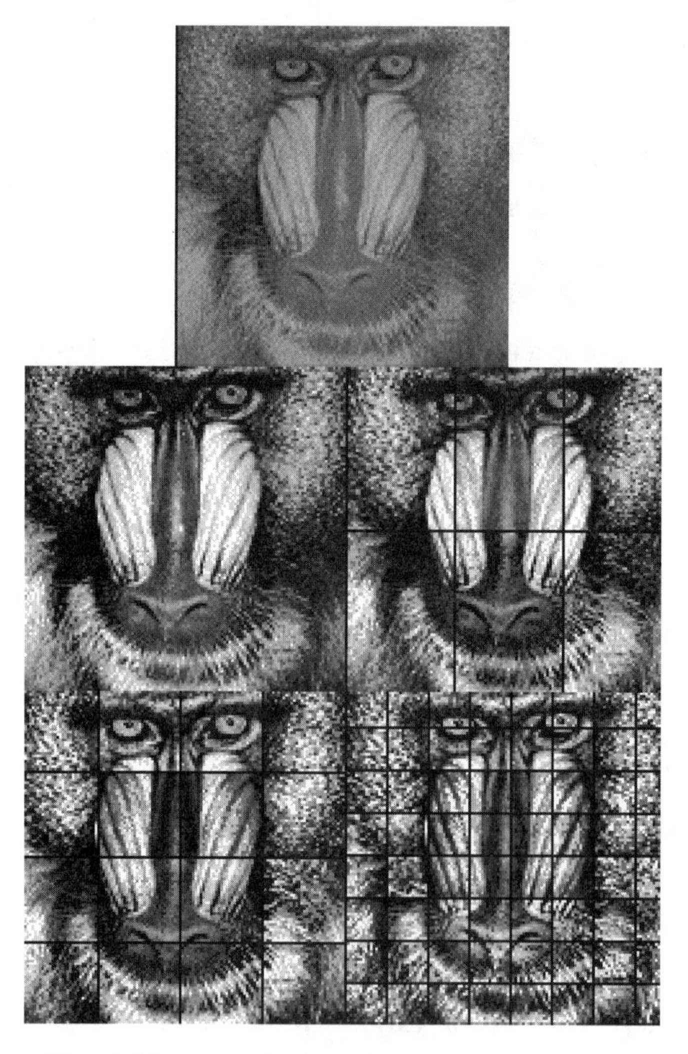

Fig. 4-22. Adaptive Histogram Equalization

Going from left to right, top to bottom, Fig. 4-22 shows 5 images. The first is the original mandrill image. The second is the result of applying a uniform non-adaptive histogram equalization. Fig. 4-23 shows the histograms for the original image, the NAUHE and the AUHE, assuming a 4x4 subdivision on the image.

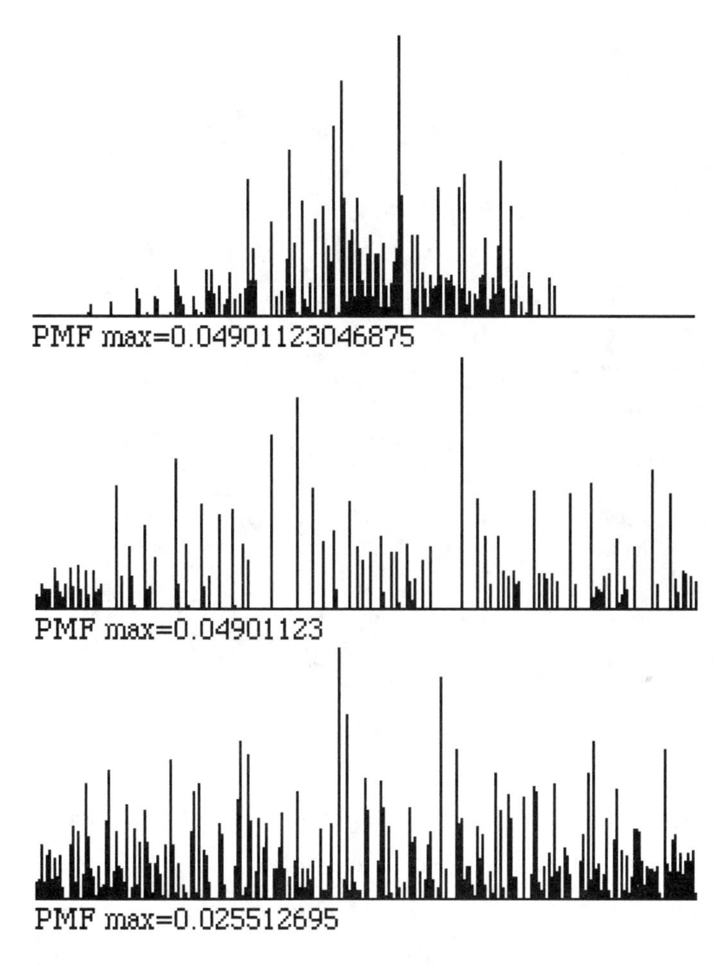

Fig. 4-23. Histograms: Original, NAUHE and AUHE

The images that follow show progressively finer subdivisions of the mandrill image. The original mandrill image is 128x128 pixels. When the image subdivision process is unable to divide the original image into an integral number of regions, the resulting image is truncated. Lines are drawn on the image where subdivision occurs. This is done for illustration purposes only.

Fig. 4-24. Adaptive Uniform Histogram Equalization

Fig. 4-24 shows the adaptive uniform histogram equalization, with and without the inscribed subdivision markings. The boundaries of the histogram subdivisions are visible in the left-hand side of Fig. 4-24. One technique that might help remove them is to employ some overlap on the boundary regions. This is a topic for future research.

4.3. The TransformTable Class

The *TransformTable* class resides in the *gui* package. The purpose of the class is to provide a simple look-up table, a container for an array of *short*.. The primary service of the TransformTable is to provide an object-oriented implementation for the printing of the contained array. Typically, this array is pasted into a graphing program (e.g., Microsoft Excel) for the purpose of visualization of data.

4.3.1. Class Summary

```
package gui;

public class TransformTable {

  public TransformTable(int size)
  public short[] getLut()
  public void setLut(short lut[])
  public void print()
 }
```

4.3.2. Class Usage

A TransformTable is instanced using the constructor:

```
public TransformTable(int size)
```

The *size* argument is used to allocate an array of *short*. A reference to the internal array of *short* is passed using the accessor methods:

```
public short[] getLut()
public void setLut(short lut[])
```

4.4. The NegateFrame Class

The *NegateFrame* class resides in the *gui* package. It provides the services of performing histogram equalization, creating and drawing tilings of an image and deriving and assembling subimages. One of the helpful features of the NegateFrame is that it can display and process the subframes as a part of its processing. Thus we can watch the progress of the image processing as subframes are created and displayed. In some ways, this could be viewed as an animated version of the adaptive histogram equalization algorithm and, as such, becomes a good tool for teaching (and debugging!) such systems.

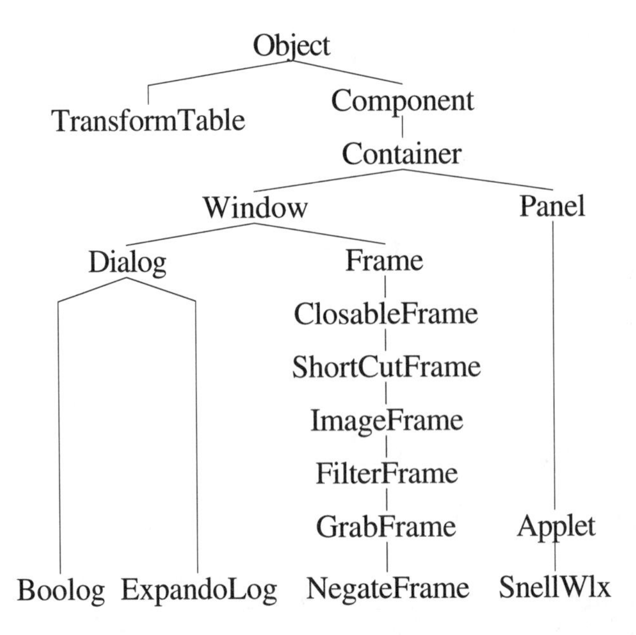

Fig. 4-25. The Class Summary Showing NegateFrame

Fig. 4-25 shows the *NegateFrame* as it relates to the other classes discussed so far. The strategy is to continue to pile the classes deeper as we add more features to the image processor. Java's lack of multiple inheritance does not permit us to adopt a more flexible development strategy. For example, it would be really nice if we could generate a low-overhead version of the *NegateFrame*, that contains only the short array data structures and the means to process them. Lack of multiple-inheritance is a deep problem with the Java language, one that is beyond the scope of this chapter.

4.4.1. Class Summary

```
package gui;
public class NegateFrame extends GrabFrame {
   public double getBBar()
   public double getGBar()
   public double getRBar()
   public double[] average(
         double a[], double b[], double c[])
   public double[] getAverageCMF()
   public int getMax()
   public int getMin()
   public NegateFrame subFrame(int x1, int y1, int w, int h)
   public NegateFrame(
         short _r[][], short _g[][], short _b[][],
         String title)
   public NegateFrame(String title)
   public short linearMap(short v,
   public static void main(String args[])
   public void add10()
   public void applyLut(short lut[])
   public void assembleMosaic(NegateFrame nf, int x1, int y1)
   public void computeStats()
   public void drawMosaic()
   public void drawMosaic(int blocksHigh, int blocksWide )
   public void enahe(double alpha)
   public void eponentialLog()
   public void histogram()
   public void linearLog()
   public void linearTransform(double c, double br)
   public void auhe()
   public void auhe(int blocksHigh, int blocksWide )
   public void negate()
   public void powImage(double p)
   public void printPMFb()
   public void printPMFg()
   public void printPMFr()
   public void printPMFs()
   public void printStats()
   public void printTT()
   public void rayleighLog()
   public void rnahe(double alpha)
   public void unahe()
}
```

4.4.2. Class Usage

This sections documents Java code in the NegateFrame using some locally scoped set of variables. For example, there is no attempt to make the *N* of this section match the *N* of any other. In the case where the variables are locally scoped and unclear, they are defined. Otherwise, we have tried to hold true to the notation of the rest of the book. While the JavaDoc program is probably useful for most applications when generating Java documentation automatically, we have found it to be impoverished when documenting code that is best described in standard math notation.

```
public double getBBar()
public double getGBar()
public double getRBar()
```

The methods:

```
public double getBBar()
public double getGBar()
public double getRBar()
```

Use the internal arrays of *short* to compute the average. The name, *Bar*, is used to denote the appearance of the notation. For example:

$$\bar{r} = \frac{1}{N}\sum_{i=1}^{N} r_i = getRBar();$$
$$\bar{g} = \frac{1}{N}\sum_{i=1}^{N} g_i = getGBar(); \quad (4.38)$$
$$\bar{b} = \frac{1}{N}\sum_{i=1}^{N} b_i = getBBar();$$

where

$$N = \text{ total number of pixels} \quad (4.39)$$

To get the average of three double precision arrays, all of the same length, and return a new array, also of the same length, use

```
public double[] average(double a[], double b[], double c[])
```

In other words, average returns:

$$\bar{x}_i = (a_i + b_i + c_i)/3 \\ i \in [0 \ldots N] \tag{4.40}$$

To get the Cumulative Mass Function (CMF) as an average of the CMFs for the red, green and blue color planes, use

```
public double[] getAverageCMF()
```

In math, we write:

$$\overline{P}_V(v) = [P_R(r) + P_G(g) + P_B(b)]/3 = getAverageCMF(); \tag{4.41}$$

After the *computeStats* method is invoked, there are internally held data structures that have accessor methods. The methods:

```
public int getMax()
```

and

```
public int getMin()
```

will return correct and updated values only if the *computeStats* method has been invoked since the image has changed.

In math, we write:

$$V_{\max} = \text{getmax}(); \\ V_{\min} = \text{getmin}(); \tag{4.42}$$

in order to extract a sub-frame from the NegateFrame instance, the method:

```
public NegateFrame subFrame(int x1, int y1, int w, int h)
```

This method will start at position *x1,y1* and proceed *w* pixels across and *h* pixels down, copying the pixels into a new array. The new array is used to make a new instance of the NegateFrame. The new instance is displayed and has a full-fledged interface with its own data structures. This carries a lot of overhead and can cause out of memory exceptions when abused.

To make a new instance of a NegateFrame, starting with three two-dimensional color planes, use

```
public NegateFrame(
   short _r[][], short _g[][], short _b[][], String title)
```

The typical constructor for the NegateFrame class assumes that you do not have the color planes available. This one merely asks for a string to title the frame:

```
public NegateFrame(String title)
```

A linear brightness and contrast look-up function is available; it is typically applied to a loop-up table. Clipping is performed when the argument is out of range. The syntax:

```
public short linearMap(short v, double c, double b)
```

Has the following math equivalence:

$$f(v) = \begin{cases} cv + b & \text{if } v \in [0...255] \\ 0 & \text{otherwise} \end{cases} = \text{linearMap}(v, c, b) \tag{4.43}$$

There is a *main* method that may be invoked directly; it makes an instance of the NegateFrame and is typically used for testing:

```
public static void main(String args[])
```

To add the constant, 10, to every pixel in the image, use

```
public void add10()
```

Such a method is good as an example of how to write elementary programs in the NegateFrame class.

To apply a look-up table to every pixel in an image use

```
public void applyLut(short lut[])
```

The *applyLut* works in all color planes. In math, it has the effect of:

$$r_i \leftarrow lut(r_i)$$
$$g_i \leftarrow lut(g_i) \text{ for all } i \in [0 \dots N] \quad (4.44)$$
$$b_i \leftarrow lut(b_i)$$

The '←' operator is used to indicate that new values for the colors replace the old values. After all, expression $x=x+1$ appears more like meaningless gibberish, then mathematics.

To absorb a NegateFrame instance into an existing NegateFrame instance, use

```
public void assembleMosaic(NegateFrame nf, int x1, int y1)
```

If the resulting frame goes out of bounds, an *array out-of bounds* exception is thrown. This method is used to merge the mosaic sub images, such as those shown in Fig. 4-22.

To compute the statistics for an image, use

```
public void computeStats()
```

To make an instance of a dialog that allows for interactive specification of the parameters needed to subdivide an image into rectangular blocks, use

```
public void drawMosaic()
```

The *drawMosaic* method will draw the block and set the parameters used by an invocation of *auhe*.

To subdivide an image with truncation, so that it has *blocksHigh* rectangles and *blocksWide* rectangles, use

```
public void drawMosaic(int blocksHigh, int blocksWide )
```

Note that *drawMosaic* does not specify how many pixels the rectangles have, because this is image resolution dependent.

To perform the exponential non-adaptive histogram equalization, as described in Section 4.2.4, use

```
public void enahe(double alpha)
```

To create a dialog that inputs the value for α, use

```
public void eponentialLog()
```

In math, this is represented by (4.35) which is repeated here:

$$lut_i = V'_{\min} - \frac{1}{\alpha}\ln(1-h_i) \tag{4.45}$$

To display and compute the red, green and blue histograms, use

```
public void histogram()
```

To display the brightness and contrast dialog used to apply the *linearTransform* interactively, use

```
public void linearLog()
```

To apply the linear transformation of brightness and contrast remapping on the red-green and blue color planes, use

```
public void linearTransform(double c, double br)
```

In math, we write:

$$\begin{aligned} r_i &\leftarrow f(r_i) = \text{linearMap}(r_i, c, br) \\ g_i &\leftarrow f(g_i) = \text{linearMap}(g_i, c, br) \\ b_i &\leftarrow f(b_i) = \text{linearMap}(b_i, c, br) \end{aligned} \tag{4.46}$$

To perform an adaptive uniform histogram equalization using the parameters entered into the dialog created by *drawMosaic,* use

```
public void auhe()
```

The default values for the *auhe* method are 2x2. The effect is to create for sub-frames using the overloaded version of the *auhe method*, which appears below:

```
public void auhe(int blocksHigh, int blocksWide )
```

This version of the *auhe* method allow the programmer to subdivide an image into blocksHigh * blocksWide sub-images, each held in its own frame. Each subframe has a non-adaptive uniform histogram equalization applied.

To negate an image, use

```
public void negate()
```

Originally, this is all the NegateFrame did: negate an image!! For historical reasons, the name for the class has stuck!

To apply (4.14) to an image, use

```
public void powImage(double p)
```

In math, we write:

$$
\begin{aligned}
r_i &\leftarrow f(r_i) = 255\left(\frac{r_i}{255}\right)^p \\
g_i &\leftarrow f(g_i) = 255\left(\frac{g_i}{255}\right)^p \\
b_i &\leftarrow f(b_i) = 255\left(\frac{b_i}{255}\right)^p
\end{aligned}
\tag{4.47}
$$

To print the probability mass functions for the red, green and blue color planes, use

```
public void printPMFr()
public void printPMFg()
public void printPMFb()
```

To print all the above PMFs with a single invocation, use

```
public void printPMFs()
```

To print the statistics for an image, use

```
public void printStats()
```

To print the look-up table (which consists of an instance of the TransformTable class), use

```
public void printTT()
```

To bring up a dialog for entering the value of α used by *rnahe,* use

```
public void rayleighLog()
```

To transform an image using the Rayleigh non-adaptive histogram equalization, use

```
public void rnahe(double alpha)
```

This implements the transfer function of (4.37) and applies it to each color plane.

To perform the uniform non-adaptive histogram equalization, use

```
public void unahe()
```

This implements the algorithm described in section 4.2.4.

4.5. Summary

In this chapter we explored two methods of altering the appearance of an image: the direct methods and the indirect methods.

The direct methods employ a mapping function, such as the linear, exponential or square root functions. These mapping functions are applied directly to the color planes of an image in order to perform homogeneous point processing.

The indirect method made use of the statistical properties of the image, as contained in the image's histogram. These are broadly classified into two types as well: adaptive and non-adaptive.

The adaptive image enhancement techniques were implemented by subdividing an image into smaller parts. Each subdivision was processed and then reassembled into a new image. The coarse adaptive subdivision left blemishes in the final image where the sub-images were respliced. An object oriented implementation permitted a new instance of a frame to be created for each subimage. This was educational, but it turned out to be an inefficient way to perform the image processing. A fast technique is to perform window sliding by increasing an index. Such a procedure is described in [Gonzalez and Woods] and in [Umbaugh]. For more transfer functions for look-up tables, see [Pratt]. For another approach to the derivation of the histogram equalization transfer functions, see [Castleman]. For further discussion on adaptive contrast enhancement, see [Yu et al.].

5. Digital Image Processing Fundamentals

There's more to it than meets the eye.

– 19th century proverb

Digital image processing is electronic data processing on a 2-D array of numbers. The array is a numeric representation of an *image*. A *real* image is formed on a sensor when an energy emission strikes the sensor with sufficient intensity to create a sensor output. The energy emission can have numerous possible sources (e.g., acoustic, optic, etc.). When the energy emission is in the form of electromagnetic radiation within the band limits of the human eye, it is called visible light [Banerjee]. Some objects will reflect only electromagnetic radiation. Others produce their own, using a phenomenon called *radiancy*. Radiancy occurs in an object that has been heated sufficiently to cause it to glow visibly [Resnick]. Visible light images are a special case, yet they appear with great frequency in the image processing literature.

Another source of images includes the *synthetic* images of computer graphics. These images can provide controls on the illumination and material properties that are generally unavailable in the real image domain.

This chapter reviews some of the basic ideas in digital signal processing. The review includes a summary of some mathematical results that will be of use in Chapter 15. The math review is included here in order to strengthen the discourse on sampling.

5.1. The Human Visual System

A typical human visual system consists of stereo electromagnetic transducers (two eyes) connected to a large number of neurons (the brain). The neurons process the input, using poorly understood emergent properties (the mind). Our discussion will follow the eye, brain and mind ordering, taking views with a selective focus.

The ability of the human eye to perceive the spectral content of light is called color vision. A typical human eye has a spectral response that varies as a function of age and the individual. Using clinical research, the CIE (Commission Internationale de L'Eclairage) created a statistical profile of human vision called the *standard observer*. The response curves of the standard observer indicate that humans can see light whose wavelengths have the color names red, green and blue. When discussing wavelengths for visible light, we typically give the measurements in *nanometers*. A nanometer is 10^{-9} meters and is abbreviated *nm*. The wavelength for the red, green and blue peaks are about 570-645 nm, 526-535 nm, and 444-445 nm. The visible wavelength range (called the mesopic range) is 380 to about 700-770 nm [Netravali] [Cohen].

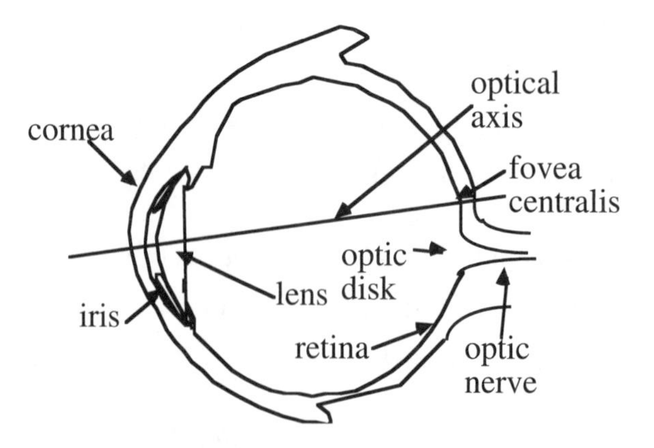

Fig. 5-1. Sketch of a Human Eye

Fig. 5-1 shows a sketch of a human eye. When dimensions are given, they refer to the typical adult human eye unless otherwise stated. Light passes through the cornea and is focused on the retina by the lens. Physiological theories use biological components to explain behaviour. The optical elements in the eye (cornea, lens and retina) form the primary biological components of a photo sensor. Muscles are used to alter the thickness of the lens and the diameter of the hole covering the lens, called the iris. The iris diameter typically varies from 2 to 8 mm. Light passing through the lens is focused upon the retina. The retina contains two types of photo sensor cells: rods and cones.

There are 75 to 150 million rod cells in the retina. The rods contain a blue-green absorbing pigment called rhodopsin. Rods are used primarily for night vision (also called the scotopic range) and typically have no role in color vision [Gonzalez and Woods].

Cones are used for daylight vision (called the photopic range). The tristimulus theory of color perception is based upon the existence of three types of cones: red, green and blue. The pigment in the cones is unknown [Hunt]. We do know that the phenomenon called adaptation (a process that permits eyes to alter their sensitivity) occurs because of a change in the pigments in the cones [Netravali]. The retina cells may also inhibit each another from creating a high-pass filter for image sharpening. This phenomenon is known as lateral inhibition [Mylers].

The current model for the retinal cells shows a cone cell density that ranges from 900 cells/mm^2 to 160,000 cells/mm^2 [Gibson]. There are 6 to 7 million cone cells, with the density increasing near the fovea. Further biological examination indicates that the cells are imposed upon a noisy hexagonal array [Wehmeier].

Lest one be tempted to count the number of cells in the eye and draw a direct comparison to modern camera equipment, keep in mind that even the fixated eye is constantly moving. One study showed that the eyes perform over 3 fixations per second during a search of a complex scene [Williams]. Further more, there is nearly a 180-degree field of view (given two eyes). Finally, the eye-brain interface enables an integration between the sensors' polar coordinate scans, focus, iris adjustments and the interpretation engine. These interactions are not typical of most artificial image processing systems [Gonzalez and Woods]. Only recently have modern camcorders taken on the role of integrating the focus and exposure adjustment with the sensor.

The optic nerve has approximately 250,000 neurons connecting to the brain. The brain has two components associated with low-level vision operations: the lateral geniculate nucleus and the visual cortex. The cells are modeled using a circuit that has an inhibit input, capacitive-type electrical storage and voltage leaks, all driving a comparitor with a variable voltage output.

The capacitive storage elements are held accountable for the critical fusion frequency response of the eye. The critical fusion frequency is the rate of display whereby individual updates appear as if they are continuous. This frequency ranges from 10-70 Hz depending on the color [Teevan] [Netravali]. At 70 Hz, the 250,000-element optic nerve should carry 17.5 million neural impulses per second. Given the signal-to-noise ratio of a human auditory response system (80 dB), we can estimate that there are 12.8 bits per nerve leading to the brain [Shamma]. This gives a bit rate of about 224 Mbps. The data has been pre-processed by the eye before it reaches the optic nerve. This preprocessing includes lateral inhibition between the retinal neurons. Also, we have assumed that there is additive white Gaussian noise on the channel, but this assumption may be justified.

Physiological study has shown that the response of the cones is given by a Gaussian sensitivity for the cone center and surrounding fields. The overall sensitivity is found by subtracting the surrounding response from the center response. This gives rise to a difference of Gaussian expression which is discussed in Chap. 10. Further, the exponential response curve of the eye is the primary reason why exponential histogram equalization was used in Chap. 4.

5.2. Overview of Image Processing

An image processing system consists of a source of image data, a processing element and a destination for the processed results. The source of image data may be a camera, a scanner, a mathematical equation, statistical data, the Web, a SONAR system, etc. In short, anything able to generate or acquire data that has a two-dimensional structure is considered to be a valid source of image data. Furthermore, the data may change as a function of time.

The processing element is a computer. The computer may be implemented in a number of different ways. For example, the brain may be said to be a kind of biological computer that is able to perform image processing (and do so quite well!). The brain consumes about two teaspoons of sugar and 20 watts of power per hour. An optical element can be used to perform computation and does so at the speed of light (and with very little power). This is a fascinating topic of current research [Fietelson]. In fact, the injection of optical computing elements can directly produce information about the range of objects in a scene [DeWitt and Lyon].

Such computing elements are beyond the scope of this book. The only type of computer that we will discuss in this book is the digital computer. However, it is interesting to combine hybrid optical and digital computing. Such an area of endeavor lies in the field of *photonics*.

The output of the processing may be a display, created for the human visual system. Output can also be to any *stream*. In Java, a stream is defined as an uninterpreted sequence of bytes. Thus, the output may not be image data at all. For example, the output can be a histogram, a global average, etc. As the output of the program renders a higher level of interpretation, we cross the fuzzy line from image processing into the field of *vision*. As an example, consider that image processing is used to edge detect the image of coins on a table. Computer vision is used to tell how much money is there. Thus, computer vision will often make use of image processing as a sub-task.

5.2.1. Digitizing a Signal

Digitizing is a process that acquires quantized samples of continuous signals. The signals represent an encoding of some data. For example, a microphone is a pressure transducer that produces an electrical signal. The electrical signal represents acoustic pressure waves (sound).

The term *analog* refers to a signal that has a continuously varying pattern of intensity. The term *digital* means that the data takes on discrete values. Let $s(t)$ be a continuous signal. Then, by definition of continuous,

$$\lim_{t \to a} s(t) = s(a) \quad \text{such that } a \in R. \tag{5.1}$$

We use the symbol R to denote the set of real numbers. Thus $R = \{x : x \text{ is a real number}\}$, which says that R is the set of all x such that x is a real number. We read (5.1) saying, in the limit, as t approaches a, such that a is a member of the set of real numbers, $s(t) = s(a)$. The expression $\{x : P(x)\}$ is read as "the set of all x's such that $P(x)$ is true" [Moore 64].

This is an *iff* (i.e., if and only if) condition. Thus, the converse must also be true. That is, $s(t)$ is not continuous iff there exists a value, a such that:

$$\lim_{t \to a} s(t) \neq s(a) \tag{5.2}$$

is true.

For example, if $s(t)$ has multiple values at a, then the limit does not exist at a.

The analog-to-digital conversion consists of a sampler and a quantizer. The quantization is typically performed by dividing the signal into several uniform steps. This has the effect of introducing *quantization noise*. Quantization noise is given, in dB, using

$$SNR \le 6b + 4.8 \tag{5.3}$$

where SNR is the signal-to-noise ratio and b is the number of bits. To prove (5.3), we follow [Moore] and assume that the input signal ranges from -1 to 1 volts. That is,

$$s(t) \in \{x : x \in R \text{ and } -1 \le x \le 1\} \tag{5.3a}$$

Note that the number of quantization intervals is 2^b. The least significant bit has a quantization size of $V_{qe} = 2^{-b}$. Following [Mitra], we obtain the bound on the size of the error with:

$$-V_{qe} \le e \le V_{qe} \tag{5.3b}$$

The variance of a random variable, X, is found by $\sigma_X^2(t) = \int_{-\infty}^{\infty} x^2 f_X(x)dx$, where $f_X(x)$ is a probability distribution function. For the signal whose average is zero, the variance of (5.3b) is

$$\sigma_e^2 = \frac{V_{qe}}{2}\int_{-1/V_{qe}}^{1/V_{qe}} \varepsilon^2 d\varepsilon = \frac{1}{3V_{qe}^2} \tag{5.3c}.$$

The signal-to-noise ratio for the quantization power is

$$SNR = 10\log\left(3 \times 2^{2b}\right) = 20b\log 2 + 10\log 3 \tag{5.3d}$$

Hence the range on the upper bound for the signal-to-quantization noise power is

$$SNR \le 6b + 4.8 \tag{5.3}.$$

Q.E.D.

In the above proof we assumed that uniform steps were used over a signal whose average value is zero. In fact, a digitizer does not have to requantize an image so that steps are uniform. An in-depth examination of the effects of non-linear quantization on SNR is given in [Gersho]. Following Gersho, we generalize the result of (5.3), defining the SNR as

$$SNR_{dB} = 20\log\left(\frac{\sigma}{< e(x)| p(x) >}\right) \quad (5.3e)$$

where

$$\sigma = \text{standard deviation}$$

and $< e|p >$ is the *mean-square distortion* defined by the inner product between the square of the quantization error for value x and the probability of value x. The inner product between e and p is given by

$$< e|p >= \int_{-\infty}^{\infty} e(x)p(x)dx \quad (5.3f).$$

where

$$e(x) = (Q(x) - x)^2 \quad (5.3g).$$

The inner product is an important tool in transform theory. We will expand our discussion of the inner product when we touch upon the topic of sampling.

We define *Q(x)* as the quantized value for x. Maximizing SNR requires that we select the quantizer to minimize (5.3f), given *a priori* knowledge of the PDF (if the PDF is available). Recall that for an image, we compute the PMF (using the *Histogram* class) as well as the CMF. As we shall see later, (5.3f) is minimized for k-level thresholding (an intensity reduction to k colors) when the regions of the CMF are divided into k sections. The color is then remapped into the center of each of the CMF regions. Hence (5.3f) provides a mathematical basis for reducing the number of colors in an image provided that the PDF is of zero mean (i.e, no DC offset) and has even symmetry about zero. That is $p(x) = p(-x)$. Also, we assume that the quantizer has odd symmetry about zero, i.e., $Q(x) = -Q(-x)$.

A simple zero-memory 4-point quantizer inputs 4 decision levels and outputs 4 corresponding values for input values that range within the 4 decision levels. When the decision levels are placed into an array of double precision numbers, in Java (for the 256 gray-scale values) we write:

```
public void thresh4(double d[]) {
  short lut[] = new short[256];
  if (d[4] ==0)
   for (int i=0; i < lut.length; i++) {
     if (i < d[0]) lut[i] = 0;
     else if (i < d[1]) lut[i] = (short)d[0];
     else if (i < d[2]) lut[i] = (short)d[1];
     else if (i < d[3]) lut[i] = (short)d[2];
     else lut[i] = 255;
     System.out.println(lut[i]);
   }
   applyLut(lut);
}
```

We shall revisit quantization in Section 5.2.2.

Using the Java AWT's Image class, we have seen that 32 bits are used, per pixel (red, green, blue and alpha). There are only 24 bits used per color, however. Section 5.2.2 shows how this relates to the software of this book.

Recall also that the digitization process led to sampling an analog signal. Sampling a signal alters the harmonic content (also known as the spectra) of the signal. Sampling a continuous signal may be performed with a pre-filter and a switch. Fig. 5-2 shows a continuous function, $f(x)$, being sampled at a frequency of f_s.

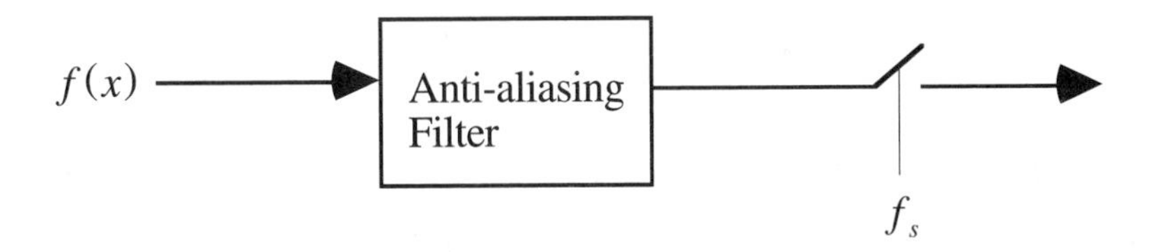

Fig. 5-2. Sampling System

The switch in Fig. 5-2 is like a binary amplifier that is being turned on and off every $1/f_s$ seconds. It multiplies $f(x)$ by an amplification factor of zero or one.

Mathematically, sampling is expressed as a *pulse train,* $p(x)$, multiplied by the input signal $f(x)$, i.e., sampling is $f(x)\,p(x)$..

To discuss the pulse train mathematically, we must introduce the notation for an impulse. The *unit impulse*, or *Dirac delta,* is a generalized function that is defined by

$$\int_{-\infty}^{\infty} \delta(x)dx = \int_{-\varepsilon}^{\varepsilon} \delta(x)dx = 1 \tag{5.4}$$

where ε is arbitrarily small. The Dirac delta has unit area about a small neighborhood located at $x = 0$. Multiply the Dirac delta by a function and it will *sift* out the values where the Dirac delta is equal to zero:

$$\int_{-\infty}^{\infty} f(x)\delta(x)dx = \int_{-\varepsilon}^{\varepsilon} f(x)\delta(x)dx = f(0) \tag{5.5}$$

This is called the *sifting property* of the Dirac delta. In fact, the Dirac delta is equal to zero whenever its argument is non-zero. To make the Dirac activate, given a non-zero argument, we bias the argument with an offset, $\delta(x - x_{offset})$. A pulse train is created by adding an infinite number of Dirac deltas together:

$$p(x) = \sum_{n=-\infty}^{\infty} \delta(x - n / f_s) \tag{5.6}$$

$$f(x)p(x) = f(x)\sum_{n=-\infty}^{\infty} \delta(x - n / f_s) \tag{5.7}$$

To find the spectra of (5.7) requires that we perform a *Fourier transform.* The Fourier transform, just like any transform, performs a correlation between a function and a *kernel.* The kernel of a transform typically consists of an *orthogonal basis* about which the reconstruction of a waveform may occur. Two functions are orthogonal if their inner product $< f|g >$ =0. Recall that the inner product is given by

$$< f|g > \equiv \int_{-\infty}^{\infty} f(x)g(x)dx \equiv \text{inner product} \tag{5.7a}$$

From linear algebra, we recall that a collection of *linearly independent* functions forms a *basis* if every value in the set of all possible values may be expressed as a linear combination of the basis set. Functions are linearly independent *iff* the sum of the functions is non-zero (for non-zero co-efficients). Conversely, functions are linearly dependent *iff* there exists a combination of non-zero coefficients for which the summation is zero. For example:

$$\begin{aligned} &c_1 \cos(x) + c_2 \sin(x) = 0 \\ &\textit{iff} \;\; c_1 = c_2 = 0 \end{aligned} \qquad (5.7b)$$

The ability to sum a series of sine and cosine functions together to create an arbitrary function is known an the *super position* principle and applies only to periodic waveforms. This was discovered in the 1800's by Jean Baptiste Joseph de Fourier [Halliday] and is expressed as a summation of sine and cosines, with constants that are called *Fourier coefficients.*

$$f(x) = \sum_{k=0}^{\infty} (a_k \cos kt + b_k \sin kt) \qquad (5.7c)$$

We note that (5.7c) shows that the periodic signal has discrete spectral components. We find the Fourier coefficients by taking the inner product of the function, *f(x)* with the basis functions, sine and cosine. That is:

$$\begin{aligned} a_k &=< f|\cos(kt) > \\ b_k &=< f|\sin(kt) > \end{aligned} \qquad (5.7d)$$

For an elementary introduction to linear algebra, see [Anton]. For a concise summary see [Stollnitz]. For an alternative derivation see [Lyon and Rao].

It is also possible to approximate an *aperiodic* waveform. This is done with the *Fourier transform.* The Fourier transform uses sine and cosine as the basis functions to form the inner product, as seen in (5.7a):

$$F(u) = \int_{-\infty}^{\infty} f(x) e^{-j2\pi ux} dx =< f|e^{-j2\pi ux} > \qquad (5.8).$$

By Euler's identity,

$$e^{i\theta} = \cos\theta + i\sin\theta \tag{5.9}$$

we see that the sine and cosine basis functions are separated by being placed on the real and imaginary axis.

Substituting (5.7) into (5.8) yields

$$F(u) * P(u) = \int_{-\infty}^{\infty} \left[f(x) \sum_{n=-\infty}^{\infty} \delta(x - n / f_s) e^{-j2\pi ux} \right] dx \tag{5.10}$$

where

$$P(u) = \int_{-\infty}^{\infty} \sum_{n=-\infty}^{\infty} \delta(x - n / f_s) e^{-j2\pi ux} dx \tag{5.11}$$

The term

$$F(u) * P(u) \equiv \int_{-\infty}^{\infty} F(\gamma) P(u - \gamma) d\gamma \tag{5.12}$$

defines a convolution. We can write (5.10) because multiplication in the time domain is equivalent to convolution in the frequency domain. This is known as the convolution theorem. Taking the Fourier transform of the convolution between two functions in the time domain results in

$$< f * h | e^{-j2\pi ux} > = < \int_{-\infty}^{\infty} f(\gamma) p(x - \gamma) d\gamma | e^{-j2\pi ux} > \tag{5.13}$$

which is expanded by (5.8) to yield:

$$= \int_{-\infty}^{\infty} \left[\int_{-\infty}^{\infty} f(\gamma) p(x - \gamma) d\gamma \right] e^{-j2\pi ux} dx \tag{5.13a}$$

Changing the order of integration in (5.13a) yields

$$= \int_{-\infty}^{\infty} f(\gamma) \left[\int_{-\infty}^{\infty} p(x - \gamma) e^{-j2\pi ux} dx \right] d\gamma \tag{5.13b}$$

with

$$P(u) = \int_{-\infty}^{\infty} p(x-\gamma)e^{-j2\pi ux}dx \tag{5.13c}$$

and

$$F(u) = \int_{-\infty}^{\infty} f(x)e^{-j2\pi ux}dx \tag{5.13d}$$

we get

$$F(u)P(u) = \int_{-\infty}^{\infty} f(x) * p(x)e^{-j2\pi ux}dx \tag{5.14}$$.

This shows that convolution in the time domain is multiplication in the frequency domain. We can also show that convolution in the frequency domain is equal to multiplication in the time domain. See [Carlson] for an alternative proof.

As a result of the convolution theorem, the Fourier transform of an impulse train is also an impulse train,

$$F(u) * P(u) = F(u) * \sum_{n=-\infty}^{\infty} f_s \delta(f - nf_s) \tag{5.15}$$

Finally, we see that sampling a signal at a rate of f_s causes the spectrum to be reproduced at f_s intervals:

$$F(u) * P(u) = f_s \sum_{n=-\infty}^{\infty} F(f - nf_s) \tag{5.16}$$

(5.16) demonstrates the reason why a band limiting filter is needed before the switching function of Fig. 5-2. This leads directly to the sampling theorem which states that a band limited signal may be reconstructed without error if the sample rate is twice the bandwidth. Such a sample rate is called the *Nyquist rate* and is given by $f_s = 2B$ Hz.

5.2.2. Image Digitization

Typically, a camera is used to digitize an image. The modern CCD cameras have a photo diode arranged in a rectangular array. Flat-bed scanners use a movable platen and a linear array of photo diodes to perform the two-dimensional digitization.

Older tube type cameras used a wide variety of materials on a photosensitive surface. The materials vary in sensitivity and output. See [Galbiati] for a more detailed description on tube cameras.

The key point about digitizing an image in two dimensions is that we are able to detect both the power of the incident energy as well as the direction.

The process of digitizing an image is described by the amount of spatial resolution and the signal -to-noise ratio (i.e., number of bits per pixel) that the digitizer has. Often the number of bits per pixel is limited by performing a *thresholding*. Thresholding (a topic treated more thoroughly in Chap. 10) reduces the number of color values available in an image. This simulates the effect of having fewer bits per pixel available for display. There are several techniques available for thresholding. For the grayscale image, one may use the cumulative mass function for the probability of a gray value to create a new look-up table. Another approach is simply to divide the look-up table into uniform sections. Fig. 5-2 shows the mandrill before and after thresholding operation. The decision about when to increment the color value was made based on the CMF of the image. The number of bits per pixel (bpp), shown in Fig. 5-2, ranging from left to right, top to bottom, are: 1 bpp, 2 bpp, 3 bpp and 8 bpp. Keep in mind that at a bit rate of 28 kbps (the rate of a modest Internet connection over a phone line) the 8 bpp image (128x128) will take 4 seconds to download. Compare this to the uncompressed 1 bpp image which will take 0.5 seconds to download. Also note that the signal-to-noise ratio for these images ranges from 10 dB to 52 dB.

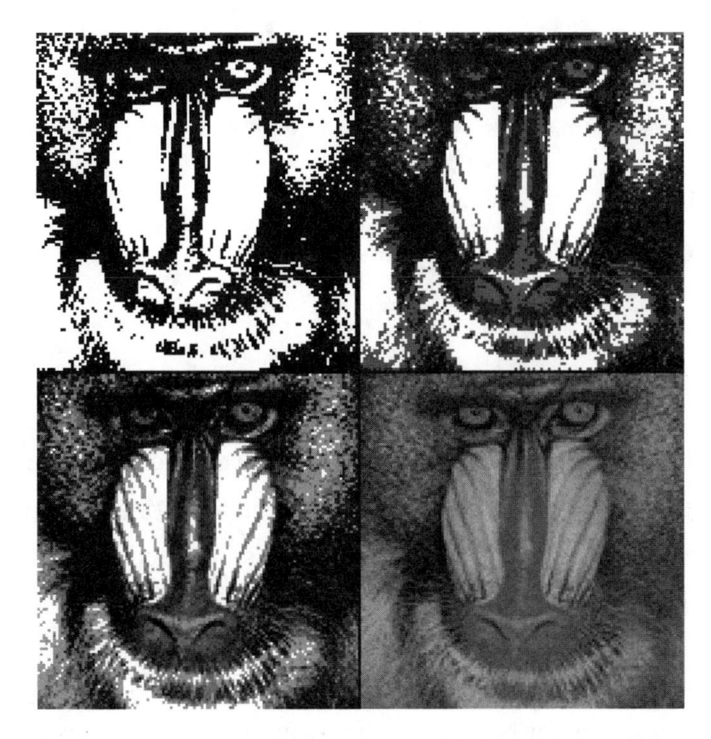

Fig. 5-3. Quantizing with Fewer Bits Per Pixel

The code snippet allows the cumulative mass function of the image to bias decisions about when to increment the color value. The input to the code is the number of gray values, *k*. There are several methods to perform the quantization. The one shown in Fig. 5-3 is useful in edge detection (a topic covered in Chap. 10). The *kgreyThresh* method follows:

```
public void kgreyThresh(double k) {
    Histogram rh = new Histogram(r,"red");
    double cmf[] = rh.getCMF();
    TransformTable tt = new TransformTable(cmf.length);
    short lut[] = tt.getLut();
    int q=1;
    short v=0;
    short dv = (short)(255/k);
    for (int i=0; i < lut.length; i++) {
        if (cmf[i] > q/k) {
            v += dv;
            q++; //(k == q+1)||
            if (q==k) v=255;
        }
```

```
            lut[i]=v;
        }
        tt.setLut(lut);
        tt.clip();
        tt.print();
        applyLut(lut);
    }
```

5.2.3. Image Display

One display device that has come into common use is the cathode-ray tube (CRT). The cathode ray tube displays an image using three additive colors: red, green and blue. These colors are emitted using phosphors that are stimulated with a flow of electrons. Different phosphors have different colors (spectral radiance).

There are three kinds of television systems in the world today, NTSC, PAL and SECAM. NTSC which stands for National Television Subcommittee, is used in North America and Japan. PAL stands for phase alternating line and is used in parts of Europe, Asia, South America and Africa. SECAM stands for Sequential Couleur à Mémorie (sequential chrominance signal and memory) and is used in France, Eastern Europe and Russia.

The gamut of colors and the reference color known as *white* (called white balance) are different on each of the systems.

Another type of display held in common use is the computer monitor.

Factors that afflict all displays include: ambient light, brightness (black level) and contrast (picture). There are also phosphor chromaticity differences between different CRTs. These alter the color gamut that may be displayed.

Manufacturers' products are sometimes adopted as a standard for the color gamut to be displayed by all monitors. For example, one U.S. manufacturer, Conrac, had a phosphor that was adopted by SMPTE (Society of Motion Picture and Television Engineers) as the basis for the SMPTE C phosphors.

The CRTs have a transfer function like that of (4.14), assuming the value, v ranges from zero to one:

$$f(v) = v^{\gamma} \tag{5.3}$$

Typically, this is termed the *gamma* of a monitor and runs to a value of 2.2 [Blinn]. As Blinn points out, for a gamma of 2, only 194 values appear in a look-up table of 256 values. His suggestion that 16 bits per color might be enough to perform image processing has been taken to heart, and this becomes another compelling reason to use the Java *short* for storing image values. Thus, the image processing software in this book does all its image processing as if intensity were linearly related to the value of a pixel. With the storage of 48 bits per pixel (for red, green and blue) versus the Java AWT model of 24 bits per red, green and blue value, we have increased our signal-to-noise ratio for our image representation by 48 dB per color. So far, we have not made good use of this extra bandwidth, but it is nice to know that it is there if we need it.

6. Input Streams

Nice thing about the Java API,
if you don't like it, just wait two minutes.

– DL

A stream in Java is an uninterpreted sequence of bytes. There are two kinds of streams in Java: *input streams* and *output streams*. They are represented by instances of the *InputStream* and *OutputStream* classes. This chapter shows how to use some of the subclasses of the InputStream and OutputStream to read and write stream data. Streams use a wide variety of data sources and data sinks (e.g., a file, the console, the WWW, etc.). Unlike C++, the lack of multiple-inheritance in Java does not permit the existence of a single input-output stream instance.

Fig. 6-1 shows a characterization of the relationships between typical input and output streams.

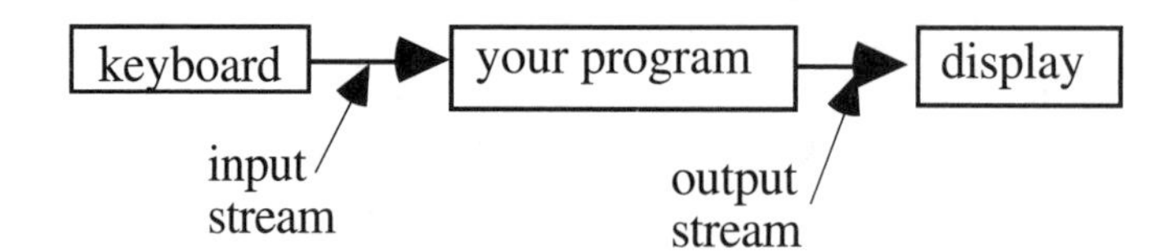

Fig. 6-1. Sketch Depicting the Relationship Between the Streams

The data entry shown by the keyboard in Fig. 6-1 is typical of the open-ended nature of a stream. The open-ended nature comes from the unknown amount of data to be processed by the stream instance. This leads to a basic question in software design, "Should the system perform character I/O or record I/O?" The answer to this question leads to a classic design trade-off: latency vs. throughput. Maximizing throughput often means increasing latency, whereas minimizing latency can mean a reduction in throughput. If the input stream comes from a human and is output to a human, as shown in Fig. 6-1, then the introduction of storage (called a *buffer*) can add to the latency in the system. In point of fact, people expect *trivial-response* time to be low for a particular action, or command. The trivial-response time is the time it takes to get an indication back from the computer. For example, when a user is typing on the keyboard, the user expects to see characters echoed on the screen quickly. Response times of less than a second lead to higher productivity [Shneiderman]. This would suggest that when human feedback is involved, unbuffered streams are needed. However, buffered streams may greatly improve a program's throughput. Thus, there is a basic difference between optimizing for throughput and optimizing for latency. Often it is the difference between a responsive program and an efficient one. For many years IBM built main-frames that use record I/O. The record used was a fixed length buffer, the size of a punch card (80 characters). With the advent of minicomputers and interactive computing, the I/O paradigm for users became character oriented. Hence, the set-up shown in Fig. 6-1 is generally performed with character I/O.

In the case when I/O is to and from files, it is much more efficient to use buffered I/O. Fig. 6-2 depicts a typical buffered I/O scenario.

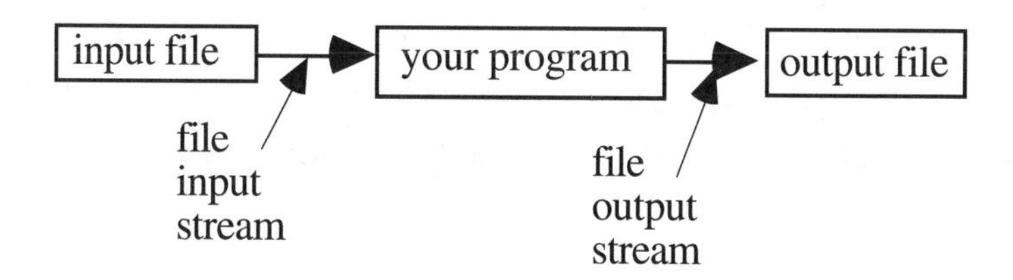

Fig. 6-2. Application for a Buffered Stream

Fig. 6-2 shows a file input stream and file output stream being used to read and write files. Such batch-style data processing is typical of the processing of image sequences and can be greatly accelerated with the introduction of buffered streams.

To further complicate the question of I/O with streams, we present the idea that a task could become I/O bound (that is, bottle-necked by low stream throughput). The most common way to avoid this is to create a thread to manage the I/O task. This is technique is described in [Lea] and [Oaks].

6.1. Getting a FileName

Java is both blessed and cursed with a rich set of classes for a programmer's use. This is a blessing because the programming environment can become richly populated with many elegant and useful classes. This is also a curse, because the number of classes, and their syntax for usage are difficult to memorize. Further, the beginner in Java is likely to be overwhelmed when confronted with the various class libraries and packages. Finally, to perform any one task, such as file I/O, probably more than one package is required. Making a *File* instance is an example of such a cross package software development task. A programmer will probably want to allow the user to select a file using a file dialog.

The *FileDialog* class resides in the *java.awt* package whereas the input and output streams reside in the *java.io* package. Thus, programs that need to perform file I/O will often import classes from both packages.

To avoid having to import the *java.awt* classes, a programmer would probably have to *embed* a file name in the Java source code. This is strongly discouraged, as file locations cannot be known if a program is moved. In fact, requiring a file to be at a specific location relative to a running application is a sure-fire way to break a program!

As a result, we incorporate a file dialog into all our file input and output examples.

There are two variants of the FileDialog class in the java.awt: the *open* dialog and the *save* dialog.

The following code snippet shows how to get a file name for *writing* a file:

```
1. String title = "Save as Java";
2. int mode =  FileDialog.SAVE;
3. Frame parent = this;
4. FileDialog fd = new
5.    FileDialog(parent, title, mode);
6. fd.show();
7. fd.hide();
8. String fn=fd.getDirectory()+fd.getFile();
9. if (fd.getFile() == null) return; // user canceled
```

Line 1 establishes the prompt that the user will see when the file dialog appears. Line 2 sets the *mode* for the dialog. Two possible modes are *FileDialog.SAVE and FileDialog.LOAD.* The *parent* Frame instance is available from within this snippet because the file dialog is being created from within a frame. In the case where the programmer does not already have a frame, line 3 may be replaced with:

```
3. Frame parent = new Frame();
```

The *getFile()* method shown on line 8 returns a string representing the file name, if the user does not cancel. Otherwise getFile returns null. The *getDirectory()* method, also shown on line 8, returns a string representing the directory in which the file resides. Neither getfile nor getDirectory should be invoked when the file dialog is visible.

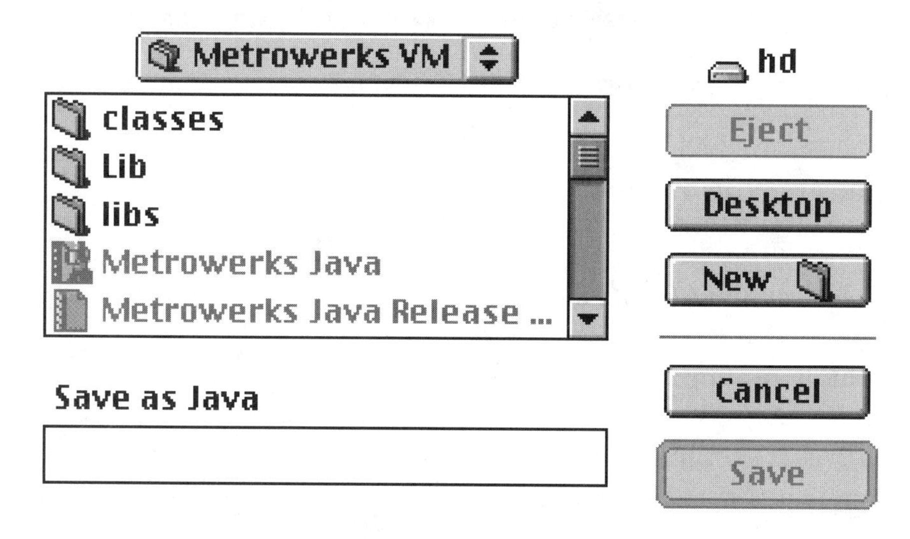

Fig. 6-3. A Sample Save File Dialog

Fig. 6-3 shows a sample save FileDialog.

To get a file name for reading, use the *FileDialog.LOAD* mode in the invocation of the FileDialog constructor. The constructor is overloaded, however, and if the *mode* is left out, the constructor defaults to using the FileDialog.LOAD mode. In fact, in the *ImageFrame* class we can place a method

```
protected String getReadFileName() {
    FileDialog fd = new
        FileDialog(this, "select file");
    fd.show();
    String file_name = fd.getFile();
    if (fd.getFile() == null) return null;
    String path_name = fd.getDirectory();
    String file_string = path_name + file_name;
    System.out.println("Opening file: "+file_string);
    fd.dispose();
    return file_string;
}
```

Note that the use of the visibility modifier, **protected**, indicates that the method may be accessed only by other methods within the class, subclass or package. This is because programmers who are outside of the package will probably want to replace:

```
FileDialog fd = new
     FileDialog(this, "select file");
```

with something like:

```
FileDialog fd = new
     FileDialog(new Frame(), "select file");
```

or a *null reference exception* will be thrown.

6.2. File Input Stream

The *FileInputStream* class resides in the *java.io* package and enables the reading of bytes from a file. A FileInputStream instance may be created from a file name. Generally, this resides in an instance of the *File* class. The File instance contains a full path-name as well as a method to extract the pathname, directory names, existence and length of the file, in bytes.

As an example, consider the following snippet, added to the ImageFrame class:

```
1. public void setFileName(String _fn) {
2. File f = new File(_fn);
3. if (f.exists()) {
4.       fileName = _fn;
5.       System.out.println(
6.       "File:"+ fileName +
7.       "\nis " + f.length() + " bytes long");
8. }
9. }
```

Line 3 shows an example of the existence of the file being tested. Line 7 gives the total length of the file, in bytes. See [Lyon and Rao] for a complete tutorial on the *File* class. The main use of the FileInputStream class is to act as a source for an uninterpreted stream of bytes. Generally, we can read from the FileInputStream one byte at a time. One use for reading the first two bytes of a file is to test the file to see what *kind* of file it is. Often, the file will be some kind of multi-media data-format file. For example, the file could be of type image, audio, video, text, etc. Within the broad category of any single type of file, there are generally an *unbound number* of subtypes. Examples of image formats include: PPM, GIF, PGM, FBM, QuickTime, Pict, pic, TIFF, etc. The same goes for audio and video formats. Chapter 7 discusses image formats in more detail. Writing a program that can read multiple formats is not easy. Java provides some tools that will *ease our pain*.

When faced with the grand unknown of the WWW, and the wondrous variety of data formats, it is wise to develop a technique for testing an input stream. One technique is to “sniff” the stream to see if it is sweet (i.e., known to be decodable). Toward this end, we have written the *StreamSniffer* class. The StreamSniffer takes an *InputStream* instance as an argument to its constructor. It then “sniffs” the bytes from the InputStream instance and returns a integer that corresponds to one of several known streams. The bytes are then *pushed back* into the FileInputStream instance, so that decoders will find the bytes they are seeking. In other words, the StreamSniffer is a non-destructive reader that inspects the stream and then covers its tracks!

Sniffing returns an integer that enables the construction of case-based switch dispatch of stream decoding. A switch dispatch is an efficient way to implement multiple decisions. The reason switch is efficient is that compilers will implement the switch statement using the so-called *computed goto*. The computed goto uses the switch argument to compute a relative address that is used for a goto statement, buried deep in the Java assembler. In fact, switch statements have a direct correspondence to the Java assembler commands, *lookupswitch* and *tableswitch* [Meyer].

The StreamSniffer class uses the InputStream instance (which may come from *any* source of bytes) and creates an instance of a *BufferedInputStream*. The BufferedInputStream internally stores bytes in a protected byte array. The StreamSniffer requires a BufferedInputStream instance in order to read a few bytes and then *reset* the BufferedInputStream to the beginning of the stream. This is a completely CPU bound task, and no additional I/O is performed during the act of a *sniff*. The methods of the BufferedInputStream require that we *mark* how many bytes we intend to read, then *reset* to the next point in the Stream.

A code snippet from the StreamSniffer follows:

```
1. package gui;
2. import java.io.*;
3. import java.util.*;
4. class StreamSniffer {
5.  private BufferedInputStream bis;
6.  private byte header[] = new byte[6];
7.  private int numberActuallyRead = 0;

8.  StreamSniffer(InputStream is) {
9.      bis = new BufferedInputStream(is);
10.      init();
11.      sniff();
12.  }
```

Lines 1-12 show the creation of a *BufferedInputStream* instance as well as the allocation of the space required for a header. After we *sniff* the buffered stream, we cover our tracks by resetting the internal stream pointer to the beginning of the stream. This enables the passing of the unmodified buffered input stream to a CODEC (COder-DECoder) without having to re-open the stream.

```
private void sniff() {
    if (! bis.markSupported()) {
        System.out.println(
            "StreamSniffer needs"+
            " a markable stream");
         return;
    }
    bis.mark(header.length);

    try {
        numberActuallyRead
            = bis.read(header);
        bis.reset();
    }
    catch (IOException e) {
        System.out.println(e);
        numberActuallyRead = -1;
    }

}
```

In order to use the StreamSniffer class, we must first make an instance of an InputStream. We let the StreamSniffer do the rest:

```
public StreamSniffer openAndSniffFile() {
  String fn = getReadFileName();
  InputStream is;
  if (fn == null) return null;
  try {
       is = new FileInputStream(fn);
  }
  catch (FileNotFoundException e) {
        return null;
  }
  StreamSniffer ss = new StreamSniffer(is);
  System.out.println("Hmm, this smells like a "+ss);
  return ss;
}
```

When an instance of the StreamSniffer is passed to *System.out.println*, the *toString* method is invoked. The StreamSniffer overrides the default *toString* method to print a string representation of what is found in the beginning of the file. For example, when the *openAndSniffFile* is invoked on one file, it prints:

```
Hmm, this smells like a GIF89a
```

This is beginning to shape up nicely. To top off our sniffer, we need a database of file types. In fact, there is no comprehensive database available, but there are some pretty good ones that can serve as a starting point (i.e., the *file* command of UNIX). It is typical to be able to tell what kind of stream you are dealing with by looking at the first few bytes. The file command of UNIX is able to use the *file name suffix* to help determine the file type. Streams on the Web do not inform a program of a file name suffix. More over, the file name suffix is often assigned by humans and may not be reliable as a source of information. The sniffer uses only the bytes found in the stream. So far, we have expanded this technique to identify forty common image and compression streams. Of course, this is only the tip of the iceberg! There are so many stream data types that no single program (that we know) can read them all. However, the StreamSniffer can help you to extend the database of known streams by giving you information about the stream. Thus, if you know what the stream is, but StreamSniffer does not, it is a straightforward matter to insert the stream into StreamSniffer's database.

The act of writing the StreamSniffer was very educational! First, let's review how Java stores numeric constants. In Java, an *octal* number (that is, one expressed in base 8) is written with a leading zero. For example,

```
0377
```

is the octal for the *int* 255. Consider that any octal digit passed as an argument is *automatically promoted to an int*! Thus, the invocation of a method for matching a series of octal constants must take ints in its parameter list. For example, to tell if a stream is a JPG stream we write:

```
if (match(0377,0330,0377,0356))
    return JPG;
```

Thus, the *match* method is invoked with four ints, each of which is denoted in its octal numeric equivalent. Octal is used primarily because the stream specification is written in octal. It is therefore a matter of being consistent relative to the specification. As an alternative, we could transform the octal numbers into their decimal equivalent.

Sometimes a specification is written using *hexadecimal* (radix 16). In Java, hexadecimal is written with a leading "0x" or "0X". A hexadecimal digit may be upper or lower case, i.e., one of {0, 1, 2, 3, 4, 5, 6, 7, 8, 9, a, b, c, d, e, f, A, B, C, D, E, F}. For example:

```
if (match(0xFF, 0xD8, 0xFF, 0xE0))
    return JPEG;
```

The hexadecimal numbers are also promoted to full ints. Thus, this use of the match method also requires a series of *int* parameters. In fact *int* is the default numeric promotion for all integer type constants.

The match method for the 4 ints follows:

```
public boolean match(
    int c0, int c1,
    int c2, int c3) {
    byte b[] = header;
    return
        ((b[0]&0xFF) == (c0&0xFF)) &&
        ((b[1]&0xFF) == (c1&0xFF)) &&
        ((b[2]&0xFF) == (c2&0xFF)) &&
        ((b[3]&0xFF) == (c3&0xFF));
}
```

The match method shows that each of the *int* values is truncated to its least significant 8 bits before a comparison is performed. Sometimes a stream specification will describe a string as being part of the header. The following snippet for the *classifyStream* method shows the header being converted into a string instance and then being compared to a prefix string:

```
public int classifyStream() {
    byte b[] = header;
    String s = new String(b);
    if (s.startsWith(".snd"))
       return SUN_NEXT_AUDIO;
.....
```

Now the reader is probably asking, “Where are these integer values stored?” The are stored as a series of public static final constants inside of the StreamSniffer class. For example:

```
public static final int TYPENOTFOUND = 0;
public static final int UUENCODED = 1;
public static final int BTOAD =2;
public static final int PBM =3;
public static final int PGM =4;
public static final int PPM =5;
.....
public static final int GZIP=34;
public static final int HUFFMAN=35;
public static final int PNG_IMAGE=38;
public static final int JPEG=39;
public static final int JPG=40;
```

A full listing of the interface to the StreamSniffer class are given in Section 6.4.1. One of the interesting problems with Java is that it lacks an unsigned byte. This is a truly perplexing issue that strikes at the heart of low-level stream programming.

Programmers who attempt to print bytes with various radices must address the lack of the unsigned byte in Java. For example:

```
byte b = (byte)128;
System.out.println(b);
```

will print:

```
-128
```

This is because the *byte* data type uses its most significant bit to represent the sign of the value. To address this problem, we created a *Ubyte* class. The Ubyte class is a *missing class* from the java.lang package because there is no corresponding scalar data type. The java.lang package contains a series of *wrapper* classes that enable the promotion from scalar data types to reference. For example, the *int, char, double* and *float* are all scalar data types that are passed by value. They correspond to class types with the following names: *Integer*, *Double*, *Character* and *Float*. These form the *wrapper* classes. One missing scalar data type is the unsigned byte and its corresponding wrapper, the *Ubyte* class. As non-language developers, we are unable to add scalar data types to Java and therefore we cannot add an unsigned *byte*. However, we can still add the unsigned *byte class*.

The *Ubyte* class resides in the *gui* package and provides a series of methods for handling bytes in an *unsigned* manner. An example of the use of the *Ubyte* class follows:

```
byte b = (byte)128;
Ubyte.printToOctal(b);
Ubyte.printToHex(b);
Ubyte.printToDecimal(b);
System.out.println(Ubyte.toString(b,8));
```

The output follows:

```
0200 0x80 128
200
```

All the Ubyte class methods are public and static. The *toString* method takes a byte and an *int* which is used to denote the radix the string should be in. The radix range is from Character.MIN_RADIX (=2) to Character.MAX_RADIX (=36). The value of the radix is always given in base 10.

The *printToOctal* method works by shifting and performing a bit-wise AND, byte 3 bits at a time. These bits are used to index into a character table, as shown in the following *printToOctal* example:

```
public static void printToOctal(byte b) {
        char oct[] = {
              '0','1','2','3','4',
              '5','6','7'};
        System.out.print("0"+
              oct[(b>>6) & 03] +
              oct[(b>>3) & 07] +
              oct[ b & 07]+" ");
   }
```

To obtain a decimal quantity from a byte, we use positional notation. Positional notation is a technique that is used to convert a number in any base (also called the *radix*) and convert it into base ten (called *decimal*). Each digit of the number in the source base is converted to its decimal equivalent. The decimal value is multiplied by the radix raised to a power that is equal to the position of the digit. For example,

$$0xDEAD = 13R^3 + 14R^2 + 10R^1 + 13R^0 \tag{6.1}$$

where R represents the radix. For R=16 the decimal for (6.1) is:

$$7853 = 4096 + 3584 + 160 + 13 \tag{6.2}$$

In general we write:

$$(N)_{10} = \sum_{i=0}^{D-1} s_i R^i \tag{6.3}$$

where

$(N)_{10}$ = the decimal result,

s_i = the decimal value of the symbol at digit i

D= the total number of digits, and

R = radix

We isolate the most significant 4 bits, multiply by 16, then add the least significant 4 bits. This yields an unsigned integer representation of the byte quantity:

```
1. public static int byteToInt(byte b) {
2.         return
3.                 ((b >> 4) & 0xF)*16 +
4.                 (b & 0xF);
5. }
```

Line 3 shows a shift to the right by 4 bits, (b >> 4). Line 4 shows a bit-wise AND. The bitwise AND is different from the boolean AND. The boolean AND is denoted '&&' and returns a *boolean* type. The bit-wise AND is denoted '&' and returns an *int* type.

Once we have an unsigned integer, we can convert it to a string that represents the integer in an radix between 2 and 36, using:

```
public static String toString(byte b,int radix) {
        return
                Integer.toString(byteToInt(b), radix);
}
```

Note that the *toString* method returns something different from the print methods. The toString method is designed to be consistent with the toString methods of the wrapper classes. The goal of the print method is to allow the StreamSniffer class to print out a header dump that can be used as an argument to the match method, thereby extending the StreamSniffer's data base of known streams. Now when we sniff an unknown stream, we get a program fragment that can be used to help identify the stream later:

```
In hex...
0x01 0x3B 0x61 0x00 0x11 0x00
in base 8...
0001 0073 0141 0000 0021 0000
 in ASCII
 ;a
if (match(1,59,97,0))
Hmm, this smells like a TYPENOTFOUND
The id is :0
```

If we sniff a known stream, we get:

```
Hmm, this smells like a PPM_RAWBITS
The id is :8
```

To map the *id* into a string that is human-readable, we use an instance of the *HashTable* class. The HashTable resides in the java.utils class and provides the ability to store and retrieve instances based on a *key*. The HashTable instance is a protected variable, so that subclasses may add new entries into the database. As an example of use, we add a series of entries into the StreamSniffer database:

```
add(TIFF_BIG_ENDIAN,"TIFF_BIG_ENDIAN");
add(TIFF_LITTLE_ENDIAN,"TIFF_LITTLE_ENDIAN");
add(FLI,"FLI");
add(MPEG,"MPEG");
```

Where the *add* and *getStringForId* methods are given by

```
protected void add(int i, String s) {
    h.put(new Integer(i), s);
}

public String getStringForId(int id) {
     return (String) h.get(new Integer(id));
}
```

6.3. The Ubyte Class

The Ubyte class resides in the *gui* package and provides printing and conversion services. It is used to represent the unsigned byte quantity.

All the public methods in the Ubyte class are static. All are intended to treat a byte as an unsigned quantity. Thus, when we convert the byte to an int, we always get a positive quantity that ranges from 0...255. The print instructions do not add a new line after they print. Their octal and hexadecimal counterparts add the “0” and the “0x” prefix to any numeric strings they print. The *toString* methods add no such prefix.

```
package gui;
public class Ubyte {
```

- To convert the *byte* to an *int* and returns the *int*.:

```
public static int toInt(byte b)
```

- To convert the byte to a string in any radix from 2-36:

```
public static String toString(byte b,int radix)
```

- To convert the byte to a string in radix 10:

```
public static String toString(byte b)
```

- To print out the byte in base 8:

```
public static void printToOctal(byte b)
```

- To print out the byte in base 10:

```
public static void printToDecimal(byte b)
```

- To print out the byte in base 16:

```
public static void printToHex(byte b)
}
```

6.4. The StreamSniffer Class

The *StreamSniffer* class resides in the *gui* package. It is used to read and classify the header of a file. Typically, a stream decoder will use the classification to dispatch the stream to the correct decoder.

6.4.1. Class Summary

```
package gui;
import java.io.*;
import java.util.*;
public class StreamSniffer {
   public boolean match(char c0, char c1)
   public boolean match(char c0, char c1, char c2, char c3)
   public boolean match(int c0, int c1)
   public boolean match(int c0, int c1, int c2, int c3)
   public BufferedInputStream getStream()
   public int classifyStream()
   public StreamSniffer(InputStream is)
   public String getStringForId(int id)
   public String toString()
   public void printHeader()
   public static final int TYPENOTFOUND = 0;
   public static final int UUENCODED = 1;
   public static final int BTOAD =2;
   public static final int PBM =3;
   public static final int PGM =4;
   public static final int PPM =5;
   public static final int PBM_RAWBITS=6;
   public static final int PGM_RAWBITS = 7;
   public static final int PPM_RAWBITS = 8;
   public static final int MGR_BITMAP =9;
   public static final int GIF87a =10;
   public static final int GIF89a =11;
   public static final int IFF_ILBM =12;
   public static final int SUNRASTER =13;
   public static final int SGI_IMAGE =14;
   public static final int CMU_WINDOW_MANAGER_BITMAP =15;
   public static final int SUN =16;
   public static final int TIFF_BIG_ENDIAN =17;
   public static final int TIFF_LITTLE_ENDIAN =18;
   public static final int FLI =19;
   public static final int MPEG =20;
   public static final int SUN_NEXT_AUDIO=21;
   public static final int STANDARD_MIDI=22;
   public static final int MICROSOFT_RIFF=23;
   public static final int BZIP=24;
   public static final int IFF_DATA=25;
   public static final int NIFF_IMAGE=26;
   public static final int PC_BITMAP=27;
   public static final int PDF_DOCUMENT=28;
   public static final int POSTSCRIPT_DOCUMENT=29;
   public static final int SILICON_GRAPHICS_MOVIE=30;
```

```
    public static final int APPLE_QUICKTIME_MOVIE=31;
    public static final int ZIP_ARCHIVE=32;
    public static final int UNIX_COMPRESS=33;
    public static final int GZIP=34;
    public static final int HUFFMAN=35;
    public static final int PNG_IMAGE=38;
    public static final int JPEG=39;
    public static final int JPG=40;
}
```

6.4.2. Class Usage

An instance of the StreamSniffer is constructed using an *InputStream* instance. The StreamSniffer internally constructs an instance of a *BufferedInput* stream which is used to scan the header in the stream for identification. Scanning and classification are performed once, automatically, at the construction of the StreamSniffer.

Suppose that the following variables are declared:

```
StreamSniffer ss;
InputStream is;
String fn;
```

To see if the stream begins with 2 unsigned bytes whose character representation is known, use

```
public boolean match(char c0, char c1);
```

To see if the stream begins with 4 unsigned bytes whose character representation is known, use

```
public boolean match(char c0, char c1, char c2, char c3)
```

To see if the stream begins with 2 unsigned byte quantities whose integer representation is known, use

```
public boolean match(int c0, int c1)
```

To see if the stream begins with 4 unsigned byte quantities whose integer representation is known, use

```
public boolean match(int c0, int c1, int c2, int c3)
```

To get an *untainted* instance of the buffered input stream, use

```
public BufferedInputStream getStream()
```

To get an identifier that corresponds to one of the static final constants in the StreamSniffer class, use:

```
public int classifyStream()
```

To create an instance of the StreamSniffer from a file name, *fn*, use

```
 InputStream is;
if (fn == null) return null;
try {
     is = new FileInputStream(fn);
}
catch (FileNotFoundException e) {
     return null;
}
StreamSniffer ss = new StreamSniffer(is);
```

To map the identifier returned by *classifyStream* into a string, use

```
public String getStringForId(int id)
```

To return a minimal string representation of the stream and invoke the *printHeader* method if the header is known, use

```
public String toString()
```

To print the head of the stream in three radices and create a prototype *if*-statement that can check this stream type, use

```
public void printHeader()
```

7. Image File Readers and Writers

People who live on the edge of technology
are destined to die on the edge of technology.

– Unknown

This chapter addresses several topics related to input and output. While references are made to streams (uninterpreted sequence of bytes), the basic goal is to read and write files to the local file system.

The reading and writing of image files is a surprisingly deep topic. This is particularly true for modern image file formats that include compression techniques. The primary focus is on how to save files and how to select a format. We defer discussion of exactly *what* is in the format and avoid the discussion of various compression techniques (a recurring topic).

In order to write an image file, we must focus our attention toward output streams. Output streams are typically generated by a program and are directed toward a byte channel (file, network server, array of bytes, etc.). It is typical for high-level data types (e.g., strings, integers, instances of classes, etc.) to be transformed into an output stream for transmission into the channel. Based upon the characterization of the channel (latency, reliability and bandwidth), the data may be encoded to obtain a particular *quality of service*. There are some channels which are said to be reliable and some that are unreliable. An unreliable channel (such as an instance of a *DatagramSocket*) can lose data or can cause data to arrive out of order. A reliable channel (such as an instance of a *Socket*) will always deliver the data in order and without loss, *in theory*.

In the core API of Java, the vast majority of data streams is reliable. Hence, the focus of this chapter is on the serialization of high-order data structures (images) into a reliable output stream represented by an instance of the *FileOutputStream* class.

7.1. Getting a Filename and Writing a PPM File

The first order of business is to write out an image to a file. This requires that we select an image file format and there are many to choose from. The *StreamSniffer* of Chapter 6 knows a small fraction of them. As we shall see, there is always room for more image file formats. In fact, this chapter shows how to write some new ones! Java provides the enabling technology to create and decode image file formats so that formats can be proliferated on the Internet with wild abandon (this can be both good and bad).

The first step is to get a save file name. The code for getting a file name follows:

```
public String getSaveFileName(String prompt) {
        FileDialog fd = new
        FileDialog(this, prompt, FileDialog.SAVE);
    fd.setVisible(true);
    fd.setVisible(false);
    String fn=fd.getDirectory()+fd.getFile();
    if (fd.getFile() == null) return null; //usr canceled
    return fn;
}
```

The *getSaveFileName* method appears in the *SaveFrame* class, which is in the *gui* package. The *SaveFrame* class is documented in Section 7.5. The *vs.WritePPM* class is described in Section 7.10. The *prompt* string, passed as an argument to the *SaveFrame* class, appears in the dialog used to enter the save file name. Fig. 7-1 shows a save file dialog.

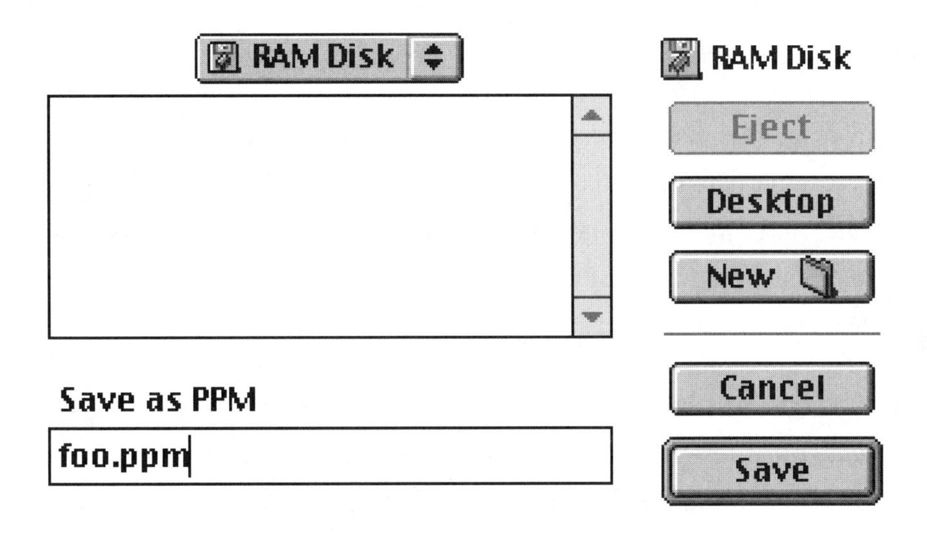

Fig. 7-1. Save File Dialog

The code used to create the save file dialog shown in Fig. 7-1 follows:

```
public void saveAsPPM() {
    String fn = getSaveFileName("Save as PPM");
    if (fn == null) return;
    vs.WritePPM.doIt(r, g, b, fn);
}
```

Thus, the *saveAsPPM* method is an elementary method that is used to invoke the *doIt* method in the *WritePPM* class, which resides in the *vs* package. We note that *doIt* takes the two-dimensional arrays, *r, g, and b,* directly as its arguments, writing an uncompressed, 24-bit-per-pixel image out to disk.

In the following section, we show how to write a method that writes an image to Java source code in a file.

7.2. Writing a Program That Writes a Program

In this section we assume that the reader has access to the API documentation for the *java.io* package. In particular, we assume that the reader can look up the usage of the *PrinterWriter* and *FileWriter* classes. In general, their usage may be inferred from the context of the program, but the reader may like to read more about these classes in [Chan et al.].

Suppose that you are interested in writing a Java program that is able to carry an image, as an *integrated* resource. An integrated resource is one that becomes a part of the program so that a data file does not have to be read in. There are several excellent arguments for integrating image resources in an image processing program. If a Java program must display an image, it typically either relies on a specific network resource (e.g., a URL), a system resource (e.g., a file in a particular location), or a human resource (e.g., a pop-up dialog box that says, "Get me an image file that I know how to read"). It may not always be easy to find a network-based resource (some computers are not on the net all the time). It may not be easy to find a specific file (programs move, files change relative position, etc.). Starting a program that requires the user to locate an image *every time* it starts is both rude and painful.

It is much better if a program can create internal data structures that can store an image. Once this is done, the image becomes a part of the Java program and is compiled into a class file. Thus, one interesting image format is the Java programming language itself! In fact, the image format, if human readable, can enable the human recognition of patterns that can lead to an insight into the interpixel coherence of the image. A code snippet for a gray-scale version of the mandrill image follows:

```
package gui;

public class NumImage {
    public  static short gray[][] = {

{ 83,  62,  63,  85,  98,  57, 116, 109, 126,  96,  94, 142,
       94,  94, 127, 142, 126, 142,  96, 109, 126,  86, 130,
       99,  86, 116,  99, 141, 156, 141, 119, 166, 156, 112,
       92, 141, 132, 107, 104,  83,  92, 112,  85, 116,  99,
      141, 120, 120, 141, 140, 156, 109, 120, 120, 101, 146,
      132,  67, 104, 146, 120, 104, 146, 107,  82, 104, 135,
      135, 116, 157, 116, 166, 132, 132, 137, 107, 120, 119,
      135, 138, 138, 146, 104, 137, 137, 120, 120, 120, 101,
      107, 120, 120, 138, 146, 137,  70, 140, 166, 138, 157,
      132, 141, 119, 112, 119, 112, 109, 112, 174, 141, 141,
      166, 180, 130, 164, 148, 148, 164, 164, 141, 141, 116,
      164, 142,  85,  96, 104, 104},
{ 96,  82,  85,  62,  62,  57, 114, 109, 109, 133, 126, 160,
      127,  94, 109, ....
```

The code for opening, writing and closing a Java file follows:

```
1.    public void saveAsJava()  {
2.    String fn = getSaveFileName("Save as Java");
3.     if (fn == null) return;
4.    saveAsJava(fn);
5.    }
```

Note that if the file name returned by the *getSaveFileName* method (shown on line 2) is **null,** then the user has canceled and the method returns. The save methods in the *SaveFrame* are overloaded so that batch processing can be done with greater ease. The overloaded version of the *saveAsJava* method appears below:

```
1.    public void saveAsJava(String fn)  {
2.     try {
3.         FileWriter fw = new FileWriter(fn);
4.         PrintWriter pw = new PrintWriter(fos);
5.         printJavaImage(pw);
6.         pw.flush();
7.         fw.close();
8.    }
9.     catch (Exception e)  {}
10.   }
```

An instance of the *FileWriter* class (created on line 3) is used to channel the bytes into a file. To perform the formatting needed for the data, it is useful to use an instance of the *PrintWriter* class (see line 5). The purpose of a *PrinterWriter* instance is to write out string representations of data elements. The *printJavaImage* method makes use of a *ProgressFrame* instance in order to keep the user updated with the progress of the image writing process. The *ProgressFrame* is described in Section 7.8. The *printJavaImage* method follows:

```
1.     public void printJavaImage(PrintWriter pw) {
2.       int g;
3.       ProgressFrame pb =
4.             new ProgressFrame("writing java...");
5.         pb.setVisible(true);
6.       pw.println("package gui;\n"
7.         +"\nclass NumImage {\n"
8.         +"\tpublic static short gray[][] = {\n");
9.       for (int y=0; y< height; y++) {
10.           pw.print("{");
11.           for (int x=0; x < width-1; x++) {
12.                 g = r[x][y];
13.                 if (g <10) pw.print("  ");
14.                 else if (g <100) pw.print(" ");
15.                 pw.print(g+", ");
16.           }
17.           pw.println(r[width-1][y]+"},");
18.           double percent = (y*1.0/height);
19.           if ((y % 5) == 0)
20.                 pb.setRatioComplete(percent,
21.                       "Percent done:"+(int)(percent*100)+"%");
22.      }
23.       pw.println("};}");
24.       System.out.println("Done writing image");
25.        pb.setVisible(false);
26.    }
```

Lines 13-15 are primarily concerned with formatting the numbers for readability. When the gray value, *g*, falls below 10, it is padded with two spaces:

```
13.                 if (g <10) pw.print("  ");
```

When the value falls between 10 and 100 it is padded with one space:

```
14.                 else if (g <100) pw.print(" ");
```

Finally, the numeric value is printed, followed by a comma and a space:

```
15.                 pw.print(g+", ");
```

The *printJavaImage* method treats an instance of the *PrintWriter* as if it were a *System.out* instance. That is, the methods *print* and *println* act like *System.out*. The only difference is that *PrintWriter* prints to a file and does not flush the output stream buffer when a line terminator is encountered. Line 17:

```
17.            pw.println(r[width-1][y]+"},");
```

shows that the only value being saved to the output is the value of the red channel. This is because we are writing a grayscale image. Lines 3 and 4:

```
3.        ProgressFrame pb =
4.             new ProgressFrame("writing java...");
```

introduce a new class called the *ProgressFrame*. The *ProgressFrame* creates a *ProgressFrame* instance called *pb*. This provides feedback to the user. Updates to the *ProgressFrame* instance are performed by

```
18.            double percent = (y*1.0/height);
19.            if ((y % 5) == 0)
20.                 pb.setRatioComplete(percent,
21.                     "Percent done:"+(int)(percent*100)+"%");
```

The *percent* is a number that must be of type *double* and range from zero to one. It should be sampled at regular intervals in the program. The update, shown on line 20, should be called in only one place in the program. The *ProgressFrame* estimates the amount of time left based on the fraction completed.

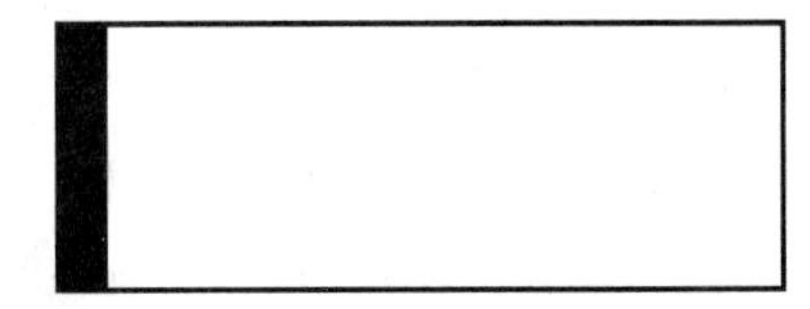

Percent done:7%

Elapsed Time: 3.25 Seconds

Estimated time left: 38.34 Seconds

Fig. 7-2. The *ProgressFrame*

Fig. 7-2 shows a sample of the *ProgressFrame* as it appears during a long save operation.

7.3. Writing Lossless 24-Bit Color Image Files

The PPM (Portable Pix Map) image file format is a 24-bit image format that is supported by many applications and several powerful UNIX tools. One problem with the PPM format is that it is not compressed and therefore takes a great deal of disk space. This section shows a simple technique for creating an output stream that is compressed using a lossless compression algorithm that resides in the public domain called *GZIP* [RFC1952]. The GZIP format is file system, CPU, character set and operating system independent and is therefore suitable for data distribution on the Internet. Further, unencumbered by patents, the algorithm can be practiced freely and is decompressable via existing utilities (such as *gunzip*). GZIP uses a compression algorithm that is a combination of the LZ77 algorithm and Huffman coding [LZ77]. The compression is unencumbered by patents and is comparable to the best currently available lossless compression techniques.

The goal of the Huffman coding scheme is to use the probability of a symbol to minimize the length of binary transmissions, *on average*. Huffman coding is a kind of prefix coding that sorts a series of symbols according to their probability of appearance. The probability is estimated by statistically measuring the input data stream. A binary string is assigned to each symbol so that the most popular symbols are given the shortest length binary string. The algorithm is typically implemented by starting with the two least popular symbols. These form the deepest node of a binary tree. The symbols are added to the tree, in order of lowest probability first. Traversing left on the node of the tree corresponds to a zero. Traversing right on the node of the tree corresponds to a one. See [Standish] for a Java implementation and further explanation of Huffman coding.

The goal of the LZ77 algorithm is to use a sliding variable-size window to build a dictionary of symbols that appear often in the input stream. These symbols are replaced by two numbers. The first is the distance indicating where in the dictionary the symbol is and the second is a length of the symbol in the dictionary. For implementation details, see [RFC1951].

Compression is controlled by the *java.util.zip.Deflater* class. Compression settings permit a tradeoff of between compression ratio and CPU time. The relationship between CPU time and compression ratio is not linear and varies according to the input data. We have accepted the default settings for all experiments reported in this book. The default settings use the LZ77/Huffman coding in combination with a compression level of 6 (with the maximum being 9). At first blush, a compression setting of 9 doubles the CPU time required but increase compression by around 25%, depending on the data. An extensive and careful study is needed to further document this phenomenon.

Chapter 15 uses GZIP as a post-processor for lossy multi-resolution wavelet-based image compression. One of the most compelling features of the GZIP file format is the ease of implementation in Java. A summary for the *GZIPOuputStream* class appears in [Chan et al.]. The details for the *WritePPM* class are shown in Section 7.10. The *Timer* class is summarized in Section 7.7. The code for generating a compressed GZIP output stream follows:

```
1.      public void saveAsPPMgz() {
2.      String fn = getSaveFileName("Save as PPM.gz");
3.       if (fn == null) return;
4.      WritePPM wppm = new WritePPM(width, height);
5.      Timer t = new Timer();
6.      t.start();
7.      try {
8.           GZIPOutputStream os = new GZIPOutputStream(
9.              new FileOutputStream(fn));
10.          wppm.header(os);
11.          wppm.image(os, r, g, b);
12.          os.finish();
13.          os.close();
14.         } catch(Exception e) {
```

```
15.             System.out.println("Save PPM Exception - 2!");
16.           }
17.       t.stop();
18.       t.print(" saveAsPPMgz in ");
19.   }
```

Lines 8 and 9 show an instance of a GZIP output stream being created. Once the *finish* and *close* methods are invoked (which finishes the output stream and then closes it) the *GZIPOutputStream* instance may no longer be used. For delivering a series of files, only *finish* should be invoked. In this case, a series of images may be transmitted as a part of the same *GZIPOutputStream* instance. This is a very useful feature for a client/server application. Table 7-1 shows the save times and file sizes for several images. The times are computed for a PowerMac 8100/100 (a PowerPC 601 running at 100 MHz). The MetroWerks CodeWarrior IDE 3.0 was used with a Just-In-Time Compiler. All files were written to RAM-disk in order to eliminate the effects of disk I/O speed. All images are 24 bits per pixel.

Let

$$t_{wui} = \text{time to write an uncompressed image}$$

and

$$t_{wgz} = \text{time to write a GZIP image}$$

so that the ratio of the compression time increase is given by

$$R_t = t_{wgz} / t_{wui} \tag{7.1}$$

Further, we define the compression ratio as

$$\mathrm{R_c} = \text{uncompressed\_size/compressed\_size} \tag{7.2}$$

The information theorist will often use the term *source coding* instead of compression. For the purpose of this discussion, the two terms are synonymous.

File Name	Dimensions	size in bytes	time to write	R_c	R_t
baboon.ppm	512x512	786,447	3.257		
baboon.ppm.gz	512x512	752,459	7.693	1.05	2.4
girl.ppm	480x512	737,259	2.802		
girl.ppm.gz	480x512	610,673	7.14	1.21	2.5
lena.ppm	512x512	786,447	2.954		
lena.ppm.gz	512x512	733,863	7.638	1.07	2.6

Table 7-1. Effect of GZIP on Image Compression

Table 7-1 shows that the time to write an image (in seconds) increases by a factor of 2.4 to 2.6 times when written in the GZIP format. Yet this lossless compress scheme has saved only 4% to 17% on the image size. The results shown in Table 7-1 are image dependent.

A comparison of the *GZIPOutputStream* file size with the file size of a shareware version of GZIP showed a very small difference (0.06%), well within any reasonable tolerance for a difference in implementation.

For the transmission of an image via a phone line using a 28 kbps modem, a saving of 4% on the *baboon.ppm* file results in a 33,998 byte savings. This translates to a 271,904 bit savings and saves 9.7 seconds in download time. There is a 4.9-second savings with a 56 kbps non-compressing modem. Therefore, in the application of transmitting images using current state-of-the-art modems, GZIP compression seems generally better than none. Some modems have built-in compression, however. This is due to the incorporation of v.42 bis which uses a Ziv-Lempel compression algorithm [Ziv et al.]. However, the hardware implementation of this algorithm is characterized by limited memory resources, and this prevents optimal implementation of the compression algorithm. Using the *girl.ppm* and *girl.ppm.gz* files, an upload was performed via a phone line. The modem (v.32 bis) was non-compressing and the trial used file transfer protocol (ftp) with a binary mode. The upload time was 36 seconds faster with the *girl.ppm.gz* than with the *girl.ppm* file, even though the average transfer rate on the *girl.ppm.gz* was slower (24.2 kbps for the *girl.ppm* vs.

the 23.6 kbps of the *girl.ppm.gz* file). The GZIP format is a clear winner for this type of image and modem combination (4 seconds of CPU time saved 36 seconds of transfer time).

For the purpose of storage, the GZIP compression does add time to the saving of an image. However, with the advent of much faster CPUs, this should not be a pressing issue. To guide the selection of the criterion of optimization, we attempt to predict the growth of CPU speed and disk space over the next 10 years.

Assume that Ram, clock speed and disk size double every 18 months [Hennessy et al.]. In 1998, a 32 MB RAM, 200 Mhz machine with 1 GB of disk is common. In 10 years we would expect about 2048 MB RAM in 12 GHz machines with 64 GB of disk, all for around $1800 (assuming 3% inflation compounded over a 10-year period with a $1500 initial cost). A chart of RAM and clock speed as a function of year is shown in Fig. 7-3.

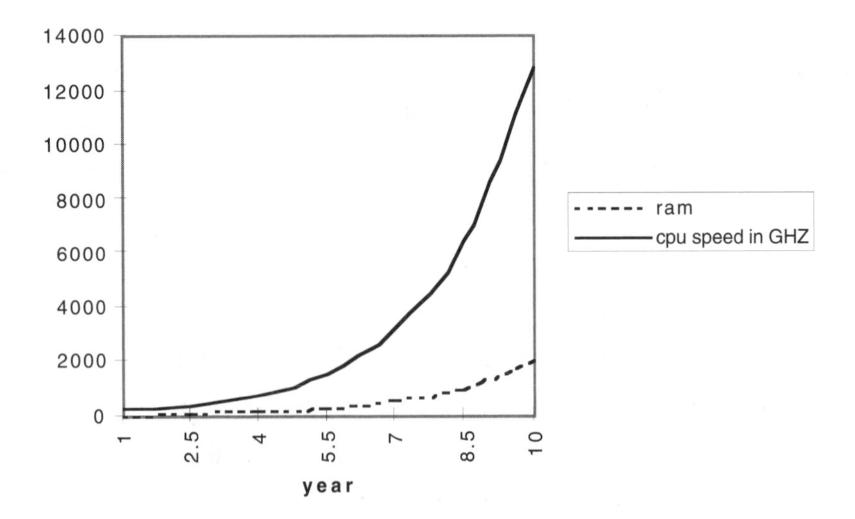

Fig. 7-3. Chart of RAM and CPU Speed for the Next Ten Years.

When we look at modem rates since 1962, we see a nearly linear growth in speed for modems using POTS (Plain Old Telephone Service). Table 7-2 shows the modem model, the year of introduction and the bit rate [Forney].

Model	Year	bps
Bell 201	1962	2400
Milgo 4400/48	1967	4800
Codex 9600c	1971	9600
Paradyne	1980	14400
Codex/2680	1985	19200

Table 7-2. Modem Milestones

Fig. 7-4 shows the year as a function of bit rate.

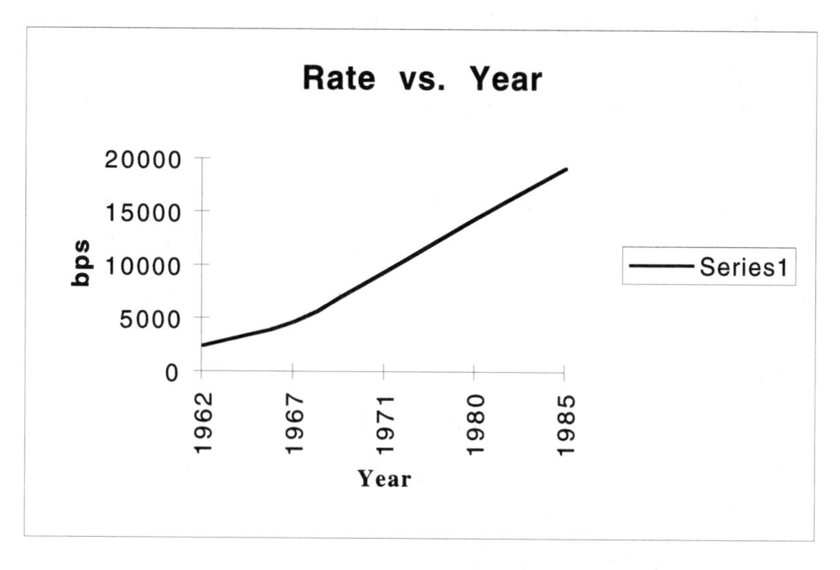

Fig. 7-4. Modem Bit Rate vs. Year

Based on the historical data, we can formulate a second order polynomial regression fit that may serve as a prediction of the POTS modem rate as a function of the year:

$$\tilde{R} = 7{,}195{,}094 - 8000\,year + 2.21\,year^2 \tag{7.3}$$

Equation (7.3), is a prediction based on past performance and does not take into account the migration toward revolutionary technologies, such as cable modems.

According to (7.3) the estimate for the 1998 POTS modem speed is on the order of 33.4 kbps. This is an underestimate because the current x2 technologies are able to perform 56 kbps in the downstream direction (although it is a pretty good estimate for the upstream direction). The rate of increase will, of course, be leap-frogged by hybrid-fibre cable systems that will be deployed within the next few years [@Home].

It is not clear how long the POTS modem will be used in the home. As long as people are using POTS-based modems to connect to the Internet, downloads will almost always be I/O bound. Thus, algorithms that use CPU and RAM to help make I/O more efficient should continue to improve the overall though-put of POTS modem transfers.

7.4. Reading GZIP Compressed PPM Files

This section shows how to decode the GZIP file created by the *saveAsPPMgz* method of Section 7.3. The *GZIPInputStream* class resides in the *java.io* package and makes the decoding of the GZIP files a simple matter. The idea is to make an instance of a *GZIPInputStream* and to read bytes from the *GZIPInputStream* instance as if they were being produced from an uncompressed source.

The first order of business is to determine what kind of file is trying to be read. The *OpenFrame* class uses the *StreamSniffer* (described in Section 6.4) to dispatch the appropriate decoding method. The central method is called *openImage*:

```
1.     public void openImage() {
2.     String fn = getReadFileName();
3.     if (fn == null) return;
4.     File f = new File(fn);
5.     if (!f.exists()) return;
6.     setFileName(fn);
7.     InputStream is= null;
8.     try {
9.          is = new FileInputStream(fn);
10.     }
11.     catch (FileNotFoundException e) {
12.          e.printStackTrace();
13.     }
```

Line 2:

```
   2.      String fn = getReadFileName();
```

creates a dialog box that allows the user to either select a file or to cancel. If the user cancels, the *fn* String instance is set to *null* and the method returns. Line 4 makes an instance of a *File* and tests to see if it exits. Line 5:

```
   5.      if (!f.exists()) return;
```

returns if the file does not exist. Lines 7-13 make an instance of an *InputStream* which is suitable for use by the *StreamSniffer*. We make an instance of the *StreamSniffer* in order to test the stream for readability.

```
        StreamSniffer ss = new StreamSniffer(is);
        int id = ss.classifyStream();
```

Once we have a classification identifier (called *id*), we can dispatch to the proper stream decoder. The following code snippet shows the use of a *switch* statement to perform the dispatch:

```
   1.      switch (id) {
   2.          case StreamSniffer.PPM:
   3.              setFileName(fn);
   4.              openPPM(fn);
   5.               break;
   6.          case StreamSniffer.PPM_RAWBITS:
   7.              setFileName(fn);
   8.              openPPM(fn);
   9.               break;
   10.          case StreamSniffer.GIF87a:
   11.              openGif(fn);
   12.               break;
   13.          case StreamSniffer.GIF89a:
   14.              openGif(fn);
   15.               break;
   16.          case StreamSniffer.GZIP:
   17.              openPPMgz(fn);
   18.               break;

   19.           default: {
   20.               System.out.println("Can not open "+ss+" as
         image");
   21.          }

   22.     }
   23.    }
```

Lines 16 and 17 show the dispatch of interest here, the opening of a gzipped PPM type image. Once inside the *public* method *openPPMgz,* we are able to make an instance of the *GZIPInputStream* using:

```
GZIPInputStream in = null;

try {
  in = new GZIPInputStream(
      new FileInputStream(fn));
  } catch(Exception e)
   {e.printStackTrace();}
```

We then read in all the bytes from the *GZIPInputStream* instance using a sequence of private methods whose full source code implementations are on the CD-ROM .

7.5. The SaveFrame Class

The *SaveFrame* class resides in the *gui* package. It provides a series of methods that enable the menu items under the *save* menu. The formats that have been implemented so far include GIF, PPM, PPM.GZ and the Java source code. Keep in mind, the GIF format will work only with 8-bit color images. PPM is a 24-bit color image format and PPM.GZ is the GZIP compressed version of the PPM format.

7.5.1. Class Summary

```
public class SaveFrame extends NegateFrame {
public static void main(String args[])
public String getSaveFileName(String prompt)
public void saveAsPPM()
public void saveAsPPM(String fn)
public void saveAsPPMgz()
public void saveAsPPMgz(String fn)
public void saveAsGif()
public void saveAsGif(String fn)
public void saveAsJava()
public void saveAsJava(String fn)
public void printJavaImage(PrintWriter pw)
}
```

7.5.2. Class Usage

To test the SaveFrame class, you may invoke the static *main* method directly:

```
gui.SaveFrame.main(args);
```

This creates an instance of the *SaveFrame* class and shows the frame. To make an instance of the *SaveFrame* class, a title for the frame is required. An example that automatically displays an image from internal resources follows:

```
public static void main(String args[]) {
     SaveFrame sf = new SaveFrame("save frame");
     sf.show();
}
```

Suppose that *sf* is a *SaveFrame* instance. To save the image in a GIF image file format and prompt the user for the filename, use

```
sf.saveAsGif();
```

Suppose that *fn* is a suitable output filename, and this is already known. To save the image in a GIF image file format, without prompting the user, use

```
sf.saveAsGif(fn);
```

Suppose that *prompt* is an instance of a *String* class. To get a file name, suitable for writing a file, use

```
String fileName = sf.getSaveFileName(String prompt);
```

If *fileName* is null, then the user has canceled.

To save the image in a PPM image format file and prompt the user for the filename, use

```
sf.saveAsPPM();
```

To save the image as PPM, without prompting the user, use

```
sf.saveAsPPM(fn);
```

To save the image in a GZIP PPM image file format and prompt the user for the filename, use

```
sf.saveAsPPMgz();
```

To avoid the user prompt, use

```
sf.saveAsPPMgz(fn);
```

To save the image as a Java source file that may be compiled into a Java program and prompt the user for the filename, use

```
sf.saveAsJava();
```

To avoid the user prompt, use

```
sf.saveAsJava(fn);
```

To save the image as a Java source file that may be compiled into a Java program identified by a *PrinterWriter* instance *pw,* use

```
sf.printJavaImage(pw);
```

7.6. The OpenFrame Class

The *OpenFrame* class resides in the *gui* package. It is used to centralize the methods for opening a variety of file formats. The *OpenFrame* extends the *SaveFrame* of the previous section. Thus programmers who make use of the *OpenFrame* will have access to the methods from all the super classes. A class tree is shown in Fig. 7-5.

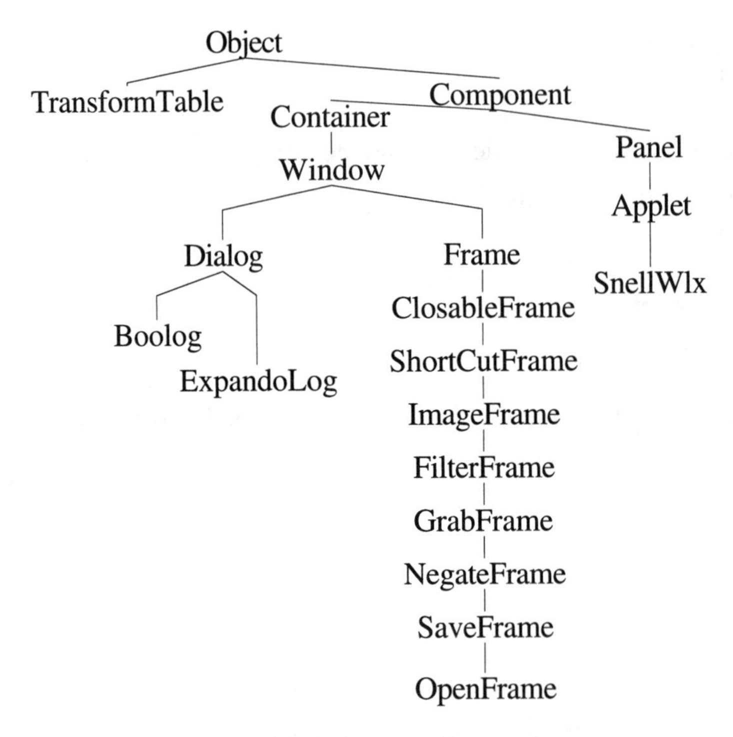

Fig. 7-5. Class Tree Showing *OpenFrame*

The *OpenFrame* class makes use of the *StreamSniffer* to read the *InputStream* instance and decide how to dispatch the opening of the file. Programmers who want to add formats for reading which the *StreamSniffer* does not know will have to modify both the *StreamSniffer* and the *OpenFrame* classes.

7.6.1. Class Summary

```
public class OpenFrame extends SaveFrame {
public OpenFrame(String title)
public void openImage()
public static void main(String args[])
public void openPPMgz(String fn)
public void openPPM(String fn)
public void openPPM()
public StreamSniffer openAndSniffFile()
}
```

7.6.2. Class Usage

To test the *OpenFrame* class, you may invoke the static *main* method directly:

```
gui.OpenFrame.main(args);
```

This creates an instance of the *OpenFrame* class and shows the frame. To make an instance of the *OpenFrame* class, a title for the frame is required. An example that displays an image from internal resources follows:

```
public static void main(String args[]) {
    OpenFrame of = new OpenFrame("Open frame");
    of.show();
}
```

Suppose that a series of variables are defined as follows:

```
OpenFrame of = new OpenFrame("Open frame");
String prompt = "select an image file";
String fn = of.getSaveFileName(prompt);
```

To open an image file in a known format and prompt the user for the file, use

```
of.openImage();
```

To open an image file in the GZIP compressed PPM format, use

```
of.openPPMgz(fn);
```

To open an image file in the PPM format, given a *String* instance, *fn*, as a valid input file name, use

```
of.openPPM(fn);
```

To open an image file in the PPM format, asking the user to select the file, use

```
of.openPPM();
```

To make an instance of the *StreamSniffer*, classify the file, print the result of the classification, and prompt the user to select the input file, use

```
StreamSniffer ss=of.openAndSniffFile();
```

7.7. The Timer Class

During the course of program development, it is often useful to time the execution of various methods. This has been born of necessity, even though optimization of reading and writing file formats is a distraction. We must have a way to examine the trade-off of compression ratios vs. CPU time. Clearly, the issue of bench marking is a deep one, and an entire books have been dedicated to bench marking techniques [Hockney]. There is so much variation in Java virtual machine technologies that bench marking can be counter-productive as it will almost always be virtual machine specific. Even so, the *Timer* class represents a simple technique for timing the execution of various methods and is too useful to ignore.

The *Timer* class resides in the *gui* package and provides elementary methods for timing method invocations.

7.7.1. Class Summary

```
public class Timer {
        public void start()
        public void stop()
        public double getTime()
        public void print(int N, String message)
        public void print(String message)
}
```

7.7.2. Class Usage

Suppose that a *Timer* instance is created as follows:

```
Timer t = new Timer();
```

To start the *Timer* use

```
t.start();
```

To stop the *Timer* use

```
t.stop();
```

To get the elapsed time, in seconds (computed internally using milliseconds), since the *Timer* instance was started and to stop the timer, use

```
double time = t.getTime();
```

Suppose the following variables are defined as follows:

```
int aNumberOfSomething = 1000;
String message = "multiplies";
```

To stop the timer, print a message, then restart the timer, use

```
public void print(aNumberOfSomething, message);
```

This will print using the following:

```
System.out.println(
        message
        +" "
        + getTime() + " seconds "
        + N/getTime()+ "  per second");
```

The idea is that the rate at which the *aNumberOfSomething* is being performed is computed by the *Timer* instance.

To print a message followed by the number of seconds, use

```
t.print(message);
```

7.7.3. Class Example: Loop Interchange

Bench marking is an art. The results of a benchmark in Java are generally virtual machine dependent. In the following example, we reproduce the code in the *LoopInterchange* class, which resides in the *gui* package.

To find out how to design two-dimensional array processing programs in Java. This is a deep-seated performance issue, and the results of this section have changed the way in which the author writes image processing programs. Following Hennessy and Patterson [Hennessy et al.], we formulate an experiment based on a simple idea: By exchanging the nesting of a loop so that data in memory is accessed sequentially, we can improve performance.

Accessing data sequentially improves performance because we can increase the chance that we will access CPU cache. Cache is a kind of memory that is generally small in size, relative to the main memory, yet is much faster for the CPU to access. By increasing the locality of memory access, we are able to reduce the number of cache misses. One way to write a loop is to index more quickly along the row. If indexing along the row is faster, then the array is said to be stored in row-major order. A loop that accesses an array in row-major order follows:

```
for (j=0; j < jmax; j++)
  for (i=0; i < imax; i++)
       x[i][j] *= 2;
```

Another loop that accesses an array in column-major may be written as

```
for (i=0; i < imax; i++)
  for (j=0; j < jmax; j++)
       x[i][j] *= 2;
```

Note that the order in which the loops are accessed has been *interchanged*. A third way to perform the same computation is to *simulate* a two-dimensional array with a one-dimensional array. This appears cumbersome because the technique for accessing the two coordinates involves a multiplication and an addition in the index:

```
for (i=0; i < imax; i++)
  for (j=0; j < jmax; j++)
       y[j+i*imax] *= 2;
```

At first glance, this appears to be a much slower technology for implementing a two-dimensional array. Indeed, the use of y[j+i*imax] is slower under some circumstances. Yet Java source code is generally written to simulate two-dimensional arrays because the Java language specification does not indicate if memory will be stored in row or column major order [Gosling et al.]. The following code uses the *Timer* class to determine the fastest way to write two-dimensional array accesses:

```
package gui;
public class LoopInterchange {
    int jmax = 700;
    int imax = 700;
    int i, j;

    float x[][] = new float [imax][jmax];
```

```
float y[] = new float [imax*jmax];

public void test() {
    double t1,t2,t3,speedup;

    t1 = testLoop1();
    System.out.println("loop1:"+t1+" seconds");
    t2 = testLoop2();
    System.out.println("loop2:"+t2+" seconds");
    t3 = testLoop3();
    System.out.println("loop3:"+t3+" seconds");
}

double testLoop3() {
    Timer t = new Timer();
    t.start();
    for (j=0; j < jmax; j++)
        for (i=0; i < imax; i++)
            x[i][j] *= 2;
    t.stop();
    System.out.println("first index faster");
    return t.getTime();
}
double testLoop1() {
    Timer t = new Timer();
    t.start();

    for (i=0; i < imax; i++)
        for (j=0; j < jmax; j++)
            y[j+i*imax] *= 2;
    t.stop();
    System.out.println("multiply access");
    return t.getTime();
}

double testLoop2() {
    Timer t = new Timer();
    t.start();
    for (i=0; i < imax; i++)
        for (j=0; j < jmax; j++)
            x[i][j] *= 2;
    t.stop();
    System.out.println("second index faster");
    return t.getTime();
}

public static void main(String args[]) {
    LoopInterchange li = new LoopInterchange();
    for (int i = 0; i < 1; i++)
```

```
                li.test();
        }
    }
```

The *LoopInterchange* class is seen now, not just as a simple class example, but as a means of determining the fastest way to write Java array processing programs *for one particular virtual machine implementation.* A sample of program output for the MetroWerks Java virtual machine version 1.2.1 (as distributed with CodeWarrior Pro 3.0) follows:

```
multiply access
loop1:8.916 seconds
second index faster
loop2:7.018 seconds
first index faster
loop3:7.464 seconds
```

Times shown are for a PowerMac 8100 (PowerPC 601 running at 100 MHz). A PowerPC 601 has a 32K cache that hold both code and data. Such a cache is known as a unified cache. Cache size and architecture are machine specific. For example, the PowerPC 603 and 604 implement a separate cache for code and another for data. Such a cache is known as a Harvard architecture. Such considerations are important because we want to be sure to process an array that is larger than the cache. This will enable the code to generate some cache misses as well as cache hits. The size of the array is 700*700*4 = 1,960,000 bytes, which is about 60 time larger than the PowerPC 601 cache.

When the cache is missed, the processor must use the system bus in order to fetch memory. A system bus is typically slower than the CPU's clock speed. For example, a PowerMac 9500 has a 132 MHz CPU with a 44 MHz system bus. Thus, the PowerMac 9500 must wait several cycles to fetch memory across the bus [Thompson].

Loop Name	Dimensions	MW 1.2.1 (sec.)	MW JIT (sec.)
first index faster	700x700	7.464	1.184
multiply access	490000	8.916	0.45
second index faster	700x700	7.018	0.562

Table 7-3. Times for the Loop Interchange

Table 7-3 shows the times (in seconds) for the loop interchange benchmarks, using the MetroWerks 1.2.1 version of the Java virtual machine. The Just-In-Time compiler (JIT) was tried both on and off. As we change from the non-JIT to the JIT compilers, the multiply access appears to go from being slowest to being fastest. In fact it is almost 20x faster than the non-JIT version of the loop and 2.6 times faster than row-major order of access. Recall that in line 49 of the *main* method of the *LoopInterchange* class, we use a for-loop. If the limit of the for-loop is increased, we may repeat the test for many iterations. When this occurs, meaningful statistics may be derived. We must compute the results of Table 7-3 over several trials before we can form any conclusions.

For example, after 10 trials the following was output:

```
multiply access
loop1:0.511 seconds
second index faster
loop2:0.409 seconds
first index faster
loop3:1.011 seconds
```

This indicates that the multiply access is somewhat slower than the column-major order access. Of course, on some hardware, there may be a multiply-accumulate instruction that is a built-in part of the CPU. In such a case, multiply access would be consistently faster.

After 10 trials on an unloaded Sparc 5 workstation, we obtained the following:

multiply access

```
loop1:0.567 seconds
second index faster
```

```
loop2:0.442 seconds
first index faster
loop3:0.664 seconds
```

In fact, for the Sparc 5 running JDK1.1.3, the loop using column major order is consistently faster. Another observation is that the JDK 1.1.3 on the Sparc 5 runs at about the same speed (for this trivial benchmark) as the MetroWerks JIT on the PowerMac 8100/100.

The approach taken by this book assumes that arrays are stored in column-major order and that multiply access is a less clear way to write a doubly-nested for-loop. We have found that multiply access can be faster on some hardware-software combinations, slower on others. It really depends on the JIT and the availability of multiply-accumulate instructions that fill the CPU pipeline.

Bench marking in Java is a fascinating topic! Java enables us to dissociate a benchmark from its compiler. We need to compile the code only once, then we may run it on any Java machine. This makes a compelling argument for bench marking in Java, with the biggest variation being in the Java machine implementation. For a Web page that lists Java benchmarks, see
http://www.webfayre.com/pendragon/cm2/index.html.

7.8. The ProgressFrame Class

The *ProgressFrame* class extends the *Frame* class and resides in the *gui* package. It is used to display an estimate of the amount of time it takes to complete an operation, and it provides an animated progress bar for user feedback.

7.8.1. Class Summary

```
public class ProgressFrame extends Frame {
        public ProgressFrame(String title)
        public void setRatioComplete(double d, String id
        public static void main(String args[])
}
```

7.8.2. Class Usage

Suppose the following variables are predefined such that:

```
String title = "test";
ProgressFrame pb = new ProgressFrame(title);
```

Then to show the *ProgressFrame* use

```
pb.setVisible(true);
```

To update the *ProgressFrame*, use a double precision variable that ranges from 0 to 1, inclusive, and invoke the *setRatioComplete* method. An example, taken from the *main* method of the *ProgressFrame* class follows:

```
public static void main(String args[]) {
    ProgressFrame pb =
      new ProgressFrame("test prog");
    pb.show();
    for (double d=0; d < 1; d = d + 0.1) {
      pb.setRatioComplete(d,
              "Percent done:"+(int)(d*100)+"%");
              try {
                  Thread.sleep(500);
          }
          catch (Exception e) {};
      }
}
```

Invocation of *ProgressFrame.main* results in an animated display, shown in Fig. 7-6.

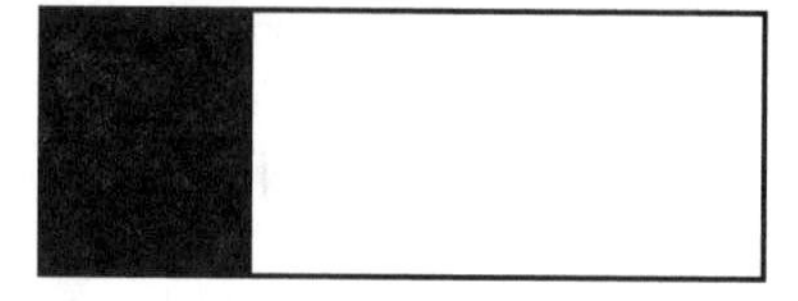

Percent done:30%

Elapsed Time: 5.0 Seconds

Estimated time left: 12.53 Seconds

Fig. 7-6. The ProgressFrame

The progress frame should be updated at regular intervals, from one place in the method being sampled. This should give a reasonable estimate of the time left to perform the process. The estimate is based on the amount left to process and the amount of time it took to process the amount done so far. Thus, the estimate assumes that the future rate of processing will be the same as the past rate of processing. Such an assumption may be unwarranted, however, and so the estimate should be used for non-time critical ventures (i.e., a cup of java!).

7.9. The WriteGif Class

The *WriteGif* class resides in the *vs* package. It is designed to take an 8-bit-per-pixel image and write it out to a GIF file. GIF files may be read by most browsers without the addition of Java. On the other hand, they do not support 24-bit color and the copyright for the format of the GIF image is owned by Compuserve [Kay]. One interesting advantage of the GIF format is that it can display a sequence of overlayed images. It also has the ability to store interleaved image data. This can make the reading of GIF format a challenge. Perhaps the most compelling reason to use GIF is that the Java *awt* package supports the reading of it. Unfortunately, there is no built-in API support for writing GIF images.

Since the GIF file format is encumbered by a patent, the writing of GIF format is unimplemented in Kahindu Vision (the source code that comes with this book). To implement the code, you must set the class variable:

```
weAreLegal = false;
```

to

```
weAreLegal = true;
```

It then becomes the reader's responsibility to contact the UNISYS lawyers (LZW_INFO@UNISYS.COM) and work out a license agreement.

Writing GIF images is difficult for three reasons. First, you need to get a license for an electronic implementation, secondly GIF includes an implementation of the Ziv-Lempel algorithm [Murray] and this complicates the code somewhat. Finally, GIF images are limited to 256 colors and so a color reduction process is typically required when writing 24-bit images. Both the Ziv-Lempel algorithms and the color reduction algorithms are complex topics in their own right. As a result, their discussion is deferred until Chapter 13.

7.9.1. Class Summary

```
public class WriteGIF {
```

```
    public static void DoIt(Image img, String fname) throws
        Exception
}
```

7.9.2. Class Usage

Typically, the programmer will obtain a valid output file name and *Image* instance. The procedure is to invoke the static *WriteGIF.DoIt* method, assuming that the image has already been quantized to 256 colors. If the image has not been color reduced, an exception will be thrown. A class example follows:

```
public void saveAsGif() {
    String fn = getSaveFileName("Save as GIF");
    if (fn == null) return;
        try {
            vs.WriteGIF.DoIt(
                getImage(),
                fn);
    }
    catch(Exception e) {
        System.out.println(
                "Save GIF Exception! 24 bit images not
        handled");
    }

}
```

7.10. The WritePPM Class

The *WritePPM* class resides in the *vs* package. It is typically invoked to write out a 24-bit image. There is no need to implement a PGM file format (Portable Gray Map) or PPM (Portable Pix Map) which are 8 and 1 bit-per-pixel formats. The reason is that files may be saved as GZIP, and GZIP will automatically exploit the redundancy in the uncompressed stream. More importantly, we need not support anything other than a 24-bit-per-pixel image (which greatly simplifies the code).

7.10.1. The PPM File Format

The PPM file format written to disk by the *WritePPM* class is known as a 24-bit-per-pixel binary format. It consists of a header followed by binary image data. The header is in ASCII. The header always consists of:

```
P6
width height
255
```

where the newlines are forced with an '\n' and the width and height are represented by ASCII numeric strings. For example, a 120x230 image is represented by

```
P6
120 230
255
```

The "255" at the end of the header indicates that the maximum value is 255 for any given color plane. The rest of the information consists of raw binary data in the form: RGB where R, G and B are the red, green and blue color bytes.

7.10.2. Class Summary

```
public class WritePPM {
   public static void doIt(
       short r[][], short g[][], short b[][],
       String fn)
}
```

7.10.3. Class Example

The *WritePPM* class is invoked with a known good save file name and three two-dimensional arrays of *short*. The arrays contain image data and must have the same dimensions or an ArrayOutOfBoundsException will be thrown.

An example from the *SaveFrame* that includes a timing function follows:

```
public void saveAsPPM() {
    String fn = getSaveFileName("Save as PPM");
    if (fn == null) return;
    Timer t = new Timer();
    t.start();
    vs.WritePPM.doIt(r, g, b, fn);
```

```
        t.print(" saveAsPPM in ");
    }
```

7.11. Summary

The writing of image files is generally an ill-posed problem. For every image file format supported, a surprising amount of code must be written. To learn more about image file formats, the reader is directed to [Gunter] [Kay] [Kientzle] and [Murray].

To convert from a given image file format into the PPM format used in this book, the reader may be interested in some of the shareware programs that are the CD-ROM . One good commercial program is [Debabelizer], and another is [DataViz], but there are many more.

To learn more about the GZIP file format, see [RFC1952]. To read the specification for the compression data format used by GZIP, see [RFC1951]. The GZIP file format is based upon a Huffman encoding of the data and hence is unencumbered by patents [Huffman].

7.12. Projects

1. One file format, called *PNG* is used for storing images. The source code for this format is available under GNU license at: <http://www.visualtek.com/PNG>. Download this software and integrate it into the Kahindu package. Test your program with several PNG type images.

2. Another popular file format is called BMP. The code for the BMP file format is available from [Stevens]. Get this code and integrate it into the Kahindu package. Be sure to test your program with several BMP type images

8. Direct Convolution

The first company to bring a book to market
gets 80 percent of the market,
the second gets 20 percent, and
the third is left standing there saying
"Damn, I wish I had an idea"

– D. L. 1998

The dictionary defines "convolution" as a twisting, coiling or winding together. In image processing, convolution is an operation taken between two images. One image is typically smaller than the other image. The smaller image is called the kernel of the convolution. The convolution operation consists of a weighted sum of the area surrounding an input pixel. In fact, the convolution is the correlation between the image and a reversed kernel.

We study convolution because it is so useful. Convolutions can edge detect, smooth, and crispen [Crane] [Myler]. In fact, as pointed out in Chapter 5, the images constructed with the eye, camera and any transducer are all the result of a convolution between a point source and the scene being imaged [Bracewell].

Convolution is a primary technique for *spatial filtering.* A spatial filter takes the local area around a pixel in an image and performs an operation on each of the pixels in order to create a new pixel in an output image. The dimensions of the area about the pixel of interest are given by the convolution mask and form the outline for a *convolution window*. The convolution window is moved so that it centers about each pixel in the input image. After centering, a new pixel is computed using an arithmetic sum of products.

8.1. The Mathematical Basis of Convolution

We define the discrete one-dimension convolution as a correlation between two signals that have been reversed in time. In the continuous domain, we write:

$$h(x) = f(x) * g(x) = \int_{-\infty}^{\infty} f(u)g(x-u)du \qquad (8.1)$$

where f is the input signal, u is a dummy variable of integration, and g is called the kernel of the convolution. Other names for the kernel include the window kernel, mask, window, window mask, filter template, impulse response and PSF (Point Spread Function). This book uses the term, *kernel*. Equation (8.1) slides the kernel across the input signal as it multiplies the input signal and kernel together and sums the result. The *region of support* is defined as that area of the kernel which is non-zero. Typically, we show the region of support of the kernel and assume that the kernel has a zero value outside of the region of support. The discrete form of (8.1) is given by:

$$h(x) = f(x) * g(x) = \sum_{-\infty}^{\infty} f(u)g(x-u) \qquad (8.2)$$

When the input signal and kernel have infinite extent and the kernel has a finite region of support, then the convolution has a finite region of support and is said to be a *linear convolution*. A function that has a finite region of support is said to have *compact support*.

When the input signal and kernel are periodic sequences with the same period, then the convolution is said to be a *cyclic convolution* (in some circles, this is called a *circular convolution*). Typically, the kernel is smaller than the input signal and is repeated in order to get a sequence that is the same length as the input signal. The resulting convolution is a sequence that is the same length as the input signal. The convolution is also periodic with the same period as the input signal.

Fig. 8-1 shows an example of a discrete one-dimensional convolution. Each element of the resulting convolution is formulated as an arithmetic sum of products between the shifted kernel and the input signal.

g(u)	f(x)	g(0-u)	h(0)	g(1-u)	h(1)	g(2-u)	h(2)	g(3-u)	h(3)	h(x)
		3								
	1	2	4	3						4
1	2	1		2	10	3				10
2	3			1		2	16	3		16
3						1		2	9	9
								1		

Fig. 8-1. Example of a One-Dimensional Convolution

The continuous convolution is defined in two dimensions as

$$h(x,y) = f * g = \int_{-\infty}^{\infty} \int_{-\infty}^{\infty} f(u,v) g(x-u, y-v) du dv \tag{8.3}$$

In the discrete domain, (8.3) may be written as

$$h(x,y) = f * g = \sum_{u=0}^{u_{max}-1} \sum_{v=0}^{v_{max}-1} f(u,v) g([x-u],[y-v]) \tag{8.2}$$

where g is the periodic kernel of the convolution, f is the periodic two-dimensional input image, h is the periodic two-dimensional cyclic output convolution, u_{max} is the width of the kernel, v_{max} is the height of the kernel and

$$\begin{aligned} [x-u] &= (x-u) \bmod u_{max} \\ [y-v] &= (y-v) \bmod v_{max} \end{aligned} \tag{8.3}$$

An alternative form of (8.2), which avoids the mod operation of (8.3), is given by

$$h(x,y) = f * g = \sum_{u=-u_c}^{u_c} \sum_{v=-v_c}^{v_c} f(x-u, y-v) g(u+u_c, v+v_c) \tag{8.4}$$

Where (u_c, v_c) lies at the center of the kernel [Mitra]. While (8.4) is a more efficient form for the convolution, it assumes that the center of the kernel may be found easily. Typically, this imposes a constraint that the kernel must have odd dimensions [Jähne]. This is a reasonable assumption, because the center of the kernel must be placed over each pixel in the input image in order to correctly compute the convolution.

Suppose that the width of the input image is denoted by *width* and the height of the input image is denoted by *height*. One of the problems with (8.4) is that the range of x and y must be constrained such that $x \in [u_c ... width - 1 - u_c]$ and $y \in [v_c ... height - 1 - v_c]$. For example, if $h(0,0)$ is computed, then $x-u$ and $y-v$ will be negative and out of range for f. In order to condition the arguments of (8.4), we add circulant index functions, c_x, c_y to (8.4) and write:

$$h(x,y) = f * g = \sum_{u=-u_c}^{u_c} \sum_{v=-v_c}^{v_c} f(c_x(x-u), c_y(y-v)) g(u+u_c, v+v_c) \tag{8.5}$$

where

$$c_x(x) = \begin{cases} x - width + 1 & \text{if } x > width - 1 \\ width + x & \text{if } x < 0 \\ x & \text{otherwise} \end{cases} \tag{8.6}$$

and

$$c_y(y) = \begin{cases} y - \text{height} + 1 & \text{if } y > height - 1 \\ height + y & \text{if } y < 0 \\ y & \text{otherwise} \end{cases} \tag{8.7}$$

An implementation of (8.6) and (8.7) follows:

```
public int cx(int x) {
     if (x > width -1)
          return x - width + 1;
```

```
        if (x < 0)
            return width + x;
        return x;
    }
    public int cy(int y) {
        if (y > height - 1)
            return y - height + 1;
        if (y < 0)
            return height + y;
        return y;
    }
```

Substituting *k* for the kernel variable *g* (*g* was taken by the green channel in the code), a brute force implementation of (8.5) may be written as

```
public short[][] convolveBrute(short f[][], float k[][]) {
    int uc = k.length/2;
    int vc = k[0].length/2;

    short h[][]=new short[width][height];
    double sum = 0;

    for(int y = 0; y < height; y++) {
      for(int x = 0; x < width; x++) {
        sum = 0.0;
        for(int v = -vc; v <= vc; v++)
                for(int u = -uc; u <= uc; u++)
                sum += f[cx(x-u) ][cy(y-v)] * k[ u+uc][v+vc];
        if (sum < 0) sum = 0;
        if (sum > 255) sum = 255;
        h[x][y] = (short)sum;
      }
    }
    return h;
}
```

As an example of how (8.6) and (8.7) process the arguments to *f*, suppose that the image has the dimensions of (*width,height*) = (12,11) and that the kernel is a 3x3 matrix. Then the center of the kernel is at (1,1) and the computation of *h*(0,0) will cause (8.5) to reduce to:

$$h(0,0) = f * g = \sum_{u=-1}^{1}\sum_{v=-1}^{1} f(c_x(-u), c_y(-v))g(u,v) \tag{8.6}$$

The arguments to the circulant index functions will then become:

$$\left(c_x(-1), c_y(-1)\right) = (11,10) \tag{8.7}$$

which is the diagonally opposite corner.

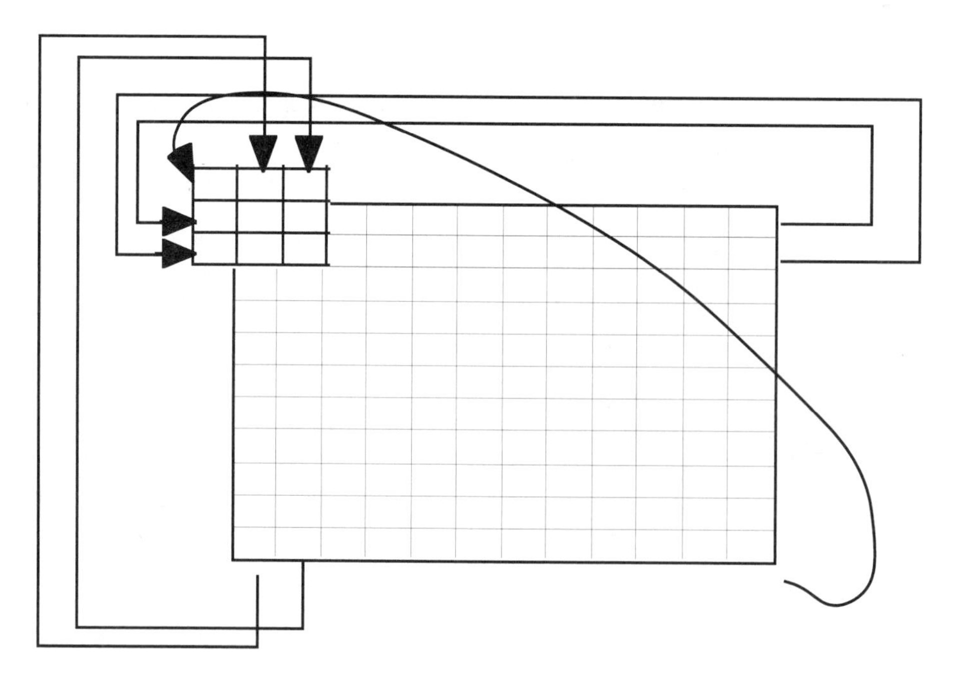

Fig. 8-2. Cyclic Convolution with Corner Kernel

Fig. 8-2 shows a diagram depicting where the image data comes from when a 3x3 kernel is convolved with the corner of an image. As another example, suppose we want to compute $h(11,10)$. In this case, the arguments to the circulant index functions are out of range at $\left(c_x(12), c_y(11)\right) = (0,0)$.

The circular nature of the borrowing of data from the corners and sides is probably the reason for the term cyclic convolution. There are several alternatives to using (8.6) and (8.7), including zero padding, ignoring the edges, extrapolation, and image replication. With zero padding, the edges beyond the image are set to zero. This is easy to implement, but it generally produces wrong results. Ignoring the edges is also easy to implement, but again, the convolution will be wrong at the edges. Extrapolation filters the pixels at the edges and thereby increases the dimensions of the image. This looks better, but overweights the edges [Jähn]. Image replication will produce a correct result, but there is a significant memory and computation cost.

Fig. 8-3 shows what happens when the edges of an image are ignored during a convolution with a 13x13 unweighted Mexican hat kernel (discussed in Chapter 10).

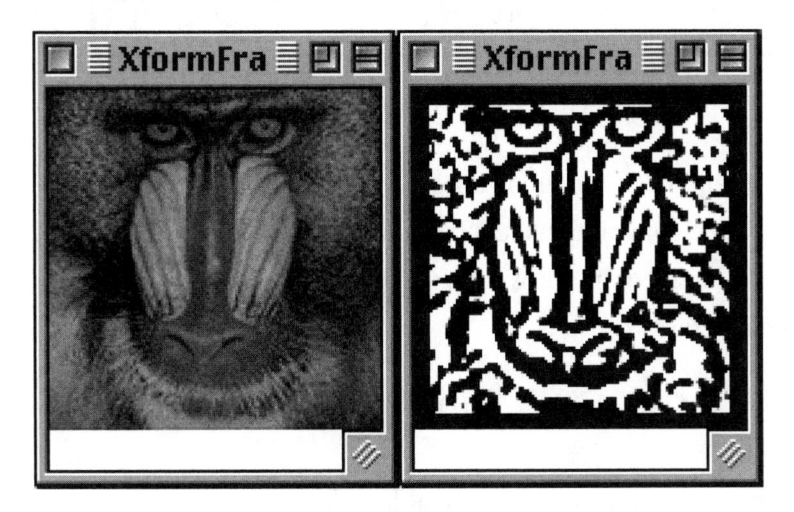

Fig. 8-3. A Convolution, Ignoring the Edges

The 13 pixel border around the edges of the image occurs because we have ignored the edges where the kernel wraps around the image. Fig. 8-4 show the same 13x13 Mexican hat kernel using the full extent of the image, given the circulant index functions of (8.5).

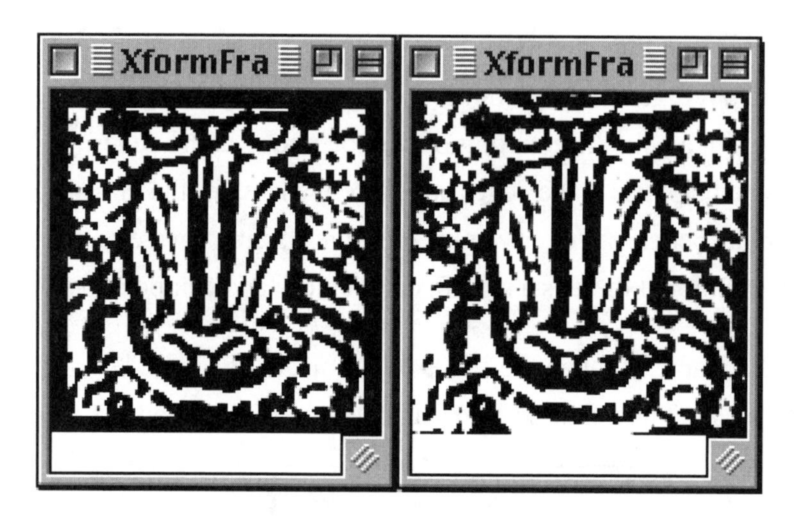

Fig. 8-4. A No-edge Convolution vs. a Cyclic Convolution

The use of the circulant index functions does save the borders of the image. But more importantly, only cyclic convolutions reduce to simple multiplication in the Fourier space! The convolution theorem (proved in Chapter 5) may be restated using the notation of this chapter as

$$\int_{-\infty}^{\infty}(f * g)e^{-j\omega t}dt = \int_{-\infty}^{\infty} f(t)e^{-j\omega t}dt \int_{-\infty}^{\infty} g(t)e^{-j\omega t}dt \tag{8.8}$$

The convolution theorem is remarkable indeed; it says that convolution in the time domain is the same as multiplication in the frequency domain. That is, the Fourier transform of the convolution is the same as the product of the Fourier transforms.

8.2. The Cyclic Convolution, Correctness and Speed

The last section showed that the cyclic convolution gave an image that filled a frame to the edges. Because the edges are included and because of the equivalence to the Fourier transform, cyclic convolution is a correct way to perform convolution.

Correctness does not come cheap! A brute force implementation, as suggested by (8.5) invokes (8.6) and (8.7) as a part of the reference to the image, *f*.

The brute force implementation takes 22 seconds for the 13x13 kernel on the 128x128 image (R, G and B planes being processed on a PowerMac 8100/100). If we dispense with the edges, the time drops to 8 seconds. Consider that a 128x128x3 image has 49,152 pixels to process. The sub-image is (128-13)x(128-13)x3 = 39,675 pixels to process. Thus, the sub image is 20% smaller, yet the process time is 8/22 *100 = 36% smaller. The key element is that the repeated invocation of two circulant index functions in the kernel of the convolution cuts the performance by 2.75 times!

Also, the circulant index functions are needed for 20% of the time (less for smaller kernels and/or larger images), yet the brute force implementation invokes them 100% of the time. One way to optimize the computation is to break up the image into 5 parts: top, bottom, left, right and center. These are shown in Fig. 8-5.

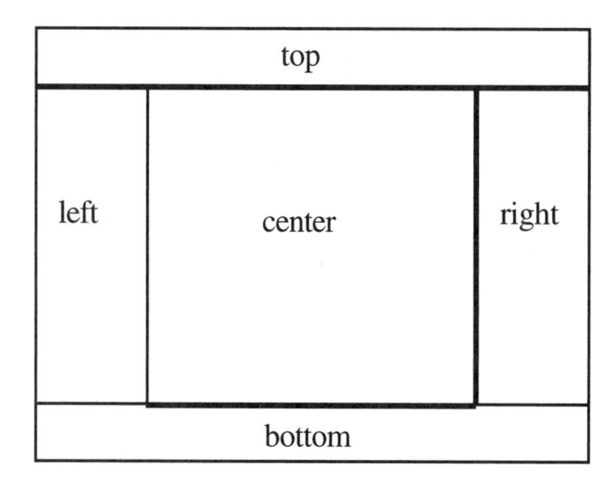

Fig. 8-5. The Five Parts of the Image

Note that in the left and right sub-images, $c_y = y$. In fact, we could subdivide the image into 9 parts (top, top-left, top-right, bottom, bottom-left, etc.). This would yield even more optimization, but at a cost of increased implementation complexity. In any case, the optimization of reducing the convolution into subparts cut's execution time by 50%, to only 11 seconds. For further optimization, we will employ different algorithms. A new Java API, called Java Advanced Imaging (JAI), is designed to accelerate the convolution operation. As a result, the implementation of the convolution will change to detect the existence of the JAI package. If the JAI package is available, it is reasonable to expect the convolution implementation to be faster than the one given here. The API specification to code in this book will be left unchanged, wherever possible.

8.3. Generating Convolution Kernels - The Mat Class

Often we are presented with a procedure and equations which, when sampled, will generate values for a convolution kernel. Once we are given the means to formulate the kernel, we must enter it into source code. Typing in kernels is both tedious and error-prone task.

The *Mat* class resides in the *gui* package and is intended to ease the burden of typing in kernels. The programmer invokes the *Mat.printKernel*(k) and the kernel is printed to the console, embedded in Java source code. The kernel is then cut and pasted into the Java source file, for speed. An example of *printKernel*'s output follows:

```
public void laplacian9(){
	float k[][] = {
	{  -1,    -1,    -1,    -1,    -1,    -1,    -1,    -1, -1},
	{  -1,    -1,    -1,    -1,    -1,    -1,    -1,    -1, -1},
	{  -1,    -1,    -1,    -1,    -1,    -1,    -1,    -1, -1},
	{  -1,    -1,    -1,     8,     8,     8,    -1,    -1, -1},
	{  -1,    -1,    -1,     8,     8,     8,    -1,    -1, -1},
	{  -1,    -1,    -1,     8,     8,     8,    -1,    -1, -1},
	{  -1,    -1,    -1,    -1,    -1,    -1,    -1,    -1, -1},
	{  -1,    -1,    -1,    -1,    -1,    -1,    -1,    -1, -1},
	{  -1,    -1,    -1,    -1,    -1,    -1,    -1,    -1, -1}};
//sum=0.0
	convolve(k);
}
```

The sum of all the elements in the kernel is printed in the comments. This is used by the kernel designer to invoke scaling or normalization, as required. The sum of the

Laplacian kernel is zero, so no scaling takes place. Kernel design is a deep topic that appears in Chapter 9.

The *Mat* class is an abstract class that resides in the *gui* package and performs several matrix manipulations. Most of the manipulations are designed to assist in the generation and processing of two-dimensional kernels.

8.3.1. Class Summary

```
package gui;
import java.awt.*;
public abstract class Mat {
   public static void print(double  a[][])
   public static void print(float  a[][])
   public static double sum(double a[][])
   public static double sum(float a[][])
   public static double sum(short a[][])
   public static void normalize(double a[][] )
   public static void normalize(float a[][] )
   public static double average(double a[][])
   public static double average(float a[][])
   public static double average(short a[][])
   public static void threshold(short a[][], short thresh)
   public static void threshold(short a[][])
   public static void printKernel(float k[][], String name)
   public static void printKernel(short k[][], String name)
   public static void scale(double a[][], double k)
   public static void scale(float a[][], float k)
   public static void scale(float a[][], double k)
}
```

8.3.2. Class Usage

Consider the following predefined variables:

```
double d[][];
float f[][];
short thresh, s[][];
double sum, avg;
```

To print a two-dimensional array of *double,* use

```
Mat.print(d);
```

To print a two-dimensional array of *float,* use
```
Mat.print(f);
```

To sum all the element of a two-dimensional array of *double*, use
```
sum = Mat.sum(d);
```

To sum the elements of a two-dimensional array of *float*, use
```
sum = Mat.sum(f);
```

To sum the elements of a two-dimensional array of *short*, use
```
sum = Mat.sum(f);
```

To normalize the elements of a two-dimensional array of *double*, so that the array has a magnitude of one (scale by 1/sum), use
```
Mat.normalize(d);
```

To normalize the elements of a two-dimensional array of *float* so that the array has a magnitude of one, use
```
Mat.normalize(f);
```

To compute the average of the elements in a two-dimensional array of *double*, use
```
avg = Mat.average(d);
```

To compute the average of the elements in a two-dimensional array of *float*, use
```
avg = Mat.average(f);
```

To compute the average of the elements in a two-dimensional array of *short*, use
```
avg = Mat.average(s);
```

To change the values in a two-dimensional array of *short* so that those below *thresh* are 0 and those greater than or equal to *thresh* are 255, use
```
Mat.threshold(s, thresh);
```

To change the values in a two-dimensional array of *short* so that those below the average are 0 and those greater than or equal to average are 255, use
```
Mat.threshold(s);
```

To print a two-dimensional array of *float,* use

```
Mat.printKernel(f, "mean3");
```

To print a two-dimensional array of *double,* use

```
Mat.printKernel(d, "mean3");
```

To print a two-dimensional array of *short,* use

```
Mat.printKernel(s, "mean3");
```

For example, to print out a 9x9 averaging array, use

```
public static void main(String args[]) {
     short s[][] = new short[9][9];
     for (int i = 0; i < s.length; i++)
         for (int j= 0; j < s[0].length; j++)
           s[i][j] = 1;
     Mat.printKernel(f,"mean9");
}
```

whose output appears below:

```
public void mean9(){
    float s =(float)81.0;
    float k[][] = {
    {1/s,1/s,1/s,1/s,1/s,1/s,1/s,1/s,1/s},
    {1/s,1/s,1/s,1/s,1/s,1/s,1/s,1/s,1/s},
    {1/s,1/s,1/s,1/s,1/s,1/s,1/s,1/s,1/s},
    {1/s,1/s,1/s,1/s,1/s,1/s,1/s,1/s,1/s},
    {1/s,1/s,1/s,1/s,1/s,1/s,1/s,1/s,1/s},
    {1/s,1/s,1/s,1/s,1/s,1/s,1/s,1/s,1/s},
    {1/s,1/s,1/s,1/s,1/s,1/s,1/s,1/s,1/s},
    {1/s,1/s,1/s,1/s,1/s,1/s,1/s,1/s,1/s},
    {1/s,1/s,1/s,1/s,1/s,1/s,1/s,1/s,1/s}};

    convolve(k);
}
```

For the overloaded version of the *printKernel* that takes the *short* array, the sum of all the elements is pre-computed and declared at the head; the rest of the array is then normalized. The kernel is then human readable, machine generated and evaluated at compile time. For the other versions of the overloaded *printKernel* method, the output is rather different. For example, if the two-dimensional array is an array of *float*, then use

```
Mat.printKernel(f, "mat3");
```

For example:

```
public static void main(String args[]) {
   float f[][] = new float[3][3];
   for (int i = 0; i < f.length; i++)
         for (int j= 0; j < f[0].length; j++)
           f[i][j] = 1;
    Mat.normalize(f);
    Mat.printKernel(f,"mean3");
}
```

produces the following output:

```
public void mean3(){
   float k[][] = {
   {  0.11111111f,   0.11111111f, 0.11111111f},
   {  0.11111111f,   0.11111111f, 0.11111111f},
   {  0.11111111f,   0.11111111f, 0.11111111f}};
//sum=1.0000000074505806
   convolve(k);
}
```

This is probably less readable for large floating point arrays, yet, if the equation for generating the kernel is computationally complex, this type of pre computation can save considerable time (as well as being less error prone). The sum is shown in comments, since the assumption is that the kernel is already properly scaled. Proper scaling for a kernel typically means that the sum adds to one or zero. The comment is intended to allow the kernel designer to check this particular kernel property.

To scale a two-dimensional array of *double* by a *double* type amount, use

```
double scale = 0.1;
Mat.scale(d,scale);
```

To scale a two-dimensional array of *float* by a *float* type amount, use

```
float scale = 0.1f;
Mat.scale(f,scale);
```

To scale a two-dimensional array of *float* by a *double* type amount, use

```
double scale = 0.1;
Mat.scale(f,scale);
```

8.4. Implementing Direct Convolution - The ConvolutionFrame

The *ConvolutionFrame* class resides in the *gui* package and provides the services of performing direct convolution using several pre-defined kernels. As the Java API changes, the implementation of the *ConvolutionFrame* will be altered. However, it is fully expected that the API described in this section will be preserved. Thus, the *ConvolutionFrame* implementation represents a 100% pure Java implementation of the convolution algorithm that will isolate the programmer from further API changes (such as the introduction of the JAI, which is used, if detected).

8.4.1. Class Summary

```
package gui;
import java.awt.*;
public class ConvolutionFrame extends OpenFrame {
      ConvolutionFrame(String title)
      public int cx(int x)
      public int cy(int y)
      public short[][] convolveBrute(short f[][], float k[][])
      public short[][] convolve(short f[][], float k[][])
      public short[][] convolveNoEdge(short f[][], float
        k[][])
      public static int rand(int min, int max)
      public void convolve (float k[][])
}
```

8.4.2. Class Usage

Suppose that the following variables are pre-defined:

```
float k[][] = {
     { 0, -1,  0},
     {-1,  5, -1},
     { 0, -1,  0}
     };
```

where *k* is a kernel with odd dimensions.

Also suppose that there is a two-dimensional array of *short* that contains some image data:

```
short img[][] = new short[128][128];
```

To make an instance of the *ConvolutionFrame* and display it, use

```
ConvolutionFrame cf = new ConvolutionFrame("Convolution");
cf.setVisible(true);
```

To invoke the circulant index functions of (8.6) and (8.7), use

```
int ix = cf.cx(ix);
int iy = cf.cy(iy);
```

There are three ways in which to gain access to a convolution. To perform a direct, brute force convolution, use

```
img = cf.convolveBrute(img, k);
```

To perform a convolution using techniques described in this chapter, use

```
img = cf.convolve(img, k);
```

To perform a direct convolution, ignoring the edges, use

```
img = cf.convolveNoEdge(img, k);
```

To obtain a random integer with uniform probability density function between *min* and *max* inclusive, use

```
int min =1; int max = 100;
int randomNumberBetween1And100 = ConvolutionFrame.rand(min,
        max);
```

To perform a convolution, using the techniques described in this chapter, between a kernel and the red, green and blue color planes, replacing the color planes with the new value, and displaying the result, use

```
cf.convolve(k);
```

8.5 Summary

This chapter disclosed the theory of convolution. The implementation of the convolution method is placed in a class called the *ConvolutionFrame*. The *ConvolutionFrame* will be updated with the fastest implementation of convolution available. This will probably be the Java Advanced Imaging (JAI) API. As of this writing, the JAI is only a specification. If the JAI is detected, be assured that this code will take advantage of it. It is the intention of this author to leave the interface to the *ConvolutionFrame* unchanged.

9. Spatial Filters

Algebraic manipulation, without conceptualization,
deepens the socio-pathetic abscess
between theory and practice.

– D.L., 1998

Spatial filters operate over a local neighborhood about a pixel in an image. The last section described the theory and implementation of convolution. Convolution is a linear operation that can be undone. Convolution is also a kind of spatial filter. There are non-linear spatial filters that are typically not implemented by convolution and, by extension, are typically not accelerated by the FFT's. Such non-linear spatial filters can be useful. Thus, this chapter describes both linear and non-linear spatial filters, grouping them by their visual effect. The effects covered include blurring, sharpening, and median filtering.

9.1. Blurring

One type of blurring that can be performed on an image is called neighborhood averaging. This is equivalent to a low-pass filter (also called a *smoothing filter*). Typically, such filters consist of a weighted summation of all pixels in a neighborhood, divided by some scaling amount. The scale is intended to make the weight of all the elements in the kernel sum to one. Fig. 9-1 shows the original mandrill image, followed by four images that result from convolution with 3x3 kernels. The kernels are called *average*, *lp1, lp2* and *lp3*. The implementation of the kernels shown in this chapter is excerpted from the *SpatialFrame* class, which is summarized in Section 9.5.

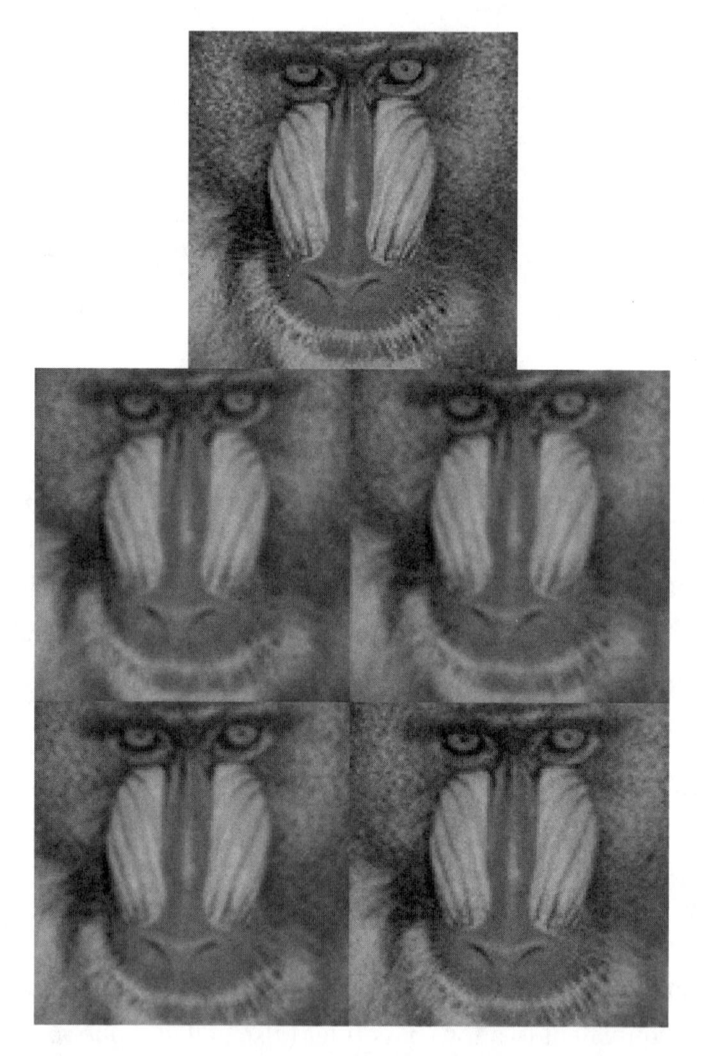

Fig. 9-1. Mandrill, *average*, *lp1*, *lp2* and *lp3*

The *average* kernel consists of an equally weighted sum of all the pixels in a 3x3 neighborhood. The method for implementing the average kernel follows:

```
public void average() {
float k[][] = {
   {1, 1, 1},
   {1, 1, 1},
   {1, 1, 1}
   };
   Mat.scale(k,1/9.0);
   convolve(k);
}
```

The *average* filter is an example of a low-pass filter. A low-pass filter retains the D.C. component in an image and removes some of the higher frequencies. The kernels of a low-pass filter are characterized by elements that are positive and sum to one. In the case of the example *average* filter, the elements sum to 9, and so we divide by 9, using *Mat.scale(k,1/9.0)*. This *normalizes* the kernel i.e., it makes the elements in the kernel sum to one. Let the gain be defined by

$$G = \sum_{i=0}^{W-1} \sum_{j=0}^{H-1} k_{ij}$$

If the gain is equal to one, the filter will have unity gain. If the gain is less than one, the filter will attenuate, and if the gain is greater than one, the filter will amplify [Galbiati].

We notice, from Fig. 9-1, that the image appears to be getting less blurry as we proceed from the *average* to the *lp3* filters. The reason is that the center pixel in the matrix is given more weight than the surrounding pixels, and this reduces the influence of the surrounding pixels on the newly computed center pixel. The code for the *lp1*-*lp3* filters follows:

```
public void lp1() {
 float k[][] = {
    {1, 1, 1},
    {1, 2, 1},
    {1, 1, 1}
    };
    Mat.scale(k,1/10.0);
    convolve(k);
 }
public void lp2() {
 float k[][] = {
    {1, 1, 1},
    {1, 4, 1},
    {1, 1, 1}
    };
    Mat.scale(k,1/12.0);
    convolve(k);
 }
public void lp3() {
 float k[][] = {
    {1, 1,  1},
    {1, 12, 1},
    {1, 1,  1}
    };
    Mat.scale(k,1/20.0);
    convolve(k);
 }
```

Such filters help to remove noise from an image, but they also tend to blur the edges. The kernels for the smoothing filters are typically of odd dimension (so that it is easy to find the center). Larger kernels result in more blur.

One kernel of particular interest represents the sampling of a *Gaussian PDF* (Probability Density Function). The Gaussian PDF is also called the *Gaussian density*, *normal probability density function* or just the *normal density*. In 2D, the Gaussian PDF is given by

$$gaussian(x, y, x_c, y_c, \sigma) = \frac{1}{2\pi\sigma^2} e^{-\frac{\left((x-x_c)^2-(y-y_c)^2\right)}{2\sigma^2}} \qquad (9.1).$$

where σ is the *standard deviation*. The maximum value of (9.1) occurs at $x = x_c, y = y_c$ and is given by:

$$g_{\max} = \frac{1}{2\pi\sigma^2} \qquad (9.2).$$

The *getGaussKernel* method uses (9.1) to generate an MxN kernel, and is described in the *SpatialFrame* class summary in Section 9.5.

The convolution of two Gaussian functions results in a Gaussian (see [Castleman] for a proof). Fig. 9-2 shows (9.1) with $x_c = y_c = 0$ and $\sigma = 3$.

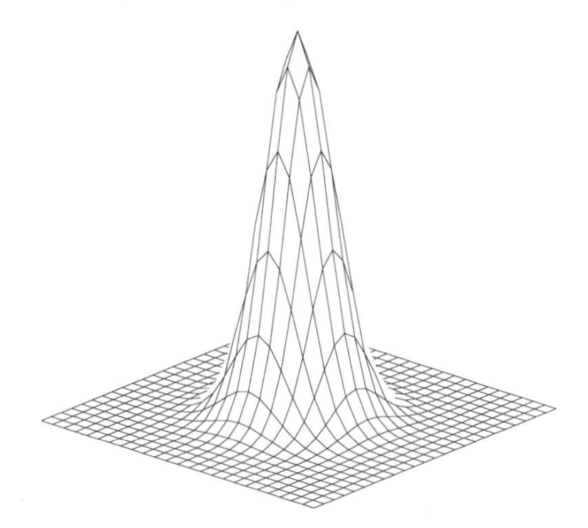

Fig. 9-2. A Gaussian Density

The integral of the Gaussian density is the *Gaussian CDF* (Cumulative Distribution Function). The Gaussian CDF is also called the *Gaussian distribution* or *normal distribution*. The Gaussian CDF is important because it is used to approximate the binomial distribution [Feller]. Also, the sum of large numbers of random numbers is normally distributed by the central limit theorem [Baker]. Recall from Chapter 5, that physiological study has shown that the response of the cones in the eye is given by a Gaussian sensitivity for the cone center and surrounding fields.

To generate a two-dimensional Gaussian kernel, we pre compute (9.1) and place the result in the Java program using the *Mat.printKernel* method described in Section

8.3.2. For an integer kernel, the dimensions of the kernel array will guide the choice for the value of the standard deviation. The kernel must be narrow enough to contain non zero coefficients [Crane]. The zeros, if they exist at all, represent a truncation error that occurs when converting from a floating point Gaussian function to an integer type kernel. Modern CPU's are fast enough so that the use of integer type kernels no longer saves CPU time (particularly in RISC systems). Even so, the long history of their use and representation makes it useful to output such kernels, for educational and check purposes only.

Example methods for implementing the 3x3 and 7x7 Gaussian kernels (entered manually) follow:

```
public void gauss3() {
  float k[][] = {
     {1, 2,  1},
     {2, 4,  2},
     {1, 2,  1}
     };
     Mat.scale(k,1/16.0);
     convolve(k);
  }
public void gauss7() {
   float mask [][] = {

     {  1, 1, 2,  2, 2, 1, 1},
     {  1, 2, 2,  4, 2, 2, 1},
     {  2, 2, 4,  8, 4, 2, 2},
     {  2, 4, 8, 16, 8, 4, 2},
     {  2, 2, 4,  8, 4, 2, 2},
     {  1, 2, 2,  4, 2, 2, 1},
     {  1, 1, 2,  2, 2, 1, 1}
     };
     Mat.normalize(mask);
     convolve(mask);
  }
```

For complex formula, for large kernels and for high precision, we are motivated to use the computer to generate the Java source that contains the kernels. This type of pre-computation is able to trade space in the Java class file for the time it take to compute the kernel.

A method for generating an MxN Gaussian kernel (where M and N are odd) follows:

```
public static void main(String args[]) {
        int M = 3;
        int N = 3;
        double sigma = 1;
        double centerPeak = 4;
        printGaussKernel(M,N,sigma,centerPeak);
}
```

Note that the *printGaussKernel* method is directed to create a *centerPeak* whose value is 4. The output produced by this invocation follows:

```
public void gauss3(){
        float s =(float)16.0;
        float k[][] = {
        {1/s,2/s,1/s},
        {2/s,4/s,2/s},
        {1/s,2/s,1/s}};
        convolve(k);
}
```

The invocation allowed the programmer to select the *centerPeak* value in advance. In fact, the *printGaussKernel* method is overloaded, so that if the *centerPeak* value is omitted, the following results:

```
public void gauss3(){
        float k[][] = {
        {  0.07511361f,    0.1238414f, 0.07511361f},
        {  0.1238414f,    0.20417994f, 0.1238414f},
        {  0.07511361f,    0.1238414f, 0.07511361f}};
        //sum=0.9999999701976776
        convolve(k);
}
```

The truncation error between 4/16 (=0.25) and 0.20417994 is 0.04582006. This represents a 4% error in the central pixel weighting. Additionally, 1% errors are made on each of the corners. Thus, there is greater than 8% truncation error in the integral representation of the Gaussian 3x3 kernel. The response to truncation error is to type in floating-point numbers. The problem with entering in floating-point numbers is that it is both a tedious and an error-prone task. On the other hand, to generate the above output we can use

```
public static void main(String args[]) {
        int M = 3;
```

```
        int N = 3;
        double sigma = 1;
        printGaussKernel(M,N,sigma);
    }
```

And to dynamically compute the Gaussian kernel and perform a convolution, use

```
    public void gauss31() {
        double sigma = 2.0;
        float k[][] = getGaussKernel(31,31,sigma);
        convolve(k);
    }
```

On a Pentium 5 running at 133 MHz, using CodeWarrior Pro 3, with the Microsoft JRE, a 128x128 color image was convolved with a 7x7 Gaussian kernel in 4 seconds. The 31x31 Gaussian kernel took 14 seconds. Clearly, a faster technique is needed to perform a convolution and this leads naturally to a discussion of the FFT. Unfortunately, the FFT can not accelerate non-linear filtering, as is discussed in Section 9.2.

Fig. 9-3 shows the mandrill image and the output after convolution with a 3x3, 7x7 and 15x15 Gaussian kernel.

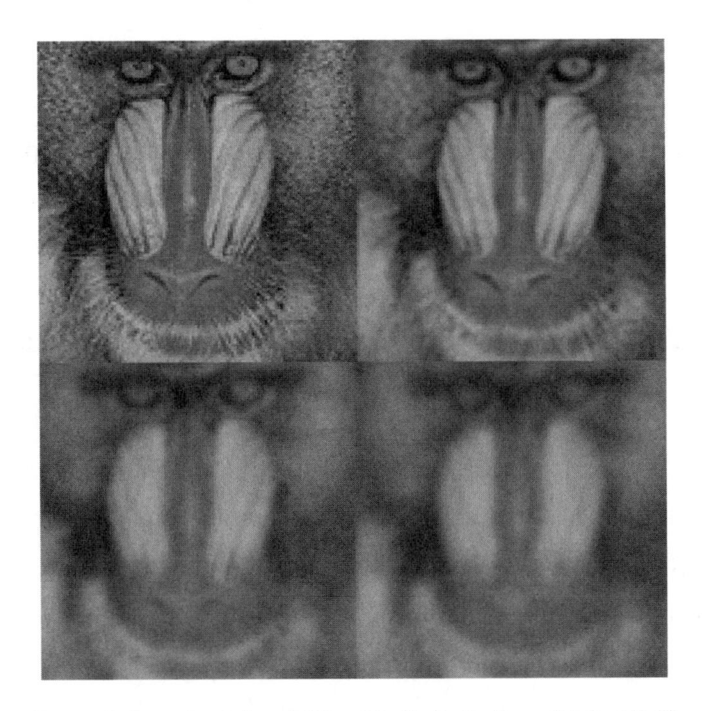

Fig. 9-3. Convolving the Mandrill with 3x3, 7x7 and 15x15 Gaussian Kernels

Low pass filtering can smooth over noise in an image. The drawback is that it can also smooth over the edges in an image. The loss of detail may be acceptable if the objective is to identify large areas of constant intensity.

9.2. Median Filtering

One kind of spatial filter is called a *median filter*. A median filter inputs a sliding window and outputs a value for the pixel located at the window's center. The new pixel value is equal to the median value of the pixels in the input window. Typically, the input window is symmetric in shape [Jain].

Median filters are not linear and their operations do not have an inverse (unlike the convolution). Also, non-linearity means that the median of the sum is not equal to the median of the medians. In comparison, average is a linear operation and the average of the sum is equal to the average of the averages.

The median (also called the *sample* median) is the middle of a list of samples that are listed in ascending order. When the number of samples in the list is odd, the median is the value of the sample at position $n/2$, where n is equal to the number of samples. When the number of samples in the list is even, the median is the value of the average of the samples at position $n/2$ and $n/2+1$. Sorting the list of samples is a time-consuming but necessary part of computing the median. To recap: The median value of a discrete, sorted, list of samples is the value of the sample in the middle of the list [Nickell] [Allen].

The reason the median is interesting is that it tends to discard statistical *outliers*. An outlier is a value that is either much smaller or much larger than the other samples in the set. Outliers tend to heavily weight the mean, despite the fact that they may represent anomalous behavior. For example, after the dog ate the student's homework, a zero was assigned. As a result, the homework grades were: 0, 85, 90, 87 and 100. The mean is 72, but the median is the middle of {0, 85, 87, 90, and 100}=87. Unfortunately, most professors don't use the median to compute the grade (did the dog really eat the homework?).

It is easy to see that sorting is the most computationally intensive part of computing the median. Thus, it is important that the sorting be done efficiently, particularly since it must be done *3 times for each pixel in a color image*.

One common technique for sorting is called the *quick sort* algorithm. The quick sort algorithm is based on the idea of *divide and conquer*. Sorting is an out-of-place topic in an image processing book, particularly in a chapter on spatial filtering. For an introduction to sorting in Java, see [Bishop].

The *SpatialFilterFrame* has a public static method called *quickSort*. It also has a method called *testQuickSort*, whose source code follows:

```
public static void testQuickSort() {
        int a[] = {1,2,3,5,4,3,2,5,6,7};
        quickSort(a);
        for (int i=0; i < a.length;i++)
```

```
        System.out.println(a[i]);
}

1
2
2
3
3
4
5
5
6
7
```

Note that the array parameter, *a*, has been altered by the invocation of the *quickSort* method. To obtain the median value, first quicksort, then select the data-point at the center of the array. The code for the *median* method follows:

```
public static int median(int a[]) {
    quickSort(a);
    int mid = a.length/2-1;
    if ((a.length & 1) == 1)
      return a[mid];
    return (int)((a[mid]+ a[mid+1]+0.5)/2);
}
```

Note that the implementation for the *median* method is not exact. Recall that, according to the definition, the computation of the median must first detect if the sample list is odd or even in length. For the odd size, the middle of the sample, *a[mid],* is correct. For the even size, take the average of *a[mid]* and *a[mid+1]*. Of course, the average between two integers should yield a floating-point type result. Thus, the source of the error in the *median* method is a truncation error that occurs for even-length sequences. Thus, it would be incorrect to use the *median* method listed here for reporting statistics. Since we are concerned only with producing integer data that can be displayed, the round-off error is considered negligible.

A public static method called *testmedian* illustrates the use of the *median* method:

```
public static void testMedian() {
      int a[] = {1,2,3,5,4,3,2,5,6,7};
      System.out.println("The median ="+median(a));
}
```

the output of which is:

```
The median =3
```

To perform the median filter, the same issues regarding circulant index functions (discussed in Chapter 8) apply. That is, *what should we do with the edges?* This is a recurrent problem, and one that will be revisited for edge detection. A cyclic convolution requires an edge wrap for correctness.

The median filter is different from a cyclic convolution. In fact, there is no set rule regarding what median filters should do with the edges of an image. A few methods have been provided with *public* visibility in the *SpatialFilterFrame* class, so that the reader may experiment with variations on the median filter edge processing. These are summarized at the end of the chapter.

The approach taken by the Kahindu program (on the CD-ROM) is to make use of the same circulant functions that appeared in Chapter 8. This is a costly mechanism, but it does preserve the dimensions of the image. Median filtering is very good at removing *salt and pepper noise*. Salt and pepper noise, also called *shot noise,* consists of randomly placed white and black pixels added to an image. Such noise may be caused by dust and lint on the optics of an image acquisition or duplication system (i.e., a scanner or a copier) [Weeks]. To add salt and pepper noise to an image, the *saltAndPepper* method is used:

```
public void saltAndPepper(int n) {
    for (int i=0;i < n; i++) {
    int rx = rand(0,width-1);
      int ry = rand(0,height-1);
      r[rx][ry] = 255;
      g[rx][ry] = 255;
      b[rx][ry] = 255;
      rx = rand(0,width-1);
      ry = rand(0,height-1);
      r[rx][ry] = 0;
      g[rx][ry] = 0;
      b[rx][ry] = 0;
    }
    short2Image();
}
```

Note that for a given argument, *n*, the *saltAndPepper* method adds 100 pixels of white and 100 pixels of black. To implement a median filter that examines pixels in a 3x3 cross about a center point, use

```
public void medianCross3x3() {
 short k[][] = {
    {0, 1, 0},
    {1, 1, 1},
    {0, 1, 0}
    };
    median(k);
}
```

Note that the *medianCross3x3* uses a kernel whose elements are all of *short* type. The kernel elements should be set to 0 if the pixel that appears in this part of the window is to be ignored. The elements should be set to 1 if the pixel in this part of the window is to appear in a list that is used to compute the median. The *median* method is overloaded and is used on the red, green and blue color planes. This is why the number of times quicksort is invoked is 3 times the number of pixels in the image. An implementation of the *median* method follows:

```
public void median (short k[] []) {
      printMedian(k,"color median");
      Timer t = new Timer();
      t.start();
      r = median(r,k);
      g = median(g,k);
      b = median(b,k);
      t.print("Median filter time");
      short2Image();
}
```

The full class summary for the *SpatialFilterFrame* appears in Section 9.4.1.

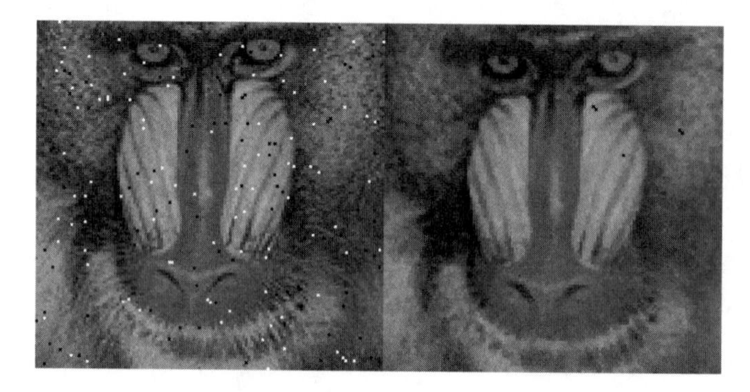

Fig. 9-4. Original and 3x3 Cross Median Filter

Fig. 9-4 shows the mandrill image on the left after an invocation of *saltAndPepper(100)*. The mandrill on the right shows the image after the application of *medianCross3x3*. Note that small black pixels still appear after the *medianCross3x3*. To filter out the black pixels, we require a more aggressive median filtering kernel. Toward that end, we devise:

```
public void medianSquare3x3() {
 short k[][] = {
      {1, 1, 1},
      {1, 1, 1},
      {1, 1, 1}
      };
      median(k);
}
```

Fig. 9-5 shows a before and after the application of the *medianSquare3x3* filter. We note that the image appears to be blurred around the edges, but at least the salt-and-pepper noise is removed.

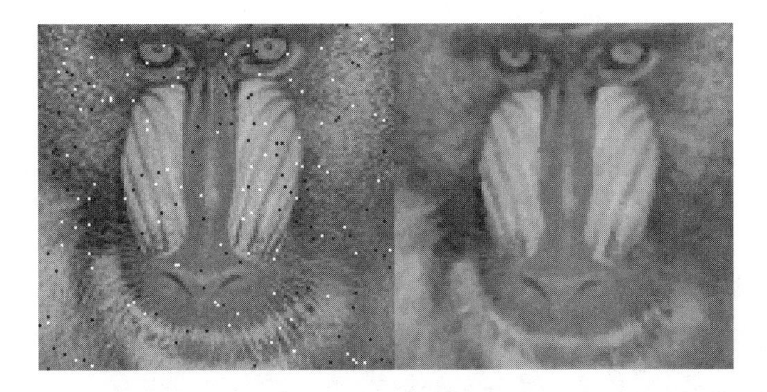

Fig. 9-5. Before and After *medianSquare3x3*

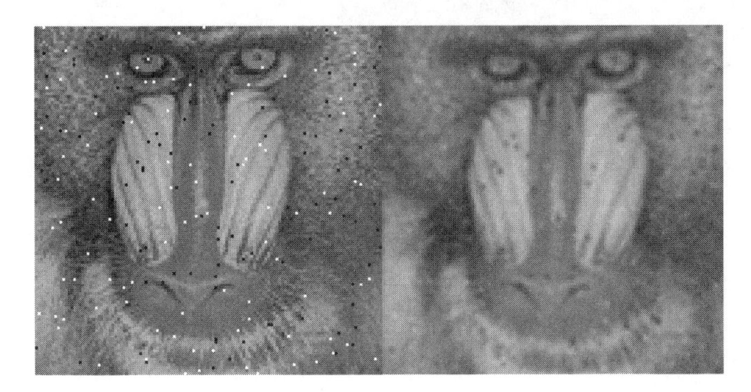

Fig. 9-6. Before and After *gauss3x3*

Fig. 9-6 compares the output of the *medianSquare3x3* with the *gauss3x3*. We see that the image is both blurry and noisy. The *gauss3x3* has been only able to blur the noise, not to remove it. To check to see how robust the median filter is, we try a series of images with *saltAndPepper(1000)*. The 3x3 median filter is unable to remove all the noise, even when performed several times, but a larger kernel can. As a result, we design a series of filter methods that permit a trade-off between noise reduction and blur in the image. Fig. 9-7 show the effect of the *medianSquare5x5* method invocation:

```
public void medianSquare5x5(){
```

```
        short k[][] = {
        {  1, 1, 1, 1, 1},
        {  1, 1, 1, 1, 1},
        {  1, 1, 1, 1, 1},
        {  1, 1, 1, 1, 1},
        {  1, 1, 1, 1, 1}};
        median(k);
    }
```

on a *saltAndPepper(1000)* image. This represents salt and pepper noise on 12% of the 128x128 image.

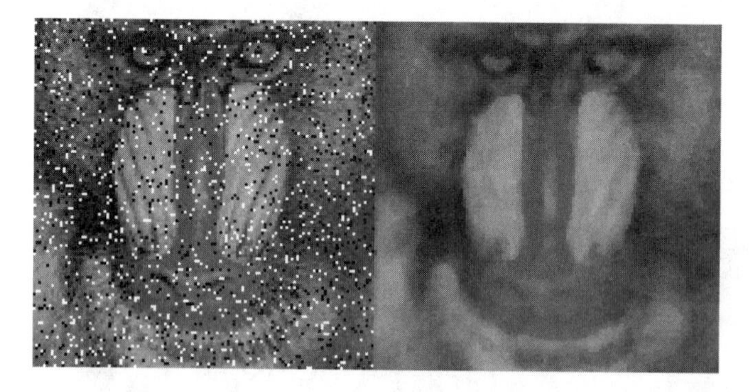

Fig. 9-7. Before and After *medianSquare5x5*.

We note that the salt-and-pepper noise is gone but that the image blur has increased. Further, the number of gray levels has been reduced and the image appears to be posterized. With the increased blur in the image, much of the detail is lost. For example, the mandrill's pupils appear to be considered a part of the salt-and-pepper noise! As a result, we seek a less aggressive median filter called the *medianOctagon5x5*:

```
public void medianOctagon5x5() {
      short k[][] = {
      {  0, 1, 1, 1, 0},
      {  1, 1, 1, 1, 1},
      {  1, 1, 1, 1, 1},
      {  1, 1, 1, 1, 1},
      {  0, 1, 1, 1, 0}};
      median(k);
}
```

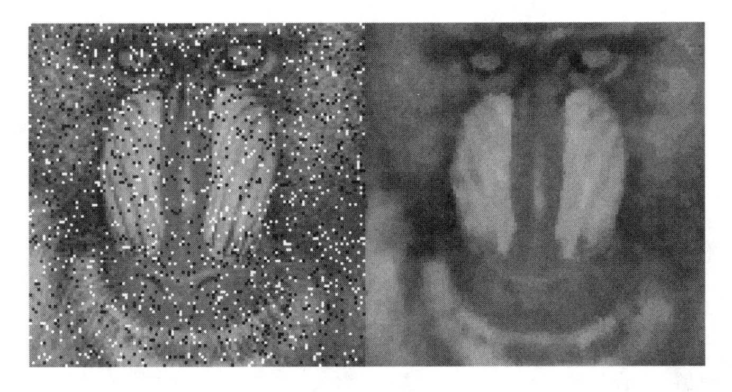

Fig. 9-8. Before and After *medianOctagon5x5*

The *medianOctagon5x5* filter has removed the noise, but blurring and posterization is still present. To make 25% of the pixels in a 128x128 image into salt-and-pepper noise, use

```
saltAndPepper(4000);
```

A *medianOctagon5x5* invocation will not clean such an image. In order to defeat the blurring artifacts, the first filter attempt is the 7x7 cross filter:

```
public void medianCross7x7(){
      short k[][] = {
      { 0, 0, 0, 1, 0, 0, 0},
      { 0, 0, 0, 1, 0, 0, 0},
      { 0, 0, 0, 1, 0, 0, 0},
      { 1, 1, 1, 1, 1, 1, 1},
      { 0, 0, 0, 1, 0, 0, 0},
      { 0, 0, 0, 1, 0, 0, 0},
      { 0, 0, 0, 1, 0, 0, 0}};
      median(k);
}
```

Fig. 9-9 shows the 25% salt-and-paper noise image before and after the application of the *medianCross7x7* filter.

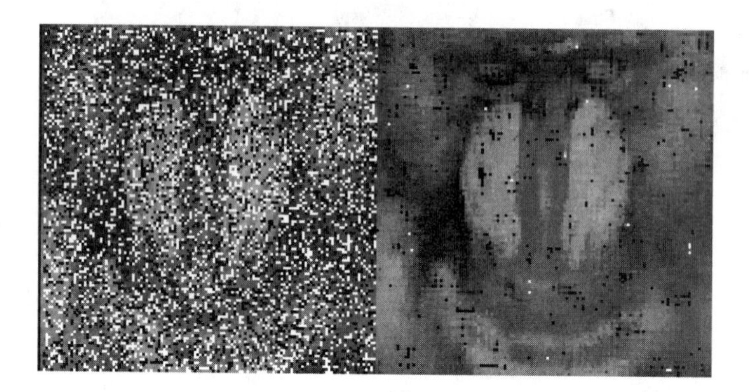

Fig. 9-9. Before and After Application of the *medianCross7x7* Filter

To process an image this noisy, we try a more aggressive filter called the *medianDiamond7x7* filter:

```
public void medianDiamond7x7(){
      short k[][] = {
      { 0, 0, 0, 1, 0, 0, 0},
      { 0, 0, 1, 1, 1, 0, 0},
      { 0, 1, 1, 1, 1, 1, 0},
      { 1, 1, 1, 1, 1, 1, 1},
      { 0, 1, 1, 1, 1, 1, 0},
      { 0, 0, 1, 1, 1, 0, 0},
      { 0, 0, 0, 1, 0, 0, 0}};
      median(k);
}
```

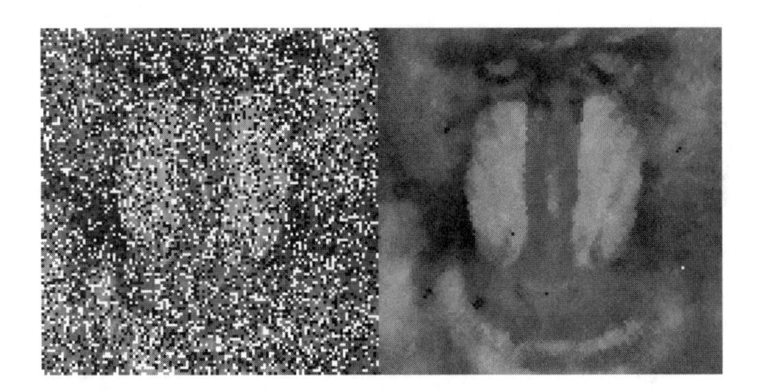

Fig. 9-10. Before and After Application of the *medianDiamond7x7* Filter

We note that Fig. 9-10 shows that there is still salt-and-pepper noise, but that it has been greatly reduced. We are guided by the size of the noisy groups of pixels and note that they are smaller than 5x5 pixels. Therefore, we attempt a *medianDiamond7x7* followed by a *medianSquare5x5.* This is called a *hybrid-median filter* [Russ].

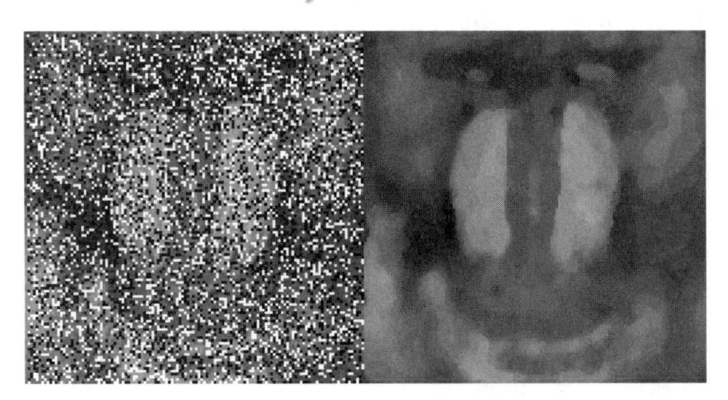

Fig. 9-11. Before and After the Hybrid-Median Filter

Fig. 9-11 shows that the salt-and-pepper noise has been removed, but that posterization artifacts have increased. It is also interesting that the edges have not moved, but detail is now gone (e.g., the pupils in the eyes have been removed). Fig. 9-12 shows the mandrill after a uniform, non-adaptive histogram equalization (UNHE) on the left, followed by the addition of 25% shot noise, hybrid-median filtering and UNHE on the right.

Fig. 9-12. UNHE Emphasis of Hybrid-Median Filtering on Posterization

The median filter has eliminated the salt-and-pepper noise. The cost is measured in image degradation. It remains to be determined where in the image a filter should be applied. Determining where and how much to filter an image is called *adaptive* filtering. One way in which adaptive filtering can be performed is to measure a local region within an image in order to determine if filtering is needed. One parameter that can be measured in a sample set is called the *sample variance.*

The *sample variance* is the sum of the squared deviations from the sample mean divided by the number of samples. That is,

$$\sigma^2 = \frac{1}{n}\sum_{i=0}^{n-1}(a_i - \bar{a})^2 \tag{9.3}$$

where

$\bar{a}$ = mean

a_i = ith sample

σ^2 = sample variance and

n = number of samples

To decide when to perform a median filtering, we devise an outlier detector. We call it the *dog-ate-my-homework* (DAMH) detector. The DAMH detector uses the coefficient of variation, defined by

$$\hat{C} = \sigma / \bar{a} \tag{9.4}$$

assuming that the mean is not zero.

Suppose that there is a poor dog that ate a student's homework, the grades are: {0,85,87,90,100}. Using (9.4), we can detect the presence of a sick dog at home:

```
public static void testOutlier() {
      int a[] = {0,85,87,90,100};
      int b[] = {95,85,87,90,100};
      System.out.println(
        "dog ate my homework ={0,85,87,90,100}"
        + outlierHere(a));
      System.out.println(
        "dog ate my homework ={95,85,87,90,100}"
        + outlierHere(b));
}

public static boolean outlierHere(int a[]) {
      return ( coefficientOfVariation(a) > .4);
}
```

Whose output appears as

```
dog ate my homework ={0,85,87,90,100}true
dog ate my homework ={95,85,87,90,100}false
```

Fig. 9-13 shows the before and after image, with outlier detection on a 3x3 median filtered image.

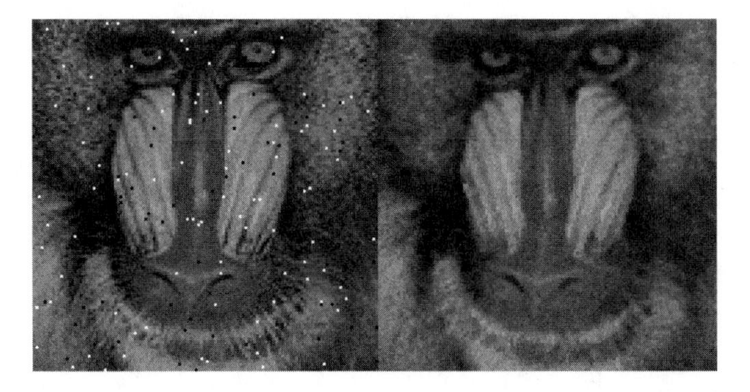

Fig. 9-13. Before and After 3x3 Median Filtering with Outlier Detection.

With the coefficient of variation > 0.1, an outlier is contained in the window and the median filter is performed. Using this technique, the number of outliers is 28,941 down from 3x128x128= 49,152. This could be further reduced by determining if the center pixel in the window is the outlier, rather than just detecting the presence or absence of an outlier. It is just as hard as determining if a pixel is part of an edge, a topic covered in Chapter 10. If the *outlierHere* method triggers on too high a value for the coefficient of variation, then some salt-and-pepper noise will get through and the number of pixels processed is reduced.

9.3. High-Pass Filtering

High-pass filtering (also known as sharpening edge crispening or crispening) will bring out the details in an image. Clinical trials indicated that edge crispening can be generally judged as more subjectively pleasing than undistorted images [Pratt]. As a side effect, sharpening will also emphasize the noise in an image.

Direct convolution may be used to perform high-pass filtering by the use of a kernel whose elements sum to one and which have both positive and negative signs. One 3x3 filter is implemented using a method called *hp1*:

```
public void hp1() {
 float k[][] = {
      { 0, -1,  0},
      {-1, 10, -1},
      { 0, -1,  0}
      };
      Mat.normalize(k);
      convolve(k);
 }
```

Fig. 9-14 shows the effect of processing the mandrill with the high-pass filter.

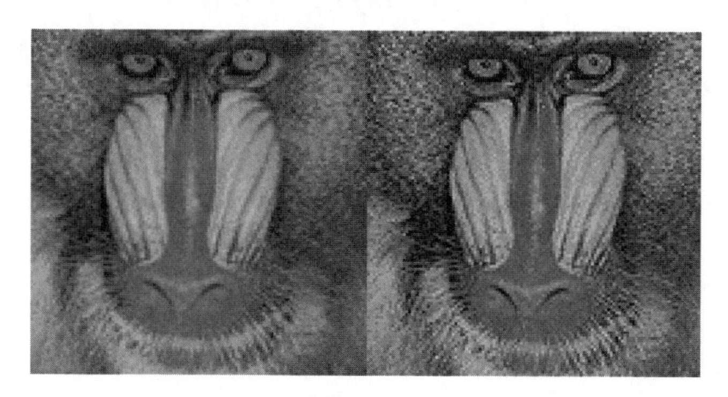

Fig. 9-14. Before and After *hp1*

To observe the effect of the center number on the high pass filter's output, we formulate a new filter called *hp2*:

```
public void hp2() {
 float k[][] = {
      { 0, -1,  0},
      {-1,  8, -1},
      { 0, -1,  0}
      };
      Mat.normalize(k);
      convolve(k);
 }
```

Fig. 9-15 shows *hp1* vs. *hp2*.

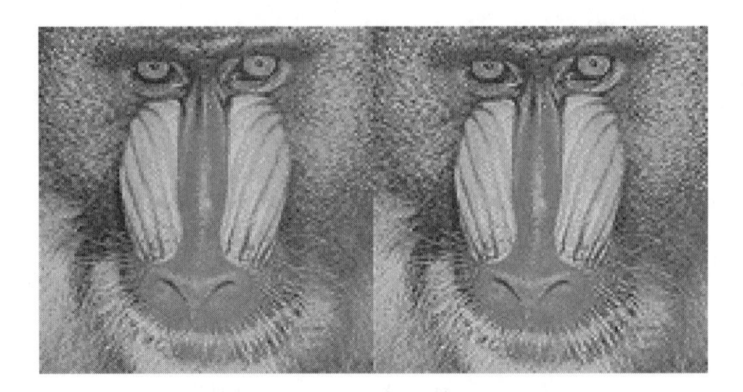

Fig. 9-15. *Hp1* vs. *hp2.*

We see from Fig. 9-15 that lowering the center of the kernel value appears to be sharpening the edges more. To test this theory, we lower the kernel center again, using *hp3:*

```
public void hp3() {
 float k[][] = {
      { 0, -1,  0},
      {-1,  5, -1},
      { 0, -1,  0}
      };
      Mat.normalize(k);
      convolve(k);
 }
```

Fig. 9-16 compares *hp2* vs. *hp3*.

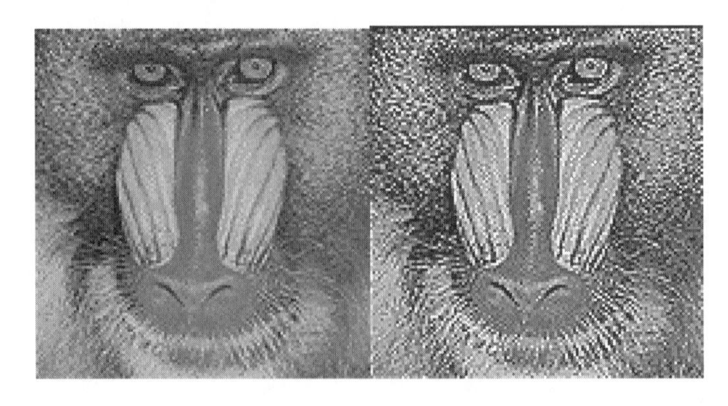

Fig. 9-16. *Hp2* vs. *hp3.*

Fig. 9-16 shows that the *hp3* filter has brought out a great deal of noise from the image (i.e., too much texture enhancement).

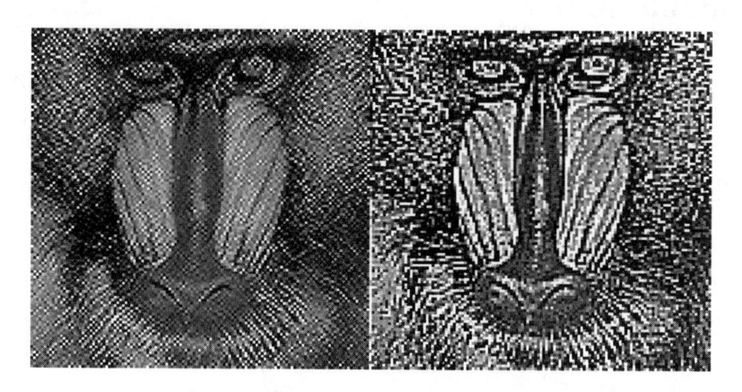

Fig. 9-17. *Hp4* vs. *hp5*

Two other types of sharpening kernels are given in methods *hp4* and *hp5*. They are compared in Fig. 9-17 and their kernels follow [Crane]:

```
public void hp4() {
 float k[][] = {
      { 1, -2,  1},
      {-2,  5, -2},
      { 1, -2,  1}
      };
      convolve(k);
 }

public void hp5() {
 float k[][] = {
      {-1,  -1,  -1},
      {-1,   9,  -1},
      {-1,  -1,  -1}
      };
      convolve(k);
 }
```

9.4. The SpatialFilterFrame

The *SpatialFilterFrame* is a class that resides in the *gui* package. It provides the services of low-pass, high-pass and median filtering. It also provides several utilities that are needed by these operations.

9.4.1. Class Summary

```
package gui;
import java.awt.*;
import java.util.*;
public class SpatialFilterFrame extends ConvolutionFrame
public SpatialFilterFrame child = null;
        public void makeChild()
        public short[][] copyArray(short a[][])
        public void subtractChild()
        public void subtract(SpatialFilterFrame f)
        public void outlierEstimate()
        public SpatialFilterFrame(String title)
        public void clip()
        public void saltAndPepper(int n)
        public void average()
        public void hp1()
        public void hp2()
        public void hp3()
        public void hp4()
        public void hp5()
        public void lp1()
        public void lp2()
        public void lp3()
        public void mean9()
        public void mean3()
        public void gauss3()
        public void gauss7()
        public void gauss15()
        public void gauss31()
        public static double gauss(
        public static void printGaussKernel(
          int M, int N,
          double sigma, double centerMax) {
        public static void printGaussKernel(
               int M, int N,
               double sigma)
        public static float [][] getGaussKernel(
```

```
          int M, int N,
          double sigma)
      public static void testQuickSort()
      public int numberOfNonZeros(short k[][])
      public short getMax(short a[])
      public short getMin(short a[])
      public static void quickSort(int a[])
      public void copyRedToGreenAndBlue()
      public void medianSquare3x3()
      public void medianSquare5x5()
      public void medianOctagon5x5()
      public void medianDiamond7x7()
      public void medianCross7x7()
      public static void printMedian(
        short k[][], String name)
      public void medianSquare7x7()
      public void medianCross3x3()
      public void median (short k[] [])
      public void medianBottom(
        short f[][], short k[][], short h[][])
      public void medianLeft(
        short f[][], short k[][], short h[][])
      public void medianRightAndTop(
        short f[][], short k[][], short h[][])
      // median, optimze the edges
      public short[][] median(short f[][], short k[][])
      public static double mean(int a[])
      public static double variance(int a[])
      public static double coefficientOfVariation(int a[])
      public short[][] medianNoEdge(short f[][], short k[][])
      public static short median(Vector v)
      public void testMedian()
      public static void testVariance()
      public static void testCoefficientOfVariation()
      public static void testOutlier()
      public static boolean outlierHere(int a[])
      public int median(int a[])
      public short[][] medianSlow(short f[][], short k[][])
      public static void printMaple(float a[][])
}
```

9.4.2. Class Usage

The *SpatialFilterFrame* extends the *ConvolutionFrame*. This adds one more class to the list of classes descended from the *ClosableFrame*. The current class hierarchy is shown in Fig. 9-18. Operations (e.g., subtraction, etc.) operate on all three color planes.

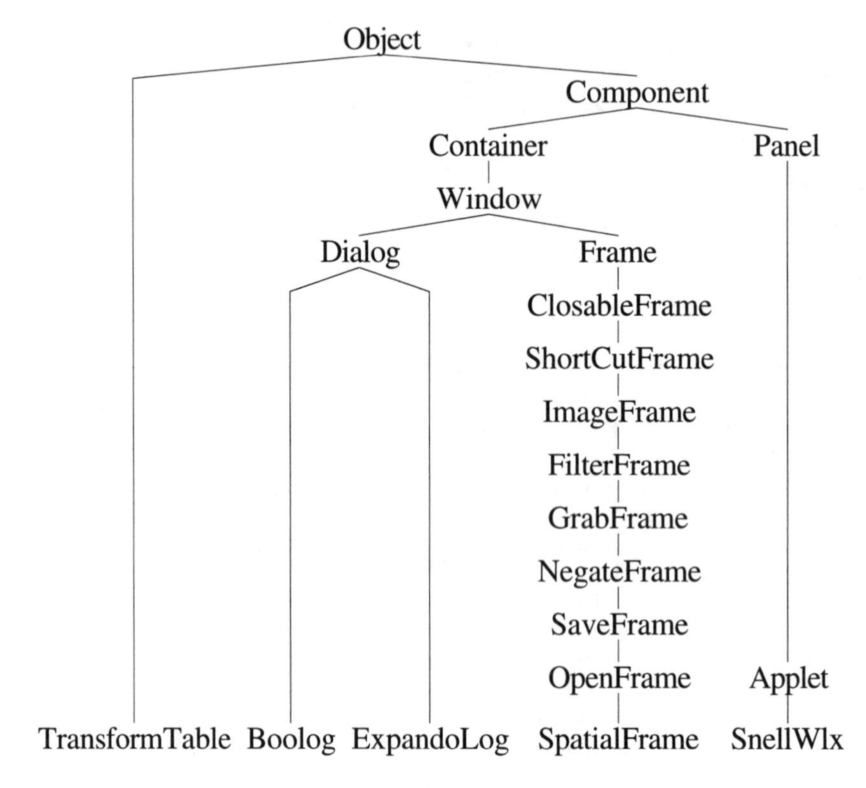

Fig. 9-18. Imaging API, Including the *SpatialFrame*

The *SpatialFilterFrame* is the first frame able to spawn child frames. The child does not have a reference to the parent frame (i.e., the frame from which the child was spawned). This may be added later. A class variable is available for the child frame and is held as a public instance. It is typically set by an invocation of *makeChild*. For example, to create a *SpatialFilterFrame* instance use

```
String title = "test";
SpatialFilterFrame sff = new SpatialFilterFrame(test);
```

Then, to create a child whose image data is that of the *sff* instance variables, use

```
sff.makeChild();
```

To copy a two-dimensional array of *short* and return a new array, use

```
short red[][] = new short[128][128];
short a[][] sff.copyArray(red);
```

To subtract the pixels in the child frame from a *SpatialFilterFrame* instance, use

```
sff.subtractChild();
```

To subtract the pixels in the *sff* from those in another *SpatialFilterFrame,* use:

```
String title2 = "test2";
SpatialFilterFrame sff2 = new SpatialFilterFrame(test2);
sff.subtract(sff2);
```

To turn outlier estimation on or off using a boolean toggle (it is on by default), use

```
sff.outlierExtimate();
```

The outlier estimation will select only those pixels that lie in a median filter window and are outliers (according to the *outlierHere* method).

To make an instance of a *SpatialFilterFrame,* given a title for the frame, use

```
String title2 = "test2";
SpatialFilterFrame sff2 = new SpatialFilterFrame(test2);
```

To hard clip intensity values that lie above 255 or below 0 to 255 and 0, use

```
sff.clip();
```

To add *n* pixels of salt noise and *n* pixels of pepper noise to an image, use

```
int n = 100;
sff.saltAndPepper(n);
```

To perform a low-pass filtering by convolving the image with:

$$\begin{bmatrix} .11111111 & .11111111 & .11111111 \\ .11111111 & .11111111 & .11111111 \\ .11111111 & .11111111 & .11111111 \end{bmatrix}$$

use

```
sff.average();
```

To perform a high-pass filtering by convolving the image with:

$$\begin{bmatrix} 0 & -.16666667 & 0 \\ -.16666667 & 1.6666667 & -.16666667 \\ 0 & -.16666667 & 0 \end{bmatrix}$$

use

```
sff.hp1();
```

To convolve with:

$$\begin{bmatrix} 0 & -.25 & 0 \\ -.25 & 2.0 & -.25 \\ 0 & -.25 & 0 \end{bmatrix}$$

use

```
sff.hp2();
```

To convolve with:

$$\begin{bmatrix} 0 & -1.0 & 0 \\ -1.0 & 5.0 & -1.0 \\ 0 & -1.0 & 0 \end{bmatrix}$$

use

```
sff.hp3();
```

To convolve with:

$$\begin{bmatrix} 1.0 & -2.0 & 1.0 \\ -2.0 & 5.0 & -2.0 \\ 1.0 & -2.0 & 1.0 \end{bmatrix}$$

use

```
sff.hp4();
```

To convolve with:

$$\begin{bmatrix} -1.0 & -1.0 & -1.0 \\ -1.0 & 9.0 & -1.0 \\ -1.0 & -1.0 & -1.0 \end{bmatrix}$$

use
```
sff.hp5();
```

To convolve with:

$$\begin{bmatrix} .1 & .1 & .1 \\ .1 & .2 & .1 \\ .1 & .1 & .1 \end{bmatrix}$$

use
```
sff.lp1();
```

To convolve with:

$$\begin{bmatrix} .083333336 & .083333336 & .083333336 \\ .083333336 & .33333334 & .083333336 \\ .083333336 & .083333336 & .083333336 \end{bmatrix}$$

use
```
sff.lp2();
```

To convolve with:

$$\begin{bmatrix} .05 & .05 & .05 \\ .05 & .6 & .05 \\ .05 & .05 & .05 \end{bmatrix}$$

use
```
sff.lp3();
```

To convolve with:

$$\begin{bmatrix} .0123 & .0123 & .0123 & .0123 & .0123 & .0123 & .0123 & .0123 & .0123 \\ .0123 & .0123 & .0123 & .0123 & .0123 & .0123 & .0123 & .0123 & .0123 \\ .0123 & .0123 & .0123 & .0123 & .0123 & .0123 & .0123 & .0123 & .0123 \\ .0123 & .0123 & .0123 & .0123 & .0123 & .0123 & .0123 & .0123 & .0123 \\ .0123 & .0123 & .0123 & .0123 & .0123 & .0123 & .0123 & .0123 & .0123 \\ .0123 & .0123 & .0123 & .0123 & .0123 & .0123 & .0123 & .0123 & .0123 \\ .0123 & .0123 & .0123 & .0123 & .0123 & .0123 & .0123 & .0123 & .0123 \\ .0123 & .0123 & .0123 & .0123 & .0123 & .0123 & .0123 & .0123 & .0123 \\ .0123 & .0123 & .0123 & .0123 & .0123 & .0123 & .0123 & .0123 & .0123 \end{bmatrix}$$

use

```
sff.mean9();
```

Note that starting with *mean9,* only 3 significant figures are shown, for formatting brevity. A full 32-bit *float* resolution is used internally.

To convolve with:

$$\begin{bmatrix} .111 & .111 & .111 \\ .111 & .111 & .111 \\ .111 & .111 & .111 \end{bmatrix}$$

use

```
sff.mean3();
```

To convolve with:

$$\begin{bmatrix} .0625 & .125 & .0625 \\ .125 & .25 & .125 \\ .0625 & .125 & .0625 \end{bmatrix}$$

use

```
sff.gauss3();
```

To convolve with:

$$\begin{bmatrix} .00714 & .00714 & .0143 & .0143 & .0143 & .00714 & .00714 \\ .00714 & .0143 & .0143 & .0286 & .0143 & .0143 & .00714 \\ .0143 & .0143 & .0286 & .0571 & .0286 & .0143 & .0143 \\ .0143 & .0286 & .0571 & .114 & .0571 & .0286 & .0143 \\ .0143 & .0143 & .0286 & .0571 & .0286 & .0143 & .0143 \\ .00714 & .0143 & .0143 & .0286 & .0143 & .0143 & .00714 \\ .00714 & .00714 & .0143 & .0143 & .0143 & .00714 & .00714 \end{bmatrix}$$

use

```
sff.gauss7();
```

To convolve with 15x15 and 31x31 Gaussian kernels use

```
sff.gauss15();
sff.gauss31();
```

The kernels are too big to format properly here.

To evaluate (9.1), use

```
double x, double y;
double xc, double yc, double sigma;
double g = SpatialFilterFrame.gauss(x,y,xc,yc, sigma);
```

Note that this is a two-dimensional Gaussian density with standard deviation given by *sigma* and whose center is located at *xc,yc*.

To print the Java source code needed to invoke a Gaussian convolution with an *MxN* kernel whose peak value is *centerMax* and whose standard deviation is *sigma*, in accordance with (9.1), use

```
int M = 31;
int N = 31;
double sigma = 1.0;
double centerMax = 1.0;
SpatialFilterFrame.printGaussKernel(M,N, sigma, centerMax);
```

To print the Java source code needed to invoke a Gaussian convolution with an *MxN* kernel whose peak value is computed so that the kernel's elements sum to one and have a standard deviation is *sigma*, in accordance with (9.1) use

```
int M = 31;
```

```
int N = 31;
double sigma = 1.0;
double centerMax = 1.0;
SpatialFilterFrame.printGaussKernel(M,N, sigma);
```

To get a Gaussian kernel, computed in accordance with (9.1) and normalized so that all elements sum to one, use

```
float k[][] = SpatialFilterFrame.getGaussKernel(M,N,sigma);
```

To test the internal quicksort facility for arrays of short integers, use

```
SpatialFilterFrame.testQuickSort();
```

To determine the number of non zero element in a two-dimensional array of short, use

```
short k[][] = new short[10][10;
int zeros = SpatialFilterFrame.numberOfNonZeros(k);
```

To get the minimum and maximum values in a one-dimensional array of short, use

```
short a[] = new short[100];
short s = sff.getMax(a);
s = sff.getMin(a);
```

To sort an array of integers in ascending order, use

```
SpatialFilterFrame.quickSort(a);
```

To copy the red color planes to green and blue, use

```
sff.copyRedToGreenAndBlue();
```

To perform the median filters, as described in Section 9.2, use

```
sff.medianCross3x3();
sff.medianSquare3x3();
sff.medianSquare5x5();
sff.medianOctagon5x5();
sff.medianDiamond7x7();
sff.medianCross7x7();
sff.medianSquare7x7();
```

See section 9.2 for the kernels and effects of these median filters.

To print the Java source needed to process a two-dimensional array of *short*, use

```
short k[][] = {
      {0, 1, 0},
      {1, 1, 1},
      {0, 1, 0}};
String name = "medianCross3x3"
SpatialFilterFrame.printMedian(k,name);
```

To perform a median filtering on all three color planes, given the kernel, *k*, use

```
sff.median(k);
```

There are some public median filtering utilities available that are left undocumented as they are typically unneeded. Their prototypes appear in the class summary, Section 9.4.1. To take the average of a one-dimensional array of **int** use

```
int a[] = new int[100];
double d = SpatialFilterFrame.mean(a);
```

To compute the variance of a one-dimensional array of **int** use

```
int a[] = new int[100];
double d = SpatialFilterFrame.variance(a);
```

To compute the coefficient of variation (9.4) of a one-dimensional array of **int** use

```
int a[] = new int[100];
double d = SpatialFilterFrame.coefficientOfVariation(a);
```

A series of test methods are available for median, variance, outlier and coefficient of variation computation. These methods are described in Section 9.2.

```
public void testMedian()
public static void testVariance()
public static void testCoefficientOfVariation()
public static void testOutlier()
public static boolean outlierHere(int a[])
```

To compute the coefficient of `median` of a one-dimensional array of **int** use

```
int a[] = new int[100];
double d = sff.median(a);
```

To print a two-dimensional array of float as a line of Maple (a symbolic manipulator) use

```
SpatialFilterFrame.printMaple(f);
```

For example:

```
public void convolve(float a[][]) {
    printMaple(a);
    super.convolve(a);
}
```

will print *a* before invoking the *convolve* method. The output for the *lp1* method is:

```
evalf(linalg[matrix](3,3,[
0.1,0.1,0.1,0.1,0.2,0.1,0.1,0.1,0.1]));
```

which in Maple evaluates to:

$$\begin{bmatrix} .1 & .1 & .1 \\ .1 & .2 & .1 \\ .1 & .1 & .1 \end{bmatrix}$$

This permits an automatic formatting of arrays. In Maple, using the command:

```
Digits := 3;
```

sets the number of significant figures to 3, which makes the arrays more readable.

9.5. Summary

This chapter showed how to use Java to generate kernels. Convolution kernels are typically small images. People generally type these images in by hand. The process of entering in kernels by hand gives rise to keyboard errors, tedium and truncation error. Having students type in kernels probably has diminishing educational returns and gives a generally bad impression of image processing. It is much more fun to write programs that write programs!

We touched on the problem of knowing when to filter and when not to filter. The modern image processing programs generally allow for interaction with the image in order to perform user-directed image process. The emphasis in this chapter has been to provide automatic mechanisms for performing the image processing, whenever possible.

9.6. Suggested Projects

1. Add a new method that performs a *selective* median filter. Use a technique that decides if the center pixel is the outlier and performs the median filter on that pixel only if it is the outlier. How does this compare to the *outlierHere* method? How will you tell if the center pixel is an outlier? Will sorting the pixels in the window and comparing the smallest and largest with the center help?

2. Add an interface that enables the user to type in the convolution coefficients. The user should be able to type in a kernel of any dimension.

3. Add an interface that allows the user to specify the size and variance of a Gaussian kernel. Then use these parameters to perform a convolution.

4. Add an interface that allows the user to outline a rectangular region of interest. Then perform processing only in that region.

5. One way in which a high-pass filter may be constructed is to subtract a low-pass filtered version of an image from the original image. This is called the *unsharp mask.* Write a method to implement the unsharp mask. How does this compare with high-pass filtering?

10. Convolution-based Edge Detection

Rock-a-bye business, on the tree top.
When the code is deprecated, the business will rock.
When code breaks, the business will fall.
And down goes the business, employees and all.

– DL, Ode to Code, 1998

Edge detection is a kind of image processing that is used to outline the boundaries of objects in the image plane [Kwok]. The extraction of useful and relevant edges continues to be a major problem in image processing [Vliet]. This chapter covers the first of two common techniques for edge detection: filtering, i.e., converting a mult-level image into a two-level image using convolution. The second technique, covered in Chapter 12, uses search to convert a multi-level image into a list of coordinates.

10.1. Laplace Filter

One way to formulate the kernel for an edge detection filter is to use the *Laplacian operator*. P.S. Laplace (1749-1827) was a French mathematician who contributed to the fields of astronomy, probability and mechanics. One of his contributions, called the Laplace transform, is typically studied in a first course on differential equations [Boyce]. It is a remarkable quirk of fate that the German mathematician/physicist, Karl Friedrich Gauss (1777-1855), was a contemporary of Laplace, and that both Laplace and Gauss lived to be exactly 78!

The Laplace operator (also called the *Laplacian)*, which is quite different from the Laplace transform, is studied in a first course on advanced calculus [Widder]. The Laplace operator, in two-dimensions, is the 2nd-order partial derivative taken with respect to each of the orthogonal directions. It is also called the *divergence of the gradient* and is given by

$$\nabla^2 f(x,y) = \frac{\partial^2 f}{\partial x^2} + \frac{\partial^2 f}{\partial y^2} \quad (10.1).$$

We are interested in the Laplacian because it is zero when the rate of intensity change in the image plane is zero or linear.

Some books negate (10.1) in order to favor a left-to-right and a bottom-to-top change in an image [Pratt].

For example, suppose that

$$f(x,y) = x^2 + y^2 \quad (10.2)$$

To compute (10.1) we need to take the 2nd-order partial derivative with respect to each coordinate, to obtain

$$\frac{\partial^2 f}{\partial x^2} = \frac{\partial^2 f}{\partial y^2} = 2 \quad (10.3).$$

Substituting (10.3) into (10.1) yields

$$\nabla^2 f(x,y) = 4 \quad (10.4).$$

As an exercise, try to substitute any linear function into (10.1) and compute the Laplacian. You should be able to prove to yourself that the function is always zero. For a more rigorous proof, see [Marr].

We now assert that first and second derivatives of smooth functions may be approximated via a convolution over 3 regularly spaced samples. This assertion forms the basis of most 3x3 edge detectors. The convolution is used to create the computational equivalent of the Taylor series approximation of the derivative. Following [Schalkoff], we make a two-point approximation that uses the Taylor series, which (in 1-D) is written as

$$f(x+\Delta x) = f(x) + \Delta x f'(x) + \frac{(\Delta x)^2}{2!} f''(x) + ... \qquad (10.4a).$$

With $\Delta x = 1$, we solve (10.4a) for $f'(x)$:

$$f'(x) = f(x+1) - f(x) - \frac{1}{2!} f''(x) - ... \qquad (10.4b)$$

and a two-point approximation follows:

$$f'(x) \approx f(x+1) - f(x) \qquad (10.4c)$$

A *centered* difference approximation is arrived at by assuming that *f* is smooth (i.e., *f* has a continuous first derivative) and averages the values sampled before and after *f(x)*. It is given by

$$f'(x) \approx \frac{f(x+1) - f(x-1)}{2} \qquad (10.4d).$$

Writing (10.4d) in terms of a convolution yields

$$f'(x) \approx \begin{bmatrix} -1/2 \\ 0 \\ 1/2 \end{bmatrix} * \begin{bmatrix} f(x-1) \\ f(x) \\ f(x+1) \end{bmatrix} \qquad (10.4e).$$

To get the second derivative, solve (10.4b) for $f''(x)$, ignoring higher-order terms (assuming that *f* has a continuous second derivative):

$$f''(x) = 2f(x+1) - 2f(x) - 2f'(x) \qquad (10.4f)$$

Substitute (10.4d) into (10.4f), expand and simplify:

$$f''(x) = 2f(x+1) - 2f(x) - 2\left(\frac{f(x+1) - f(x-1)}{2}\right)$$
$$f''(x) = 2f(x+1) - 2f(x) - f(x+1) + f(x-1) \qquad (10.4g)$$
$$f''(x) = f(x+1) - 2f(x) + f(x-1)$$

We can thus get the second derivative of *f(x)* (in 1-D) by

$$f''(x) \approx \begin{bmatrix} 1 \\ -2 \\ 1 \end{bmatrix} * \begin{bmatrix} f(x-1) \\ f(x) \\ f(x+1) \end{bmatrix}. \qquad (10.4h)$$

Q.E.D.

When a function is continuous (i.e., no breaks), we write that $f \in C^0$ (*f* has C-zero continuity). If a function has a continuous first derivative, then $f \in C^1$. It should come as no surprise that a well-posed Laplacian requires a function, *f*, that is twice continuously differentiable on the plane (i.e., $f \in C^2$). If *f* is discrete and/or not twice continuously differentiable, then we must pre-filter *f* to meet the continuity condition.

To understand how to apply the Laplacian to a discrete image, consider the 3x3 window centered over the pixel located at *f[x][y]*, given by

$$\begin{bmatrix} f[x-1][y+1] & f[x][y+1] & f[x+1][y+1] \\ f[x-1][y] & f[x][y] & f[x+1][y] \\ f[x-1][y-1] & f[x][y-1] & f[x+1][y-1] \end{bmatrix} \qquad (10.5).$$

Following [Weeks] we approximate the first-partial derivative in the x-direction (from right to center) as

$$\frac{\partial f_r}{\partial x} \approx f[x+1][y] - f[x][y] \qquad (10.6).$$

We can also approximate the first-partial derivative in the x-direction (from center to left) as

$$\frac{\partial f_l}{\partial x} \approx f[x][y] - f[x-1][y] \qquad (10.7).$$

The second partial derivative, with respect to *x*, is approximated by subtracting (10.6) from (10.7) to obtain

$$\frac{\partial^2 f}{\partial x^2} \approx \frac{\partial f_l}{\partial x} - \frac{\partial f_r}{\partial x} \approx 2f[x][y] - f[x-1][y] - f[x+1][y] \qquad (10.8).$$

Similarly, we can approximate the first-partial derivative in the y-direction (from top to center) using:

$$\frac{\partial f_t}{\partial y} \approx f[x][y+1] - f[x][y] \qquad (10.9).$$

We can also approximate the first-partial derivative in the y-direction (from center to bottom) with:

$$\frac{\partial f_b}{\partial y} \approx f[x][y] - f[x][y-1] \qquad (10.10).$$

Just like in the x-direction, we formulate the second partial derivative, with respect to y, by subtracting (10.9) from (10.10) to obtain

$$\frac{\partial^2 f}{\partial y^2} \approx \frac{\partial f_b}{\partial x} - \frac{\partial f_t}{\partial x} = 2f[x][y] - f[x][y-1] - f[x][y+1] \qquad (10.11).$$

We now substitute (10.11) and (10.8) into (10.1) to obtain the approximation for the Laplacian over a 3x3 discrete lattice:

$$\nabla^2 f(x,y) = \frac{\partial^2 f}{\partial x^2} + \frac{\partial^2 f}{\partial y^2}$$

$$\nabla^2 f(x,y) \approx 2f[x][y] - f[x-1][y] - f[x+1][y] + 2f[x][y] - f[x][y-1] - f[x][y+1]$$

which simplifies to

$$\nabla^2 f(x,y) \approx 4f[x][y] - f[x-1][y] - f[x+1][y] - f[x][y-1] - f[x][y+1] \qquad (10.12)$$

The approximation in (10.12) is typically performed by a convolution with the 3x3 Laplacian kernel, given by

$$\begin{bmatrix} 0 & -1 & 0 \\ -1 & 4 & -1 \\ 0 & -1 & 0 \end{bmatrix} \quad (10.14).$$

We can implement (10.14) directly, using

```
public void laplacian3() {
float k[][] = {
        { 0,  -1,  0},
        {-1,   4, -1},
        { 0,  -1,  0}
        };
        convolve(k);
    }
```

Thus, the pixel is replaced by the difference between itself and the average of its four nearest neighbors [Netravali]. This is generally followed by a thresholding operation. Fig. 10-1 shows the mandrill before and after filtering the image using *laplacian3* and thresholding the image at the average intensity.

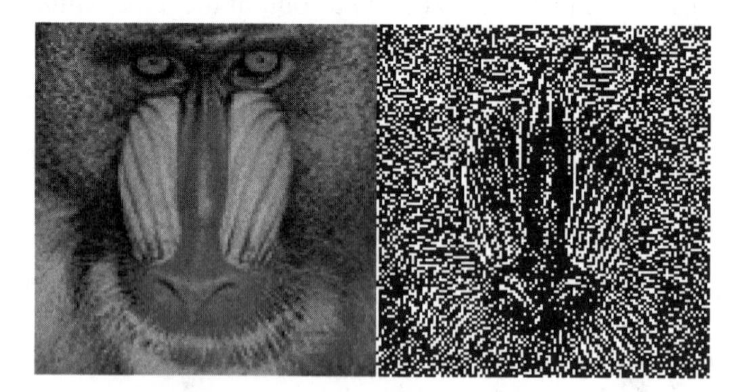

Fig. 10-1. Before and After Application of *laplacian3* and Thresholding

Fig. 10-1 shows a great deal of noise in the image. This occurs because the image was not pre-filtered (an operation we shall discuss shortly). The lack of pre-filtering causes a violation of the $f \in C^2$ condition. Alternatively, the 3x3 Laplacian may be implemented by a negated kernel, as given in [Vliet]:

```
public void laplacian3Minus() {
float k[][] = {
      { 0,  1, 0},
      { 1, -4, 1},
      { 0,  1, 0}
      };
      convolve(k);
 }
```

Fig. 10-2 shows a comparison of *laplacian3* vs. *laplacian3Minus* after thresholding. There is a subtle difference between the images as the edges appear to be shifted.

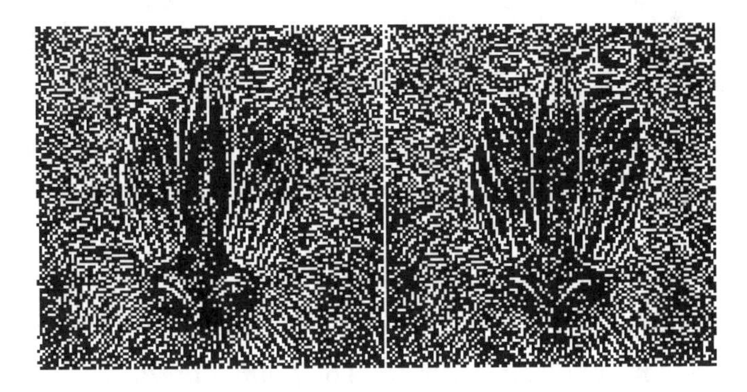

Fig. 10-2. *laplacian3* vs. *laplacian3Minus*

To reduce the noise in the edge detector, and to emphasize the difference between the edge detectors, we employ a 3x3 Gaussian low-pass pre-filter. Fig. 10-3 shows the effect of applying a 3x3 Gaussian pre-filter, then edge detecting the image with the two Laplacian edge detectors.

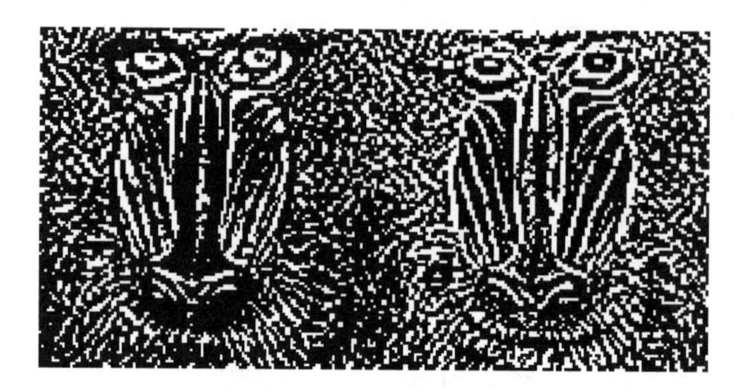

Fig. 10-3. A 3x3 Gaussian Pre-filter: *laplacian3* vs. *laplacian3Minus*

In summary, we have made three points. The first is that negating the Laplacian kernel makes a difference. The second is that numerical differentiation of images is ill-posed without a regularizing filtering operation [Torre]. The third and most important point is that pre-filtering makes the edge detection look better.

Fig. 10-4 shows *laplacian3* vs. *laplacian3Prewitt* (after a 3x3 Gaussian pre-filter and a threshold post-process). The *laplacian3Prewitt* method uses an impulse response array, after gain normalization [Pratt]. The *laplacian3Prewitt* method follows:

```
public void laplacian3Prewitt() {
float k[][] = {
        { -1,  -1, -1},
        { -1,   8, -1},
        { -1,  -1, -1}
        };
        Mat.scale(k,1/8.0);
        convolve(k);
  }
```

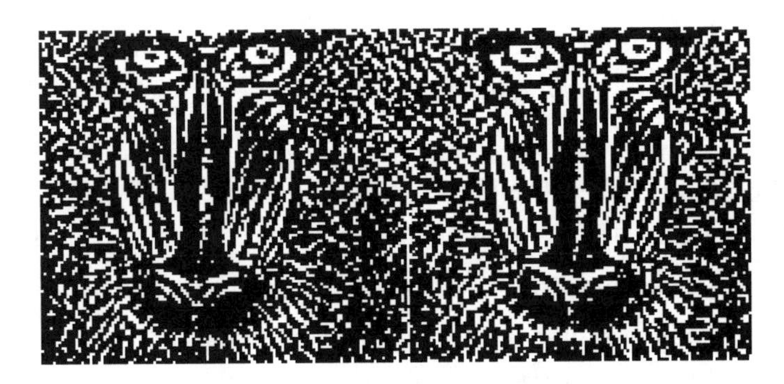

Fig. 10-4. *laplacian3* vs. *laplacian3Prewitt*

As mentioned earlier, a pre-filtering step is typically required to make the numerical differentiation of an image well-posed. As described by Marr and Hildreth [Marr], there are also physical reasons for applying a low-pass filter. One reason is to reduce the range of scales over which intensity changes take place. This leads to a filter that is smooth and band limited in frequency. The other is to group the subjectively sensed changes (i.e., illumination, orientation, and surface property changes) in an image. These changes are grouped by using a filter that has the property of *spatial localization*. This leads to a filter that is smooth and has a small variance (for spatial localization).

Canny has combined the goals of accurate edge detection and spatial localization into a single functional. He points out that we can obtain arbitrarily good localization at the expense of detection. He also shows that we can obtain arbitrarily good detection at the expense of localization. The Canny performance criterion for an edge detector requires good detection, good localization and a minimized number of responses to a single edge. Good detection means a low probability of failing to mark a real edge and a high probability of marking a correct edge. Good localization means that the marked edge should be close to the real edge.

Formulation and solution of the optimization criterion in two dimensions is a complex problem [Mehrotra]. Canny uses a criterion of optimality that shows that the

Gaussian operator is sub-optimal. But then Canny goes on to use the Gaussian operator because it can be computed with "much less effort" [Canny].

When the criterion of optimality is to arrive at a smoothing filter, with both small variance and limited bandwidth, the optimal solution is the Gaussian distribution. To put it another way, the Gaussian is the only function to minimize the bandwidth-frequency product. We may combine the Laplacian of the Gaussian (LoG) to obtain a kernel that performs the LoG filter. The reasons that this combination is possible are due to the *differentiation theorem* for convolution and the associativity of convolution. That is,

$$\frac{d}{dh}\int_{-\infty}^{\infty} f(x)I(h-x)dx = \int_{-\infty}^{\infty} f'(x)I(h-x)dx \qquad (10.15a)$$

and

$$f*(g*h) = (f*g)*h \qquad (10.15b).$$

We leave the proof of (10.15a) as an exercise for the reader. To justify (10.15b), recall that convolution in the spatial domain is equivalent to multiplication in the frequency domain. Also recall that multiplication is associative, therefore convolution must also be associative (see [Castleman] for a rigorous proof). Why do we care? We low-pass filtered the image by convolving the image with a Gaussian kernel. For edge detection, the result of low-pass filtering was then convolved with the Laplacian kernel. Now we take the Laplacian of the Gaussian, use the resulting function to generate a kernel, then take a single convolution.

Setting the center to zero, $(x_c, y_c) = (0,0)$, the Laplacian of the Gaussian (LoG) is given by

$$-\nabla^2 \frac{1}{2\pi\sigma^2} e^{\frac{x^2+y^2}{2\sigma^2}} = \frac{1}{\pi\sigma^4}\left[1 - \frac{x^2+y^2}{2\sigma^2}\right] e^{-\frac{x^2+y^2}{2\sigma^2}} \qquad (10.15c)$$

Fig. 10-5 shows a plot of (10.15c) in the range of -4...4 for x and y and with $\sigma = 1$. This is often called the Mexican hat (or LoG) kernel. The maximum of (10.15c) occurs at x=0, y=0 and is given by

$$-\left[\nabla^2 \frac{1}{2\pi\sigma^2} e^{\frac{x^2+y^2}{2\sigma^2}}\right]_{\max} = \frac{1}{\pi\sigma^4} \qquad (10.16).$$

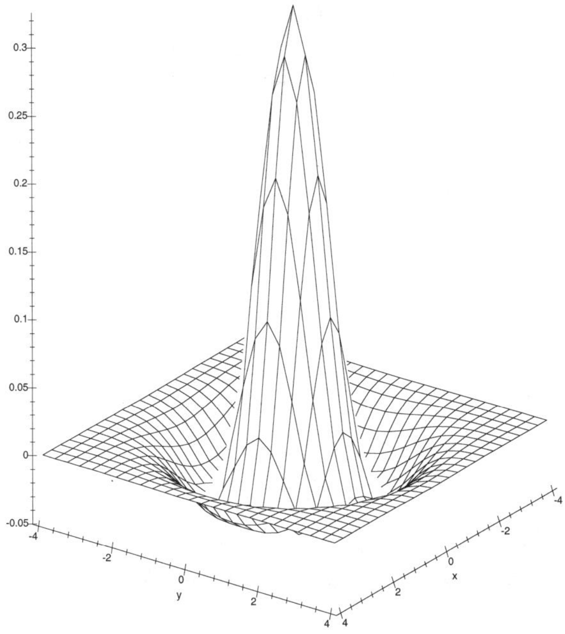

Fig. 10-5. The Mexican Hat kernel

To synthesize the Java program that generates LoG kernels, we write the following Maple statements:

```
readlib(C):
> g:=exp(-((x-xc)**2+(y-yc)**2)/(2*sigma**2))/(2*Pi*sigma**2):
> ddx:=diff(g,x,x):
> ddy:=diff(g,y,y):
> LoG := -ddx - ddy:
> C(LoG,optimized):
```

The Maple output is the basis for the Java source:

```
public static double laplaceOfGaussian(double x, double y,
        double xc, double yc, double sigma) {
         t1 = sigma*sigma;
        t2 = t1*t1;
        t5 = Math.pow(x-xc,2.0);
        t7 = Math.pow(y-yc,2.0);
        t11 = Math.exp(-(t5+t7)/t1/2);
        t13 = 1/Math.PI;
        t16 = Math.pow(2.0*x-2.0*xc,2.0);
        t18 = 1/t2/t1;
        t20 = t11*t13;
        t23 = Math.pow(2.0*y-2.0*yc,2.0);
        t26 = 1/t2*t11*t13-t16*t18*t20/8-t23*t18*t20/8;
        return t26;
}
```

The temporary variables generated by Maple require re initialization if they are locally scoped. As a result, they have been declared as class instance variables. The *EdgeFrame* class uses a method to generate the Java source code that declares the temporary variables, making them available to all sub-classes.

```
public static void tGenerator(int min, int max) {
        for (int i = min; i < max; i++)
         System.out.println(
           "static double t"
           + i + " = 0;");
}
```

The output of the *tGenerator* is shown below (there are 99 temporary variables available):

```
static double t0 = 0;
static double t1 = 0;
static double t2 = 0;...
```

10.2. Roberts

One way to detect an edge is to compute the gradient of the image and then to translate the gradient into a scalar quantity that can be used by some decision function (i.e., thresholding). The gradient is computed as the partial derivative in each of the x and y directions. The result of the gradient computation is a vector:

$$\nabla f(x,y) = \frac{\partial f}{\partial x} i + \frac{\partial f}{\partial y} j. \tag{10.17}$$

Gradients are typically studied in a course in advanced calculus [Amazigo]. The absolute value (also called the *modulus*) of the gradient is a scalar given by the square root of the sum of the squares:

$$|\nabla f(x,y)| = \sqrt{\left(\frac{\partial f}{\partial x}\right)^2 + \left(\frac{\partial f}{\partial y}\right)^2} \qquad (10.18).$$

On a two-by-two window, w, we assign pixels to a linear matrix, p:

$$w = \begin{bmatrix} p[0] & p[1] \\ p[2] & p[3] \end{bmatrix} = \begin{bmatrix} f[x][y] & f[x+1][y] \\ f[x][y+1] & f[x+1][y] \end{bmatrix} \tag{10.19}$$

We then define the difference operators on the main and secondary diagonals as

$$\Delta u = p[0] - p[3], \Delta v = p[1] - p[2] \tag{10.20}$$

The modulus is approximated, using:

$$|\nabla f(x,y)| \approx \sqrt{\Delta u^2 + \Delta v^2} \tag{10.21}$$

This is called the Roberts "cross" operator [Nadler]. From an implementation point of view, we typically start with a color image. The color image is converted into a grayscale image and edge detection is performed using (10.21). The output of the edge detection is thresholded on the average value to create a binary image. One possible process is to convert from color to gray, perform (10.21), placing the result back in the red plane, then threshold the red plane to make a binary image. Finally, copy the red plane to the green and blue planes. Knowing if, where and when to threshold an image is not easy. Fig. 10-6 shows the result of applying the *roberts2* method and thresholding on the average value.

Fig. 10-6. Before and After Application of the *roberts2* Method.

The *colorToRed* method computed the average color value at each pixel and copies it to the red color plane:

```
public void colorToRed() {
    for (int x=0; x < width; x++)
     for (int y=0; y < height; y++)
         r[x][y] = (short)
            ((r[x][y] + g[x][y]  + b[x][y]) / 3);
}
```

The *roberts2* method follows:

```
public void roberts2() {
    colorToRed();
    int p[] = new int[4];
    float delta_u = 0;
    float delta_v = 0;
    short t;
    for (int x=0; x < width-1; x++)
     for (int y=0; y < height-1; y++) {
         p[0] = r[x][y];
         p[1] = r[x+1][y];
         p[2] = r[x][y+1];
         p[3] = r[x+1][y+1];
         delta_u = p[0] - p[3];
         delta_v = p[1] - p[2];
          t = (short)
            Math.sqrt(delta_u*delta_u + delta_v*delta_v);
         r[x][y] = t;
```

```
            g[x][y] = t;
            b[x][y] = t;
        }
        short2Image();
    }
```

The *p* matrix of the *roberts2* method is set up in accordance with (10.19). The gradient is computed using (10.20) and placed into *t*. Fig. 10-7 shows the output of the *roberts2* method when thresholding is turned off. A uniform non-adaptive histogram equalization is performed before the *roberts2* is invoked.

Fig. 10-7. UNHE Mandrill Before and After *roberts2* (no Threshold).

As pointed out by [Pal], the different edge operators produce values at every location. The values are not all valid candidates for edges. The selection of a threshold value is a central issue in performance. This is due to the low intensity variation that may correspond to edges in some parts of the image. In other parts of the image, a high-intensity variation may be required. Approaches to this problem include adaptive thresholding and the use of search techniques [McGee].

10.3. Sobel and the Double Templates

During the derivation of the Roberts edge detector, we came up with a rather *ad hoc* implementation scheme that does not scale well to larger kernels. This is particularly evident when the mapping of (10.19)

$$w = \begin{bmatrix} p[0] & p[1] \\ p[2] & p[3] \end{bmatrix} = \begin{bmatrix} f[x][y] & f[x+1][y] \\ f[x][y+1] & f[x+1][y] \end{bmatrix} \qquad (10.19)$$

is extrapolated to very large kernels.

With the Sobel edge detector, we are confronted with the task of taking two convolutions for each pixel. Typically, this is for a vertical edge and a horizontal edge. We then take the square root of the sum of the squares for each of the convolution results and use this for the output image. It is typical to use a smoothing pre-filter before the Sobel edge detection (or for any edge detection, for that matter) and to use a thresholding afterward. This clouds the comparison of the edge detectors, however, and so neither task is performed, by default.

The two kernels for the horizontal and vertical Sobel edge detection for a 3x3 window are given in the *sobel3* method:

```
public void sobel3() {
 float k1[][] = {
   {-1,  -2, -1},
   { 0,   0,  0},
   { 1,   2,  1}
   };
 Mat.scale(k1,1/4.0);
 float k2[][] = {
   {1, 0,  -1},
   {2, 0,  -2},
   {1, 0,  -1}
   };
  Mat.scale(k2,1/4.0);

   templateEdge(k1,k2);
 }
```

The central processing in *sobel3* is performed in the *templateEdge* method. We start the process by converting the image to gray scale. This is done by placing the average of the red, green and blue color planes into the red plane. We then perform a convolution between kernel *k1* and the red plane, placing the answer in the green plane. The convolution between *k2* and the red plane is placed in the blue plane. The square root of the sum of the squares of the results in the green and blue planes is placed into all three color planes and is then displayed. Once again, smoothing filters have not been applied, nor has a thresholding. Thus, we are looking at the Sobel filter as a kind of high-pass filter that responds well to vertical and horizontal edges. The *templateEdge* method follows:

```
public void templateEdge(float k1[][], float k2[][]) {
      colorToRed();
      g = convolve(r,k1);
      b = convolve(r,k2);
      for (int x=0; x < width; x++)
        for (int y=0; y < height; y++) {
          r[x][y] = (short)
          Math.sqrt(g[x][y]*g[x][y]+b[x][y]+b[x][y]);
          g[x][y] = r[x][y];
          b[x][y] = r[x][y];
      }
      short2Image();
}
```

The *template* method is computing an edge strength and ignoring the edge direction. Keep in mind that *g* and *b* are place-holders that contain the results of a convolution with two orthogonal kernels. These can be used to give a vector which shows the direction of a gradient. This is left as an exercise for the reader in Section 10.6, Projects.

To understand better what happens when we perform two convolutions, we set up a 3x3 window, using (10.5):

$$\begin{bmatrix} f[x-1][y+1] & f[x][y+1] & f[x+1][y+1] \\ f[x-1][y] & f[x][y] & f[x+1][y] \\ f[x-1][y-1] & f[x][y-1] & f[x+1][y-1] \end{bmatrix} \quad (10.5).$$

Following [Galbiati] for the Sobel we convolve *k2* from *sobel3* with (10.5) to obtain

$$\Delta x = \begin{bmatrix} 1 & 0 & -1 \\ 2 & 0 & -2 \\ 1 & 0 & -1 \end{bmatrix} * \begin{bmatrix} f[x-1][y+1] & f[x][y+1] & f[x+1][y+1] \\ f[x-1][y] & f[x][y] & f[x+1][y] \\ f[x-1][y-1] & f[x][y-1] & f[x+1][y-1] \end{bmatrix} \quad (10.22).$$

Expanding (10.22) yields (recall that the "*" mean convolution, *not* multiplication)

$$\begin{aligned} \Delta x = {} & f[x-1][y+1] + 2f[x-1][y] + f[x-1][y-1] \\ & -(f[x+1][y+1] + 2f[x+1][y] + f[x+1][y-1]) \end{aligned} \quad (10.23).$$

Convolving *k1* from *sobel3* with (10.5) yields

$$\Delta y = \begin{bmatrix} -1 & -2 & -1 \\ 0 & 0 & 0 \\ 1 & 2 & 1 \end{bmatrix} * \begin{bmatrix} f[x-1][y+1] & f[x][y+1] & f[x+1][y+1] \\ f[x-1][y] & f[x][y] & f[x+1][y] \\ f[x-1][y-1] & f[x][y-1] & f[x+1][y-1] \end{bmatrix} \quad (10.24).$$

Expanding (10.24) results in

$$\begin{aligned} \Delta y = {} & f[x-1][y-1] + 2f[x][y-1] + f[x+1][y-1] \\ & -(f[x-1][y+1] + 2f[x][y+1] + f[x+1][y+1]) \end{aligned} \quad (10.25).$$

The Sobel kernels are scaled by 0.25, then the *template* method computes the square-root of the sum of the squares:

$$S = \sqrt{(\Delta x / 4)^2 + (\Delta y / 4)^2} \quad (10.26).$$

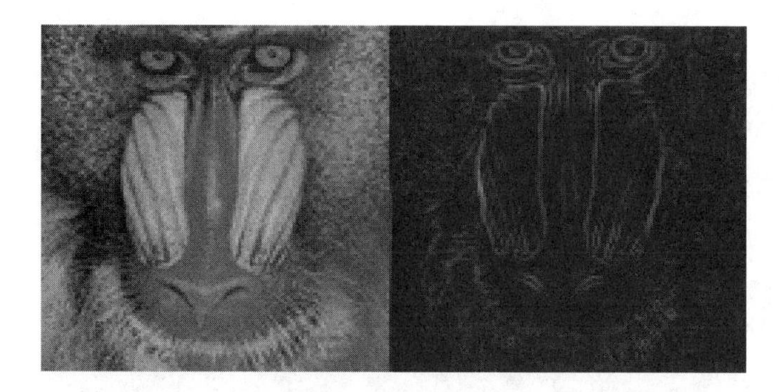

Fig. 10-8. Before and After Sobel Filtering.

Fig. 10-9. Roberts on the Left, Sobel on the Right

Fig. 10-9 shows the Roberts filter vs. the Sobel filter. No smoothing filtering or thresholding was performed. One type of thresholding uses a global average for the threshold value. Using this type of thresholding (implemented using the *thresh* method), we were able to produce Fig. 10-10, Roberts vs. Sobel, after thresholding.

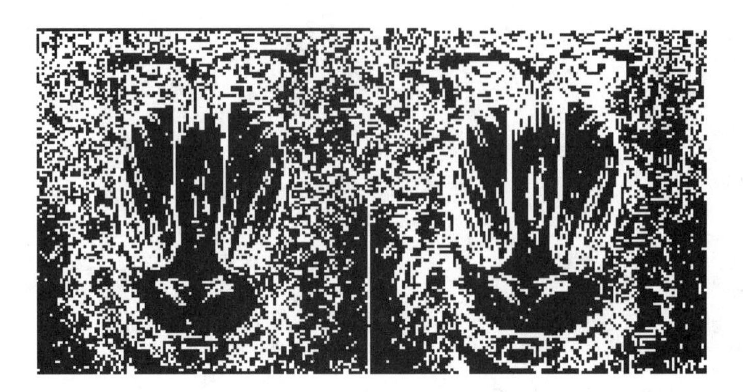

Fig. 10-10. Roberts vs. Sobel After Thresholding

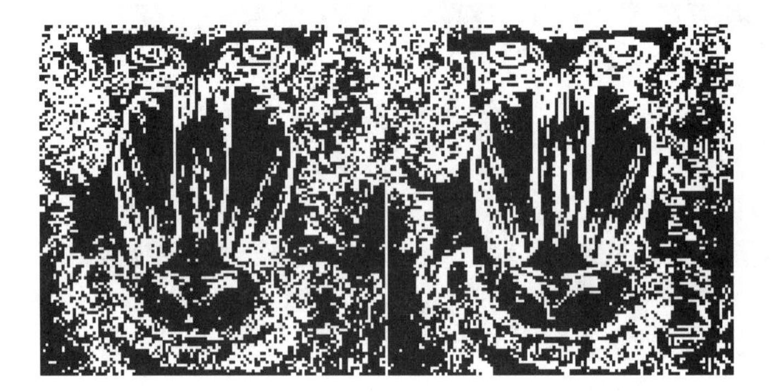

Fig. 10-11. Roberts vs. Sobel, UNAHE Pre-process, Then Thresholding

Fig. 10-11 shows that both Roberts and Sobel appear to be slightly less noisy when a non-adaptive histogram equalization (UNHE) is performed first. This is most evident in the eyebrows.

Fig. 10-12 shows the Sobel operation performed after a 7x7 Gaussian smoothing filter. It is clear that such filters can greatly alter the performance of an edge detector.

Fig. 10-12. The Sobel from 10-11 vs. a Gaussian Prefiltered Sobel

Threshold choice and pre-filter selection are both important factors in performance. Fig. 10-13 shows the peppers image before and after Roberts edge detection. The middle image uses a threshold that is selected automatically, based on the average value). The extreme right image uses a threshold that was selected manually (i.e., a tuned value for the threshold). The performance of an edge detector is a function of the kernel, threshold value and prefilter.

Fig. 10-13. Peppers Before and After Roberts Edge Detection, with No Pre-filtering

10.4. Comparing Edge Detectors

The last section raises some central issues in edge detection filter design. How do we select a prefilter? Several compelling arguments have been made for using Gaussian filters, but what happens if we try other filters anyway?

Fig. 10-14. shows two images. The left image is a 9x9 mean filtered image that has been Sobel edge-detected. The right image has been pre-filtered with a 7x7 Gaussian filter. Both images were thresholded, using the *thresh* method.

Fig. 10-14. Mean Filtering vs. Gaussian Filtering with Sobel Edge Detection

Fig. 10-14 shows that the image on the right is a clear winner, from our subjective assessment of the excellence applied to an edge detector. This begs the question: How do we formulate the criterion of an edge detector's performance? When using the LoG filter, we have been using a *rotationally invariant* filter. According to the Canny criterion, a directional derivatives edge operator is better. The primary deficiency is in the Signal-to-Noise ratio (SNR) and the localization [Mehrotra].

How do we know if the Canny criterion is correct? Formulating a criterion and showing that it is correct is a necessary, but not sufficient, step toward establishing an optimal edge detector. Two approaches to this problem are *numeric* and *symbolic*.

The numeric approach is to synthesize a test image, one in which all the edge pixels are known in an a priori fashion. A noise source is selected and added to the synthesized image, and the result is edge detected. After the edge detection, thresholding is performed. The output image contains pixels that are marked as correct (i.e., on the edge) and incorrect (i.e., not on the edge). Typically, we can use a distance from the edge as a part of the evaluation system. Unmarked edge pixels lower the performance. Choice of test image and noise source will alter the criterion of optimality. Often images are selected that have only straight edges or are uniformly textured [Ganesan]. Clearly, an edge detector optimized for one image will not necessarily work well on another. Also, shot noise is radically different from Gaussian noise.

The symbolic approach involves modeling the image with an idealized edge function. The edge function is convolved with a smoothing filter, typically a Gaussian. Such an edge function probably never exists in any real image. Optimizing for step-functions will not yield a optimal edge detector for non-step edges. The Canny approach is to compute the SNR and then to compute a function called the *localization*. The Canny functional forms the product of the SNR and the *localization*. This product is then optimized.

Subjective comparisons of edge detectors is not easy. For example, Fig. 10-15 shows an unfiltered Roberts, Sobel, and Pixel Difference edge detector, after thresholding.

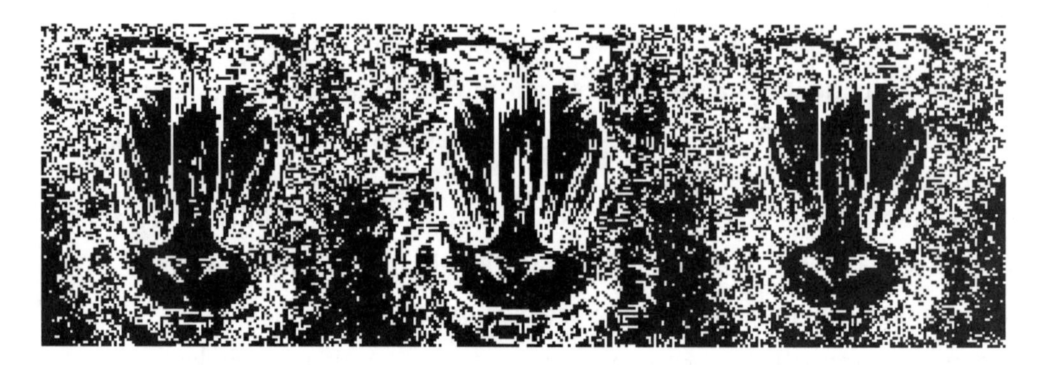

Fig. 10-15. Roberts vs. Sobel vs. Pixel Difference

The method for computing the *pixelDifference* follows:

```
public void pixelDifference() {
 float k1[][] = {
   {0,   0,   0},
   {0,   1,  -1},
   {0,   0,   0}
   };
 float k2[][] = {
   {0, -1,  0},
   {0,  1,  0},
   {0,  0,  0}
   };
   templateEdge(k1,k2);
 }
```

Based on 10-15, how do we decide which edge detector is better?

If an objective method for edge detection is found, iterative techniques can be used to *tune* various parameters. For example, once an objective function has been selected, we can search for the best level at which to threshold an image after the edge detection. Of course, this search assumes that a global thresholding technique is optimal (this is, by no means, certain). In fact, the use of global thresholding is almost certainly non-optimal, in most cases. One idea is to select a threshold by using the histogram to select a percentage of pixels that should remain active in the image [Levine]. This is an adaptive thresholding and is generally picture dependent.

In summary, there are a large number of edge detectors. Some perform well in given applications, but poorly in others [Peli]. The problem of finding an optimal edge detector is both difficult to define and to solve! Several schemes are defined by [Pratt]. It is particularly important to identify those edge detectors that have been optimized for straight lines (horizontal, vertical and diagonal) and simple circles. Thus, we need a test image that is able to incorporate several known geometric shapes. Such an image must be synthesized, so that we can control the amount of noise added to the image.

Fig. 10-16. Edge Man

Fig. 10-16 shows an image of mild-mannered *Edge Man*. Edge Man is able to illustrate how an edge detector will respond to step-edges of various shapes and sizes. Edge Man appears in the file *edgeMan.gif* on the CD-ROM that accompanies this book.

Fig. 10-17 compares 4 edge detectors using the Edge Man of Fig. 10-16. In order from left-to-right, top-to-bottom, the methods used are *pixelDifference*, *sobel*, *separatedPixelDifference* and *prewitt*. We may now compare several edge detectors. Note that the use of the italics indicates a method name. Typically, one would capitalize Sobel or Prewitt, but in Java, capitalization generally indicates a class name.

Fig. 10-17. *pixelDifference*, *sobel*, *separatedPixelDifference* and *prewitt*

Because the lines in Edge Man are exactly two pixels wide, the *pixelDifference* method has an unfair advantage over the others. The reason for this becomes clear upon examination of the kernel in the method's code. The *separatedPixelDifference* and *prewitt* methods follow:

```
public void separatedPixelDifference() {
 float k1[][] = {
   { 0,   0,  0},
   { 1,   0, -1},
   { 0,   0,  0}
   };
 float k2[][] = {
   { 0,  -1,  0},
   { 0,   0,  0},
   { 0,   1,  0}
   };
   templateEdge(k1,k2);
 }
public void prewitt() {
 float k1[][] = {
   { 1,   0, -1},
```

```
      { 1,    0, -1},
      { 1,    0, -1}
      };
    float k2[][] = {
      {-1,   -1, -1},
      { 0,    0,  0},
      { 1,    1,  1}
      };
      Mat.scale(k1,1/3.0);
      Mat.scale(k2,1/3.0);
      templateEdge(k1,k2);
    }
```

No thresholding was performed for any of the edge detection schemes.

Fig. 10-18. Roberts vs. Pixel Difference

Fig. 10-19. Pixel Difference vs. Laplacian

Fig. 10-18 shows the Roberts edge detector vs. the pixel difference filter. Fig. 10-19 shows the pixel difference vs. the Laplacian. The Laplacian used was from the *laplacian3* method, and thresholding was performed using *thresh* on both images. Pixel difference does appear to favor a positive sloped line over the negative sloped line. Further, the edges are much more uniform with the Laplacian than with pixel difference. Finally, the edges from the Laplacian are cleanly separated, as opposed to the pixel differenced image.

We are compelled to point out that the Laplacian filter dominated the other kernels because it employs rotationally invariant second-order differentiation. The Roberts 2x2 and the two-template filters favored the horizontal and vertical edges. This is particularly clear with a two-template filter like Frei-Chen.

Fig. 10-20. Before and After Frei-Chen

Fig. 10-20 shows the effect of Frei-Chen filtering on Edge Man. All the horizontal and vertical edges have been strengthened. Edge Man has a locally flat head on the top and sides, and this is emphasized by the Frei-Chen kernel. The *freiChen* method follows:

```
public void freiChen() {
  float r2 = (float) Math.sqrt(2);
  float k1[][] = {
    { 1,    0,   -1},
    { r2,   0,  -r2},
    { 1,    0,   -1}
    };

  float k2[][] = {
    {-1, -r2, -1},
    { 0,   0,  0},
```

Plate 11-1. Before and after a color dilate

Plate 11-2. Before and after a color erode.

Plate 11-3. Before and after *colorDilateErode()*

Plate 11-4. Before and after the Color Close

Plate 11-5. Before and after the Color Open

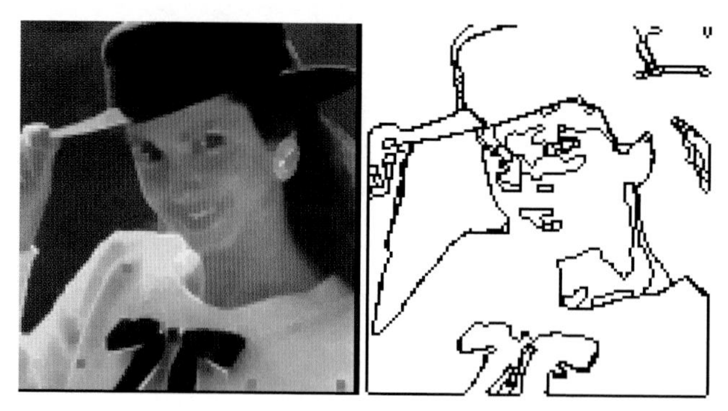

Plate 11-6. Before and after threshold and contour on Color Close.

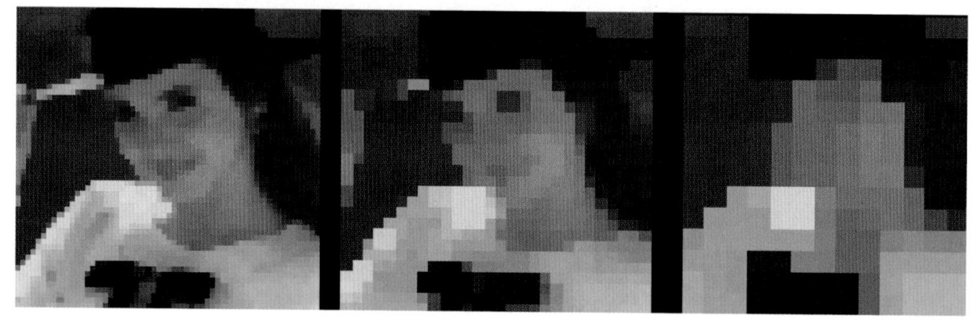

Plate 11-7. Color Morphological Pyramid Levels 1, 2 and 3

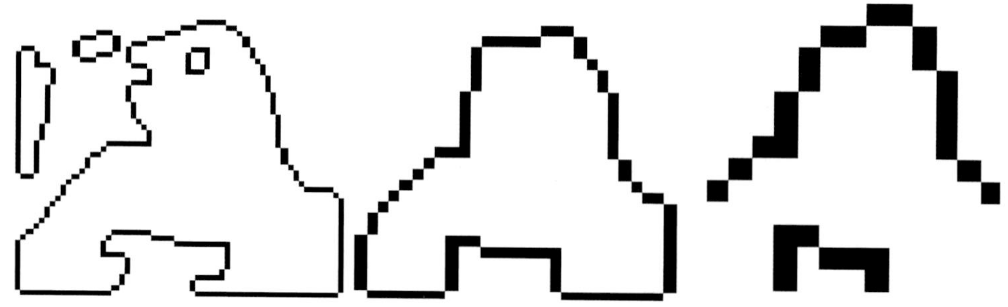

Plate 11-8. Cross kernel outline contour of thresholded Plate 11-7.

Plate 12-1. A Heuristic Search Guided by user selected points

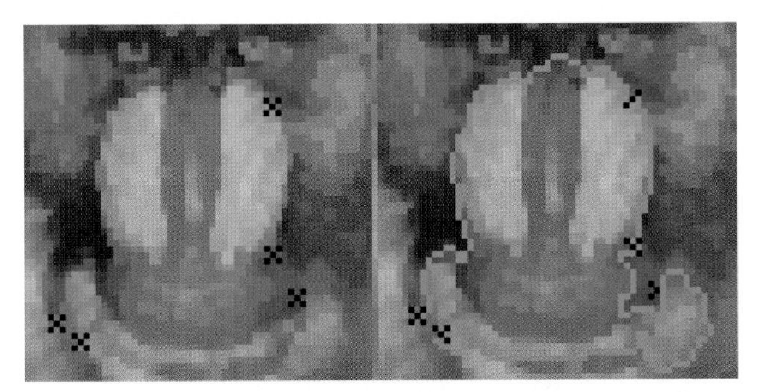

Plate 12-2. Morphological Resample and Heuristic Edge Search

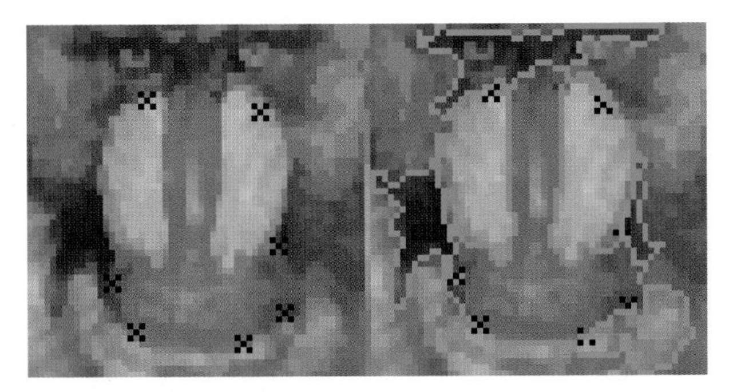

Plate 12-3. Eliminating distance from the Heuristic permits wandering

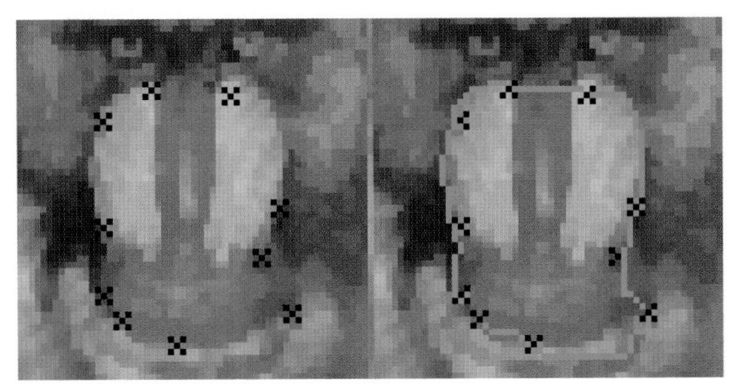

Plate 12-4. A Less Informed search takes longer

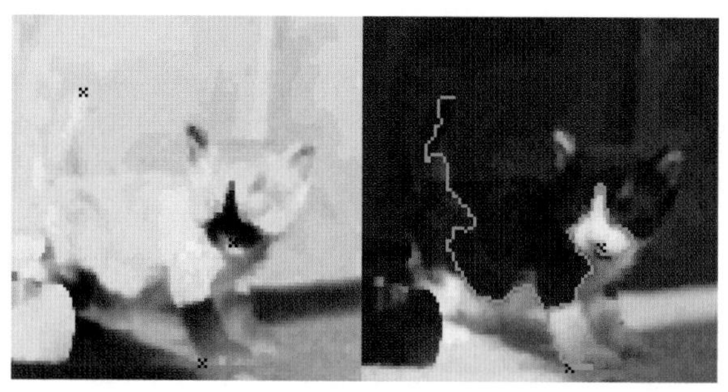

Plate 12-5. The Original *C* method for the cat image.

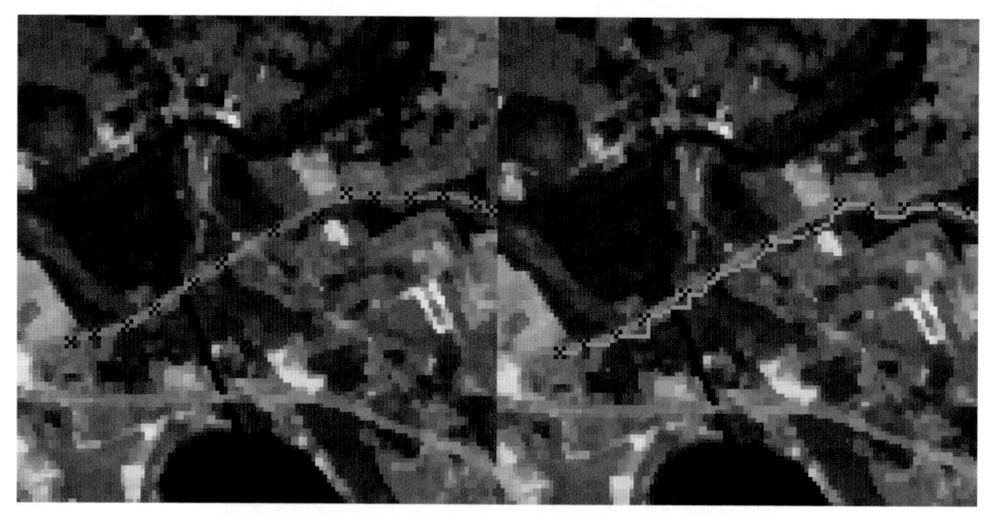

Plate 12-6. Road Following In PhotoInterpretation

Plate 12-7. River following With local Evaluations

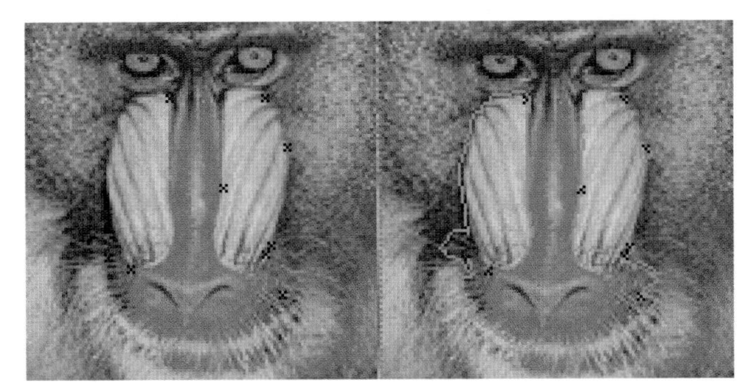

Plate 12-8. Using Roberts for Local Cost

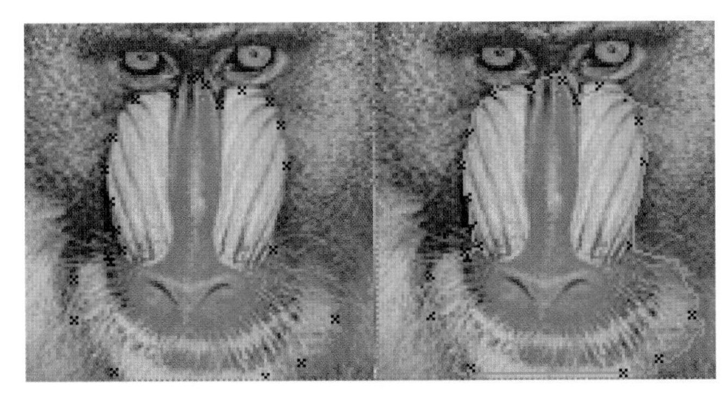

Plate 12-9. Morphological Outlining for Local Cost

Plate 12-10. Morphological Outlining for Color Images

Plate 13-1. Original Image, vs 2:1 SubSampling in Chroma

Plate 13-2. Original Image vs 4:1 SubSampling in Chroma

Plate 13-3. A Linear Cut to 15 b/p in RGB space, SNR=30dB8 b/p,

Plate 13-4. Median Cut to 16 color GIF, SNR=17dB, 1.6 b/p

Plate 14-1. Color Image Warping

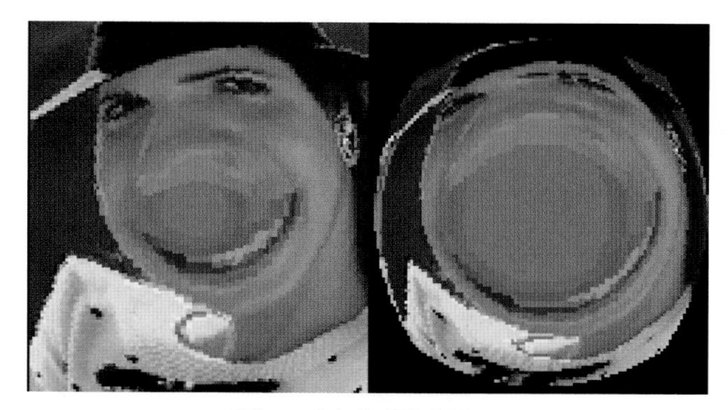

Plate 14-2. FishEye

Plate 14-3. A Polar Transform

Plate 14-4. The Sqrt Transform

Plate 14-5. 4 Point Bilinear Feedback

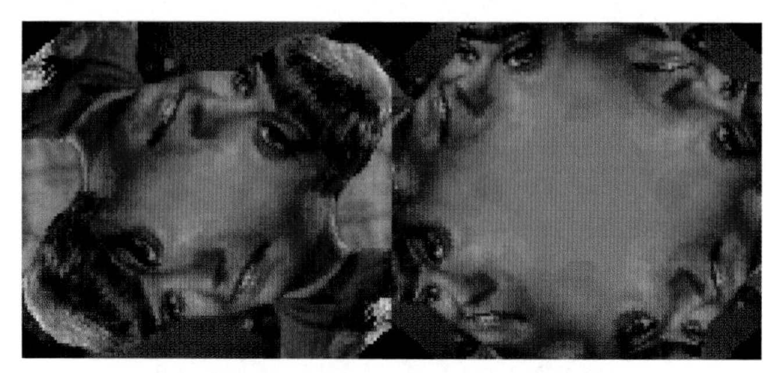

Plate 14-6. Iterative Conformal Mapping

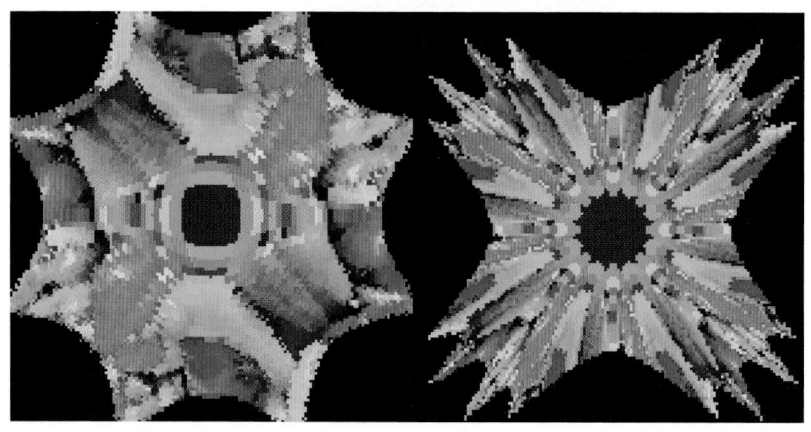

Plate 14-7. Conformal Mapping, fish eye and sqrt

```
      { 1,  r2,  1}
      };
    double s = 1/(2+r2);
      Mat.scale(k1,s);
      Mat.scale(k2,s);
      templateEdge(k1,k2);
    }
```

One of the measures of an edge detector's response is whether if it responds to edges or to isolated points. Fig. 10-21 shows Edge Man after being exposed to shot noise (*saltAndPepper(100))* and then to Roberts edge detector. Roberts has actually taken the single-pixel-width salt noise pixels and increased them in size.

Fig. 10-21. Edge Man with Shot Noise vs. Roberts

Fig. 10-22. Before and After Pixel Difference

Fig. 10-22 shows the effect of the pixel difference edge detector on salt noise. The salt has been doubled into a two-pixel long edge. Fig. 10-23 illustrates that the *laplacian3* creates a circular edge around each salt pixel.

Fig. 10-23. Before and After *laplacian3*

Fig. 10-24. 3x3 Gaussian Smoothing vs. *laplacian3* of Smoothed Image

Fig. 10-24 shows Edge Man with salt noise after a 3x3 Gaussian smoothing filter application on the left with the result of the thresholded 3x3 Laplacian edge detector on the right. Here smoothing is making the edge detector performance much worse. It smears all the edges and the salt noise, amplifying the noise around Edge Man.

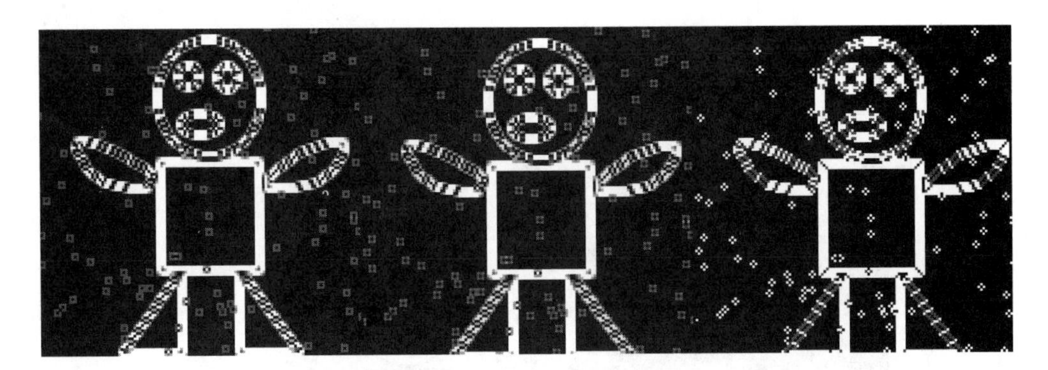

Fig. 10-25. Frei-Chen, Prewitt and Separated-Pixel Difference

Fig. 10-25 shows the Frei-Chen, Prewitt and Separated-pixel difference edge detectors applied to Edge Man + shot noise. None of the detectors discussed appears able to deal well with shot noise, and smoothing filters only smear the shot noise! All the median filters covered so far will turn Edge Man dark! This is because the edges are all treated as shot noise. In response, we narrow the median kernel to eliminate single pixel edges.

Fig. 10-26. Before and After *medianSquare2x2*

Fig. 10-26 shows the effect of a new median filter method based on a 2x2 median square. The 2x2 median square is embedded in a 3x3 matrix so that the center can be found with ease. See Section 10.6, Projects, for a suggestion on implementing even dimensioned median filters. The code for the medianSquare2x2 follows:

```
public void medianSquare2x2() {
  short k[][] = {
```

```
        {1, 1, 0},
        {1, 1, 0},
        {0, 0, 0}
        };
        median(k);
}
```

Fig. 10-27. Before and after *laplacian3*

Fig. 10-27 shows that the *median2* filter can clean the shot noise, leaving the *laplacian3* to find the edges. Unfortunately, Edge Man has been a little shot up by the shot noise. As a result, there are several discontinuities along Edge Man's perimeter. Also, edges are thicker due to the averaging property. Even worse, with a greater amount of shot noise, the median filter is considerably less robust without an increase in the kernel size. Once the size is increased, the Edge Man is judged as shot noise and is filtered out!

10.5. Summary

The edge detection operators examined in this chapter all responded to change in gray level or average gray level. All the detectors discussed responded to single points as well as to edges.

Frei-Chen, Prewitt and Separated-pixel Difference operators have a weak response to diagonal edges. Pixel Difference favored some angles more than others (left to right over right to left). The Laplacian operator has the only rotationally invariant response.

These approaches represent a *rigid template* that was selected a priori and cannot produce optimum results for all types of edges. In fact, as edges are deformed, ridge templates have a performance that worsens. One promising approach to the problem is to perform edge detection with *Snakes*. This is an advanced topic in image processing and is beyond the scope of this book; see [Lai] for more information.

Another approach to the edge detection problem is to perform boundary-type edge detection (discussed in Chapter 12). This is typically a much higher cost approach (computationally) and so is less popular that filter-based edge detection.

10.6. Projects

1. Create an interface that lets the user type in a custom kernel for convolution, without having to recompile the program. Allow the user to select the kernel size.

2. Show that the Laplacian of the Gaussian will, when convolved with an image, yield the same result as the Gaussian filtered image that has had a Laplacian performed on it.

3. Implement a fixed-template edge detector based on an operator not covered in this book (e.g., Robinson, Kirsch, boxcar, pyramid, Argyle, Macleod, etc.).

4. Remove the single pixel salt noise by writing a special case of the median filter that can handle 2x2 kernels. Hint: start by overloading the *medianNoEdge* by adding a new one in the *EdgeFrame*. A beginning prototype for the *medianNoEdge* method follows:

```
public short[][] medianNoEdge(short f[][], short k[][]) {

        if (odd(k.length) return super.MedianNoEdge(f,k);
...
}
```

To test your work, modify the *medianSquare2x2* method in the *EdgeFrame* to read:

```
public void medianSquare2x2() {
 short k[][] = {
        {1, 1},
        {1, 1}
        };
        median(k);
}
```

How does this work on the salt noise in Edge Man? What are the limits of this kind of pre-filter?

5. Review the literature; select and implement an edge detector not already implemented by the Kahindu program.

6. Recall that the *template* edge method was used to compute the *edge strength* by convolution with two orthogonal kernels. Write a method that plots scaled vectors whose magnitude is limited to a few pixels and whose direction corresponds to the arc tangent of the ratio of the outputs from the two convolutions. Consider that one convolution kernel is vertical, the other is horizontal. Be sure to identify which is which. Your program does not have to draw every vector, as this would be too cluttered.

7. Select a criterion of optimality from the literature and apply it to several edge detectors. Allow the criterion to select the optimal threshold parameters. Which edge detector is better, according to the criterion? Which one looks better according to your eye? Use a synthetic image and a digitized image to run your experiments.

8. Use a symbolic manipulation program to derive an optimal edge detector. Parameterize the type of edge you want to detect using a polynomial (at least a quintic). Repeat project 7, comparing your edge detector with others. How did you do?

10.7. EdgeFrame Class

The *EdgeFrame* class resides in the *gui* package and provides all the services described in this chapter for convolution-based edge detection. The *EdgeFrame* class also provides some simple statistical computation facilities, as well as some elementary thresholding features. *EdgeFrame* also has a new *medianSquare2x2* filter, which enables the filtering out of single pixel noise at the expense of distorting the rest of the image.

10.7.1. Class Summary

```
package gui;
import java.awt.*;
import java.util.*;
public class EdgeFrame extends SpatialFilterFrame {
public EdgeFrame(String title)
public void colorToRed()
public void medianSquare2x2()
public void roberts2()
public void shadowMask()
public void sobel3()
public void separatedPixelDifference()
public void prewitt()
public void freiChen()
public void pixelDifference()
public void templateEdge(float k1[][], float k2[][])
public void laplacian5()
public void laplacian3()
public void laplacian3Minus()
public void laplacian3_4()
public void laplacian3Prewitt()
public void laplacian9()
public void hat13()
public static double laplaceOfGaussian(double x, double y,
        double xc, double yc, double sigma)
public static void printLaplaceOfGaussianKernel(
        int M, int N, double sigma)
public static float [][] getLaplaceOfGaussianKernel(
    int M, int N,
    double sigma)
public static void tGenerator(int min, int max)
public void thresh()
public void threshLog()
public void kgreyThresh(double k)
public void thresh4(double d[])
```

```
}
```

10.7.2. Class Usage

Suppose that the following variables are defined so that

```
String title = "EdgeFrame title";
float k1[][], float k2[][];
```

To make a new instance of an *EdgeFrame,* use

```
EdgeFrame ef = new EdgeFrame(title);
```

To set the red color plane to the average of the red, green and blue color planes, use

```
ef.colorToRed();
```

To take the 2x2 median filter with a kernel of

$$\begin{bmatrix} 1.0 & 1.0 & 0 \\ 1.0 & 1.0 & 0 \\ 0 & 0 & 0 \end{bmatrix}$$

use

```
ef.medianSquare2x2();
```

To take a 2x2 Roberts gradient edge detection, use

```
ef.roberts2();
```

To convolve an image with

$$\begin{bmatrix} -2.0 & -1.0 & 0 \\ -1.0 & 0 & 1.0 \\ 0 & 1.0 & 2.0 \end{bmatrix}$$

use

```
ef.shadowMask();
```

To perform the template convolution with kernels:

$$k1 = \begin{bmatrix} -.25 & -.5 & -.25 \\ 0 & 0 & 0 \\ .25 & .5 & .25 \end{bmatrix}$$

and

$$k2 = \begin{bmatrix} .25 & 0 & -.25 \\ .5 & 0 & -.5 \\ .25 & 0 & -.25 \end{bmatrix}$$

use

```
ef.sobel3();
```

To perform the template convolution with kernels:

$$k1 = \begin{bmatrix} 0 & 0 & 0 \\ 0 & 1.0 & -1.0 \\ 0 & 0 & 0 \end{bmatrix}$$

and

$$k2 = \begin{bmatrix} 0 & -1.0 & 0 \\ 0 & 1.0 & 0 \\ 0 & 0 & 0 \end{bmatrix}$$

use

```
ef.separatedPixelDifference();
```

To perform the template convolution with kernels:

$$k1 = \begin{bmatrix} .33333334 & 0 & -.33333334 \\ .33333334 & 0 & -.33333334 \\ .33333334 & 0 & -.33333334 \end{bmatrix}$$

and

$$k2 = \begin{bmatrix} -.33333334 & -.33333334 & -.33333334 \\ 0 & 0 & 0 \\ .33333334 & .33333334 & .33333334 \end{bmatrix}$$

use
```
ef.prewitt();
```

To perform the template convolution with kernels:

$$k1 = \begin{bmatrix} .2928932 & 0 & -.2928932 \\ .41421354 & 0 & -.41421354 \\ .2928932 & 0 & -.2928932 \end{bmatrix}$$

and

$$k2 = \begin{bmatrix} -.2928932 & -.41421354 & -.2928932 \\ 0 & 0 & 0 \\ .2928932 & .41421354 & .2928932 \end{bmatrix}$$

use
```
ef.freiChen();
```

To perform the template convolution with kernels:

$$k1 = \begin{bmatrix} 0 & 0 & 0 \\ 0 & 1.0 & -1.0 \\ 0 & 0 & 0 \end{bmatrix}$$

and

$$k2 = \begin{bmatrix} 0 & -1.0 & 0 \\ 0 & 1.0 & 0 \\ 0 & 0 & 0 \end{bmatrix}$$

use
```
ef.pixelDifference();
```

To perform a template convolution, assuming that the red color plane represents the average intensity value in a gray-scale image, use
```
ef.templateEdge(k1, k2);
```

To convolve with the 5x5 Laplacian,

$$\begin{bmatrix} -1.0 & -1.0 & -1.0 & -1.0 & -1.0 \\ -1.0 & -1.0 & -1.0 & -1.0 & -1.0 \\ -1.0 & -1.0 & 24.0 & -1.0 & -1.0 \\ -1.0 & -1.0 & -1.0 & -1.0 & -1.0 \\ -1.0 & -1.0 & -1.0 & -1.0 & -1.0 \end{bmatrix}$$

use

```
ef.laplacian5();
```

To convolve with the 3x3 Laplacian,

$$\begin{bmatrix} 0 & -1.0 & 0 \\ -1.0 & 4.0 & -1.0 \\ 0 & -1.0 & 0 \end{bmatrix}$$

use

```
ef.laplacian3();
```

To convolve with the negative of the 3x3 Laplacian,

$$\begin{bmatrix} 0 & 1.0 & 0 \\ 1.0 & -4.0 & 1.0 \\ 0 & 1.0 & 0 \end{bmatrix}$$

use

```
ef.laplacian3Minus();
```

To convolve with the 3x3 Laplacian, with peak value 4,

$$\begin{bmatrix} 1.0 & -2.0 & 1.0 \\ -2.0 & 4.0 & -2.0 \\ 1.0 & -2.0 & 1.0 \end{bmatrix}$$

use

```
ef.laplacian3_4();
```

To convolve with the Prewitt version of the Laplacian,

$$\left[\begin{array}{ccc} -1.0 & -1.0 & -1.0 \\ -1.0 & 8.0 & -1.0 \\ -1.0 & -1.0 & -1.0 \end{array}\right]$$

use

```
ef.laplacian3Prewitt();
```

To convolve with the 9x9 version of the Laplacian,

$$\left[\begin{array}{ccccccccc}
-1.0 & -1.0 & -1.0 & -1.0 & -1.0 & -1.0 & -1.0 & -1.0 & -1.0 \\
-1.0 & -1.0 & -1.0 & -1.0 & -1.0 & -1.0 & -1.0 & -1.0 & -1.0 \\
-1.0 & -1.0 & -1.0 & -1.0 & -1.0 & -1.0 & -1.0 & -1.0 & -1.0 \\
-1.0 & -1.0 & -1.0 & 8.0 & 8.0 & 8.0 & -1.0 & -1.0 & -1.0 \\
-1.0 & -1.0 & -1.0 & 8.0 & 8.0 & 8.0 & -1.0 & -1.0 & -1.0 \\
-1.0 & -1.0 & -1.0 & 8.0 & 8.0 & 8.0 & -1.0 & -1.0 & -1.0 \\
-1.0 & -1.0 & -1.0 & -1.0 & -1.0 & -1.0 & -1.0 & -1.0 & -1.0 \\
-1.0 & -1.0 & -1.0 & -1.0 & -1.0 & -1.0 & -1.0 & -1.0 & -1.0 \\
-1.0 & -1.0 & -1.0 & -1.0 & -1.0 & -1.0 & -1.0 & -1.0 & -1.0
\end{array}\right]$$

use

```
ef.laplacian9();
```

To convolve with the 13x13 Mexican Hat filter (LoG),

$$\left[\begin{array}{ccccccccccccc}
0 & 0 & 0 & 0 & 0 & -1.0 & -1.0 & -1.0 & 0 & 0 & 0 & 0 & 0 \\
0 & 0 & 0 & -1.0 & -1.0 & -2.0 & -2.0 & -2.0 & -1.0 & -1.0 & 0 & 0 & 0 \\
0 & 0 & -2.0 & -2.0 & -3.0 & -3.0 & -4.0 & -3.0 & -3.0 & -2.0 & -2.0 & 0 & 0 \\
0 & -1.0 & -2.0 & -3.0 & -3.0 & -3.0 & -2.0 & -3.0 & -3.0 & -3.0 & -2.0 & -1.0 & 0 \\
0 & -1.0 & -3.0 & -3.0 & -1.0 & 4.0 & 6.0 & 4.0 & -1.0 & -3.0 & -3.0 & -1.0 & 0 \\
-1.0 & -2.0 & -3.0 & -3.0 & 4.0 & 14.0 & 19.0 & 14.0 & 4.0 & -3.0 & -3.0 & -2.0 & -1.0 \\
-1.0 & -2.0 & -4.0 & -2.0 & 6.0 & 19.0 & 24.0 & 19.0 & 6.0 & -2.0 & -4.0 & -2.0 & -1.0 \\
-1.0 & -2.0 & -3.0 & -3.0 & 4.0 & 14.0 & 19.0 & 14.0 & 4.0 & -3.0 & -3.0 & -2.0 & -1.0 \\
0 & -1.0 & -3.0 & -3.0 & -1.0 & 4.0 & 6.0 & 4.0 & -1.0 & -3.0 & -3.0 & -1.0 & 0 \\
0 & -1.0 & -2.0 & -3.0 & -3.0 & -3.0 & -2.0 & -3.0 & -3.0 & -3.0 & -2.0 & -1.0 & 0 \\
0 & 0 & -2.0 & -2.0 & -3.0 & -3.0 & -4.0 & -3.0 & -3.0 & -2.0 & -2.0 & 0 & 0 \\
0 & 0 & 0 & -1.0 & -1.0 & -2.0 & -2.0 & -2.0 & -1.0 & -1.0 & 0 & 0 & 0 \\
0 & 0 & 0 & 0 & 0 & -1.0 & -1.0 & -1.0 & 0 & 0 & 0 & 0 & 0
\end{array}\right]$$

Use
```
ef.hat13();
```

To compute the value of the LoG kernel, using an optimized version of

$$-\nabla^2 \frac{1}{2\pi\sigma^2} e^{\frac{x^2+y^2}{2\sigma^2}} = \frac{1}{\pi\sigma^4}\left[1 - \frac{x^2+y^2}{2\sigma^2}\right] e^{-\frac{x^2+y^2}{2\sigma^2}} \tag{10.15c}$$

that is centered at *xc*, *yc*, namely,

$$\left(\frac{1}{\sigma^4\pi} - \frac{1}{8}\frac{(2x-2xc)^2}{\sigma^6\pi} - \frac{1}{8}\frac{(2y-2yc)^2}{\sigma^6\pi}\right)\mathbf{e}^{\left(-\frac{1}{2}\frac{(x-xc)^2+(y-yc)^2}{\sigma^2}\right)} \tag{10.27}$$

use
```
double x, double y,  double xc, double yc, double sigma;
....
ef.laplaceOfGaussian(x,y,xc,yc,sigma);
```

To print a 10x10 LoG Kernel, with $\sigma = 2$, use
```
int M = 10;
int N = 10;
double sigma = 2;
ef.printLaplaceOfGaussianKernel(M, N, sigma);
```

To get the LoG kernel, use
```
float k[][] = ef.getLaplaceOfGaussianKernel(M, N, sigma);
```

To generate Java source code that contains temporary variables from 0 to 99, use
```
int min = 0;
int max = 99;
ef.tGenerator(min, max);
```

To threshold an image on the average value, use
```
ef.thresh();
```

To bring up a threshold dialog (shown in Fig. 10-28), use
```
ef.threshLog();
```

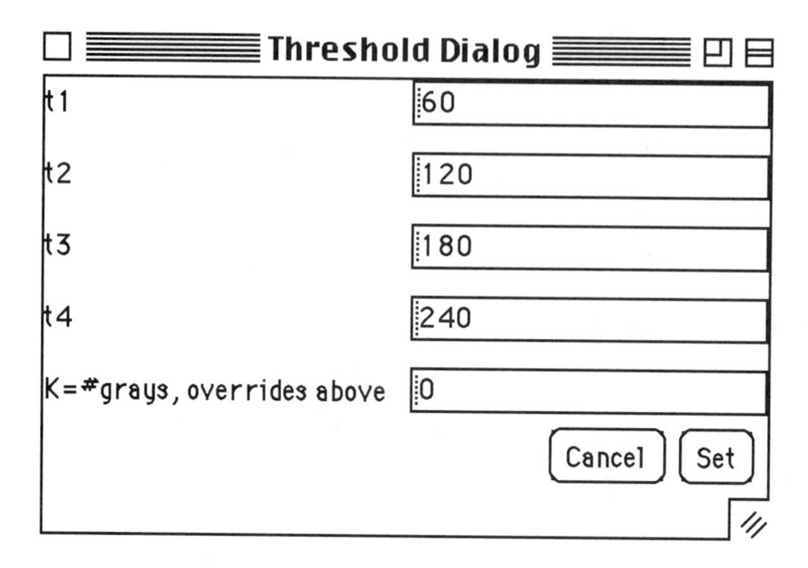

Fig. 10-28. The Threshold Dialog

Fig. 10-28 shows the threshold dialog. The threshold dialog has 4 settings for allowing the user to type in thresholds about which an image is quantized. Fig. 10-29 shows the mandrill before and after the four value thresholding.

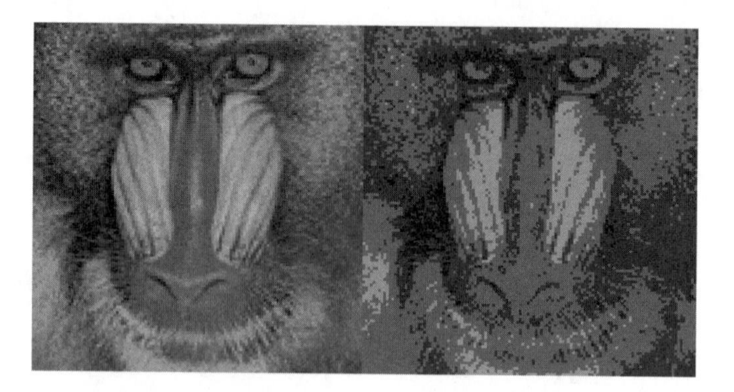

Fig. 10-29. Before and After a 4-Level Thresholding

The user can override the number of grays in an image by typing in a value for *k* in the threshold dialog. Fig. 10-30 shows the mandrill before and after a 9 level thresholding, performed using the threshold dialog.

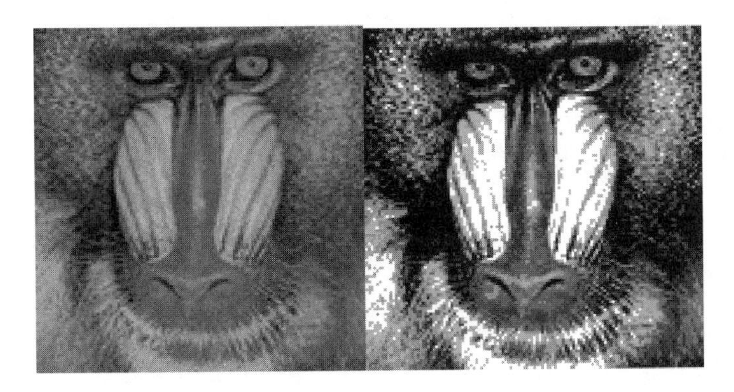

Fig. 10-30. Before and After a 9-Level Thresholding

The thresholding uses a popularity algorithm that is unsuitable for color images (it works, but the results don't look good). To specify 9 levels of gray under program control, use

```
double k = 9;
ef.kgreyThresh(k);
```

To specify an array of 4 thresholds, use

```
double d[] = {0, 120, 244, 255}
ef.thresh4(d)
```

11. Morphological Filtering

Alcohol and calculus don't mix.
Never drink and derive.

– Unknown

Chapter 10 disclosed techniques for performing edge detection using a fixed template. This served the purpose of identification of edge structures in an image. Various pre-processing techniques included median and low-pass filtering. Post-processing techniques included single and multi-level thresholding. One of the problems encountered after the thresholding of an image was that edges were not a single pixel in width. This was particularly true when pre-filtering smeared otherwise crisp step-edges.

One way to obtain a single pixel-width wide edge of good character is to use *morphological* filtering. The prefix *morph* means having to do with structure or form. The term *morphing* has gained popularity and connotes a digital image warping that is used in special effects [Wolberg]. Morphological filters manipulate the shape of an object. In image processing they typically work for either binary or gray-level images (though we have an experimental extension of morphological filtering to process color images). We defer the special effects processing associated with digital image warping until Chapter 14.

Applications for morphological image processing are not limited to finding edges. In fact, the spectrum of applications for morphological image processing includes vision, medical imaging, character recognition, inspection, etc.

11.1. Set Theory

This section presents the elementary set theory concepts for the non-theoretician or non-specialist. Readers who already know set theory should skip this section.

A set is a collection of elements, e.g., the set of all books in the library. The number of elements in a set is called its *cardinality*. The cardinality of a set can be infinite. For example, the set of all integers has an infinite cardinality, while the intersection between the set of students that skip this section and the set of students that pass the Java class has low cardinality.

Let A be a set and let x be an element contained by A. We say "x is a member of set A" and write

$$x \in A \qquad (11.1).$$

If x is not a member of set A, then we write

$$x \notin A \qquad (11.2)$$

If element x has a property, called P, then we write

$$P(x) \qquad (11.3).$$

To describe the set of all elements, x such that (11.3) is true, write

$$\{x \mid P(x)\} \qquad (11.4)$$

The *empty set* or *null set* is the set whose cardinality is zero and this is denoted as $\varnothing$.

Sets may be collected by other sets. In fact, the cardinality of $\{\varnothing, \varnothing\}$ is two. The set of all integers is denoted Z. The cardinality of Z is infinity. The cardinality of $\{Z, Z\}$ is two.

If the set B is contained by A, then B is a *subset* of A and we write:

$$B \subseteq A \tag{11.5}$$

To remember this, recall the *less-than* sign, `<`. For example, if *2* is less than *4*, we write $2 < 4$. Further, $4 > 2$ and, by analogy, we can turn (11.5) around, too:

$$A \supseteq B \tag{11.6}$$

(11.6) says that A contains B and is equivalent to (11.5). If two sets have the same elements, then they are *equal* and we write

$$A = B \tag{11.7}$$

If two sets do not have the same elements, we write

$$A \neq B. \tag{11.8}$$

A *proper subset* is both contained by the super set and not equal to the super set. That is, both (11.5) and (11.8) are true ($A \supset B$).

The *union* of all elements in set A with set B forms a new set called A union B: It is the set of all elements that are either in A or in B:

$$A \cup B = \{x \mid x \in A \text{ or } x \in B\} \tag{11.9}$$

The union operator can be mapped to all the sets in a set:

$$\bigcup\{A, B\} = A \cup B \tag{11.10}$$

The *intersection* between sets A and B yields a new set called A intersection B. It is the set of all elements that are both in A and in B:

$$A \cap B = \{x \mid x \in A \text{ and } x \in B\} \qquad (11.11)$$

The intersection operators (just like the union operator) can be mapped to all the sets in a set:

$$\bigcap\{A, B\} = A \cap B \qquad (11.12)$$

The *complement* of *B*, relative to *A*, yields a new set called the difference of *A* and *B*. These are the elements that are in *A* but not in *B*:

$$A[/]B = \{x \mid x \in A \text{ and } x \notin B\} = A \cap B^c = A[/](A \cap B) \qquad (11.13).$$

As an aside, we may now define *De Morgan's laws* using the set theoretic difference:

$$\begin{aligned} A[/](B \cup C) &= (A[/]B) \cap (A[/]C) \\ A[/](B \cap C) &= (A[/]B) \cup (A[/]C) \end{aligned} \qquad (11.13a).$$

Note that in some literature the "-" is used for the set theoretic difference. Our convention is to use the "[/]" to denote the difference between sets. The reader should be aware that some literature uses the "~" sign for the "[/]" [Moore]. In some fields, notation is a constant battle.

The *underlying set* is a set that contains all the primary elements and collections of subsets and is denoted *S*.

The *complement* of *B* relative to *S* is simply called the *complement of B*. It is written:

$$S[/]B = B^c = \{x \mid x \in S \text{ and } x \notin B\} \qquad (11.14).$$

For example, suppose *A* represents the set of students that *passes* the Java class. Suppose that *B* represents the set of students that *flunk* the Java class. *A* union *B* is the set of all students taking Java and is denoted *S*. Further, *A* intersection *B* is the null set (since a student cannot both pass and flunk). The difference of *A* and *B* is equal to A because the sets A and B do not intersect. Also, the complement of B is equal to set A. As a result, we say that sets A and B are *mutually exclusive*. For a more formal introduction to set theory, see [Nanzetta].

11.2. Erosion and Dilation

The basic idea of morphological image processing is that a structuring element is used to probe an image set. The center of the structuring element (generally an array) is moved from one pixel to the next, in the image. A set operation is used between the structuring element and that part of the image which overlaps with the element. The set operation yields structural information about the image. Hence the information obtained from morphological image processing is a function of the structuring element as well as the operation performed.

Historically, binary morphological image processing existed prior to gray-scale morphological image processing. For now, we restrict ourselves to the binary case. See [Dougherty] for a formal introduction to the gray-scale case.

To perform two-dimensional morphological image processing, we use a two-dimensional structuring element. A two-dimensional structuring element, translated across an image, reminds us of the kernel of a convolution. Morphological image processing uses set-operations rather than the multiplication and addition of convolution. As a result, morphological image processing is a kind of *non-linear* image processing.

In the binary case, the structuring element and image have only two values. The element is moved over an image and compared on an element-by-element basis. The two values are typically (0,1) or (0, 255) or (false, true). Selection of the representation is an implementation issue, which we will discuss shortly. Typically, the structuring element is formed by a two-dimensional array that has an origin. The origin is typically designated as the center of the array. In order to find the center of the array, we restrict the size of the array to be of odd dimension (e.g., 3x3, 5x5, etc.). Not all of the elements in the array need to be asserted (i.e., set equal to one). Those elements that are asserted are used to select pixels for processing.

The two most basic morphological image processing filters are *erosion* and *dilation.*

An *erosion* filter reduces the size of an object by eroding the boundary of the object. In contrast, a *dilation* filter increases the size of an object by adding to the boundary of the object. If we are given an image in a two-dimensional array, we then require a two-dimensional *structuring* element.

In the last section, everything said about set theory is generally true for all sets. In this section, we restrict ourselves to point sets in a two-dimensional integer space, Z^2 . For example:

$$A = \{a_0, a_1, ... a_{n-1}\} \qquad (11.15)$$

Each element in *A* consists of a two-dimensional point with integer coordinates. *A* contains the set of points that, when drawn, shows an image. To *translate* a point set by another point, *x*, write

$$A + x = \{a + x : a \in A\} \qquad (11.16).$$

We then define the structuring element *B*, which we use to measure the structure of *A*. We use (11.16) to translate *B* inside of *A*, performing set operations as we go. This operation is called the *Minkowski addition* .

To put it another way, the *Minkowski addition* of two sets, *A* and *B* , produces a new set, *C,* that consists of the translation of the elements in *A* by the elements in *B.* Mixing the notation of [Gonzalez and Woods] and [Schalkoff] we write

$$
\begin{aligned}
C = A \oplus B &= \bigcup (A + b : b \in B) \\
&= \left\{ c \in Z^2 : c = a + b, a \in A, b \in B \right\} \\
&= \bigcup (B + a : a \in A) \\
&= \left\{ x : (-B + x) \cap A \neq \varnothing \right\}
\end{aligned} \quad (11.17).
$$

Thus, the Minkowski addition is the union over all elements of B, of the translated points in $A+b$. Further, Minkowski addition is commutative and is called a *dilation*.

We flip the elements of a set using a *reflection*. A reflection consists of a 180-degree rotation and is denoted

$$-A \equiv \{-a \mid a \in A\} \quad (11.18).$$

The *erosion* is defined as

$$
\begin{aligned}
C = A[-]B &= \bigcap (A - b : b \in B) \\
&= \{ x : B + x \subset A \}
\end{aligned} \quad (11.19).
$$

Minkowski subtraction is given by

$$C = A[-](-B) = \bigcap (A + b : b \in B) \quad (11.19A)$$

Thus, erosion and Minkowski subtraction are equal if and only if (iff) B is symmetric (i.e., $B=-B$). Further, we define the complement of A as

$$A^c \equiv \{x \mid x \notin A\} \quad (11.20).$$

In general, the dual is given by $[\Psi(A)]^D = [\Psi(A^c)]^c$, so, following [Dougherty], erosion is the dual of dilation by $-B$:

$$C = A \oplus B = (A^c[-](-B))^c \quad (11.21)$$

and dilation is the dual of erosion by $-B$:

$$C = A[-]B = (A^c \oplus (-B))^c \quad (11.22).$$

Iff B is symmetric, then erosion is the dual of dilation. Dilation is commutative, but erosion is not.

For the purpose of implementation, we have predefined a few symmetric public structuring elements:

```
public static final
        float kv[][] = {
            {0,1,0},
            {0,1,0},
            {0,1,0}
        };
public static final
        float kh[][] = {
            {0,0,0},
            {1,1,1},
            {0,0,0}
        };
public static final
        float kCross[][] = {
            {0,1,0},
            {1,1,1},
            {0,1,0}
        };
public static final
        float kSquare[][] = {
            {1,1,1},
            {1,1,1},
            {1,1,1}
        };
```

We probe an image with the structuring elements using the *erode* method:

```
public short[][] erode(short f[][], float k[][]) {
  int uc = k.length/2;
  int vc = k[0].length/2;

  short h[][]=new short[width][height];
  short sum = 0;

  for(int x = uc; x < width-uc; x++) {
    for(int y = vc; y < height-vc; y++) {

      sum = 255;
      for(int v = -vc; v <= vc; v++)
              for(int u = -uc; u <= uc; u++)
                  if (k[u+uc][v+vc] == 1)
                   if (f[x-u][y-v] < sum)
```

```
                        sum = f[x-u][y-v];
        h[x][y] = sum;
      }
    }
    return h;
  }
```

To invoke the *erode* method, an overloaded method is used, also called *erode:*

```
    public void erode(float k[][]) {
        r = erode(g,k);
        copyRedToGreenAndBlue();
        short2Image();
    }
```

In the overloaded form, we see that the structuring element is used only on the green channel and that the output is placed in the red channel. The red channel is then copied to the green and blue channels and these are shown with the invocation of *short2Image*. The result of invoking

```
erode(kSquare);
```

on a thresholded version of the mandrill image is shown in Fig. 11-1. Thresholding is performed using an invocation of *thresh()*.

Fig. 11-1. Erosion by *kSquare*

The code for the *dilate* method is different from the *erode* method in that we perform the ">" operation:

```
public short[][] dilate(short f[][], float k[][]) {
    int uc = k.length/2;
    int vc = k[0].length/2;

    short h[][]=new short[width][height];
    short sum = 0;

    for(int x = uc; x < width-uc; x++) {
      for(int y = vc; y < height-vc; y++) {

        sum = 0;
        for(int v = -vc; v <= vc; v++)
                for(int u = -uc; u <= uc; u++)
                    if (k[u+uc][v+vc] == 1)
                     if (f[x-u][y-v] > sum)
                       sum = f[x-u][y-v];
        h[x][y] = sum;
      }
    }
    return h;
  }
```

Just like the *erode* method, we overload *dilate* to take a single structuring element, *k*:

```
public void dilate(float k[][]) {
    r = dilate(g,k);
    copyRedToGreenAndBlue();
    short2Image();
}
```

The idea is that the pixels will be valued at either 0 or 255, with each channel containing the same data. Thus we use the green channel as an input and the red channel as a convenient area to store the output. Such an implementation may be optimized in one of several ways (see the Section 11.7 for a few suggestions). The result of applying

```
dilate(kSquare);
```

to the mandrill image is shown in Fig. 11-2.

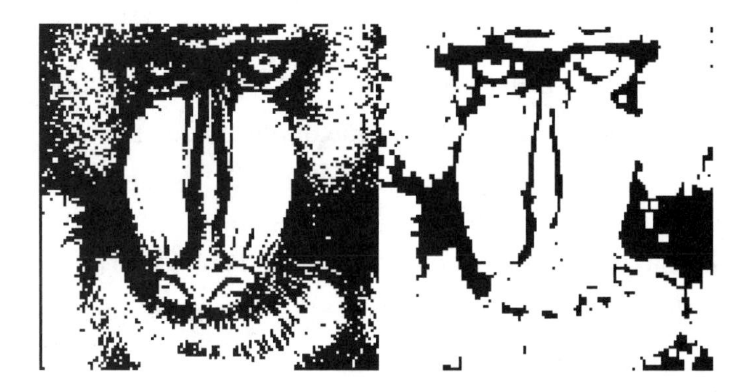

Fig. 11-2. The Dilated Mandrill

Recall that we can relate erosion to dilation using (11.21):

$$C = A \oplus B = (A^c[-](-B))^c \tag{11.21}$$

To show an example of (11.21), consider Fig. 11-3.

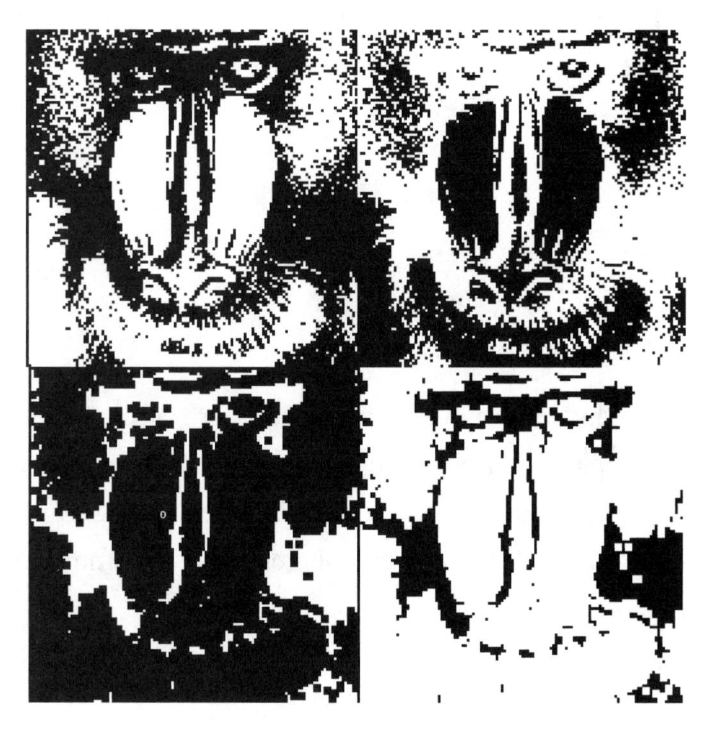

Fig. 11-3. Using (11.21) to Dilate an Image

Fig. 11-3 shows, from left to right, top to bottom, the thresholded mandrill, the negative of the thresholded mandrill, the erosion of the negated thresholded mandrill and finally the negation of the erosion of the negated thresholded mandrill. The result is the same image as the dilated mandrill shown in Fig. 11-2 (a satisfying check!).

We also define methods for performing color dilation and erosion. This permits a kind of special effect that is shown in Color Plate 11-1. We invoke the *colorDilateErode* method using

```
colorDilateErode(kSquare);
```

The code for the support methods for the color dilation and erosion follows:

```
public void colorDilateErode(float k[][]) {
  for (int i=0; i < 5; i++)
   colorDilate(k);
  for (int i=0; i < 5; i++)
   colorErode(k);
}
public void colorDilate(float k[][]) {
  r =      dilategs(r,k);
  g =      dilategs(g,k);
  b =      dilategs(b,k);
  short2Image();
}
public void colorErode(float k[][]) {
  r =      erodegs(r,k);
  g =      erodegs(g,k);
  b =      erodegs (b,k);
  short2Image();
}
```

In fact, the way in which the *dilate* and *erode* methods were defined, there is a required change for the implementation of the gray-scale dilation and erosion. Following <http://www.ph.tn.tudelft.nl/Courses/FIP/frames/fip-Morpholo.html> we define gray scale erosion and dilation using the minimum and maximum:

$$E_G(A,B) = \min_{[i,j]\in B} \{a[m+i,n+j] - b[i,j]\} \qquad (11.22a)$$

and

$$D_G(A,B) = \max_{[i,j]\in B} \{a[m-i,n-j] + b[i,j]\} \qquad (11.22b).$$

We shall restrict our selves to a symmetric structuring element that is set to zero when $k_{ij} = 1$. In this case, erosion and dilation may be defined by

$$E_G(A, B_s) = \min(A \cap B_s) \tag{11.22c}$$

and

$$D_G(A, B_s) = \max(A \cap B_s) \tag{11.22d}.$$

Equations (11.22c) and (11.22d) are easy to implement as *erodegs* and *dilategs*:

```
public short[][]
          erodegs(short f[][], float k[][]) {
  int uc = k.length/2;
  int vc = k[0].length/2;

  short h[][]=new short[width][height];
  short min = 0;
  short sum = 0;

  for(int x = uc; x < width-uc; x++) {
    for(int y = vc; y < height-vc; y++) {
     min = 255;
     sum = 0;
     for(int v = -vc; v <= vc; v++)
            for(int u = -uc; u <= uc; u++)
              if ( k[u+uc][v+vc] == 1) {
                  sum = f[x-u][y-v] ;
                  if ( sum < min)
                        min =  sum;
            }
     h[x][y] = min;
    }
  }
  return h;
}
```

```
public short[][] dilategs(
              short f[][], float k[][]) {
  int uc = k.length/2;
  int vc = k[0].length/2;

  short h[][]=new short[width][height];
  short max = 0;
  short sum = 0;
  for(int x = uc; x < width-uc; x++) {
    for(int y = vc; y < height-vc; y++) {
      max = 0;
      sum = 0;
      for(int v = -vc; v <= vc; v++)
            for(int u = -uc; u <= uc; u++)
              if (k[u+uc][v+vc] == 1) {
                  sum = f[x-u][y-v] ;
                  if ( sum > max)
                        max =  sum;
            }

      h[x][y] = max;
    }
  }
  return h;
}
```

The full gray scale modification, as described in (11.22a) and (11.22b), has been left as an exercise for the reader (see Section 11.7). As it is currently implemented, the erosion and dilation methods are going to give correct results for binary symmetric structuring elements only.

By combining a series of color dilate-erodes, we are able to arrive at a series of interesting color images. Large areas of a single color are easy to identify and this may have further application in compression and image segmentation.

```
public void colorDilateErode(float k[][]) {
  for (int i=0; i < 5; i++)
    colorDilate(k);
```

```
        for (int i=0; i < 5; i++)
         colorErode(k);
    }
```

Color plate 11-1 shows an example of color dilation, plate 11-2 shows an example of color erosion, and plate 11-3 shows the effect of *colorDilateErode*. All the structuring elements used were from *kSquare*.

Color dilation and erosion have not been found in the literature (as far as we know) and so, for lack of an available theoretical color morphological basis, we have defined it as the gray-scale dilation on the red, green and blue color planes. Since the color morphological image processing presented here is not based on any formal theoretical framework it should be treated with caution (it is, after all, only an experiment). Further study is needed to establish a theoretic framework for this approach.

11.3. Opening and Closing

Erosion filtering followed by dilation filtering is called an *opening* filter. The opening filter derives its name from the enlargement of holes that appear in an object.

Dilation filtering, followed by erosion filtering, is called a *closing* filter. The closing filter derives its name from the closing of holes that appear in an object.

Following the notation of [Dougherty], and [Schalkoff], the opening operation is denoted by

$$\begin{aligned} A \circ B &= (A[-]B) \oplus B \\ &= \bigcup \{B + x : B + x \subset A\} \end{aligned} \quad (11.23).$$

The dual operation for opening is closing. The closing operation is denoted by

$$\begin{aligned} A \bullet B &= [A \oplus (-B)][-](-B) \\ &= (A^c \circ B)^c \end{aligned} \quad (11.24)$$

In addition, because of the dual property, we can write

$$A \circ B = (A^c \bullet B)^c \qquad (11.25).$$

The binary implementations of open and close follow:

```
public void open(float k[][]) {
  r = dilate(erode(r,k),k);
  copyRedToGreenAndBlue();
  short2Image();
}
public void close(float k[][]) {
  r = erode(dilate(r,k),k);
  copyRedToGreenAndBlue();
  short2Image();
}
```

As in Section 11.4, we define the color morphological open and close being performed on the red, green and blue color planes as follows:

```
public void colorOpen(float k[][]) {
      r = dilategs(erodegs(r,k),k);
      g = dilategs(erodegs(g,k),k);
      b = dilategs(erodegs(b,k),k);
      short2Image();
}

public void colorClose(float k[][]) {
      r = erodegs(dilategs(r,k),k);
      g = erodegs(dilategs(g,k),k);
      b = erodegs(dilategs(b,k),k);
      short2Image();
}
```

Plate 11-4 shows how color close can remove the dark buttons from a white sweater. Plate 11-5 shows the result of a color open.

The result of a color morphological process is a function of the color *space* over which the operation occurs. Various color spaces are discussed in Chapter 13.

Fig. 11-4. Before and After the Close Filter

Fig. 11-4 shows a dilated version of Edge Man before and after the close filter is applied. Edge Man has had his eyes filled in by the invocation of

```
close(kSquare);
```

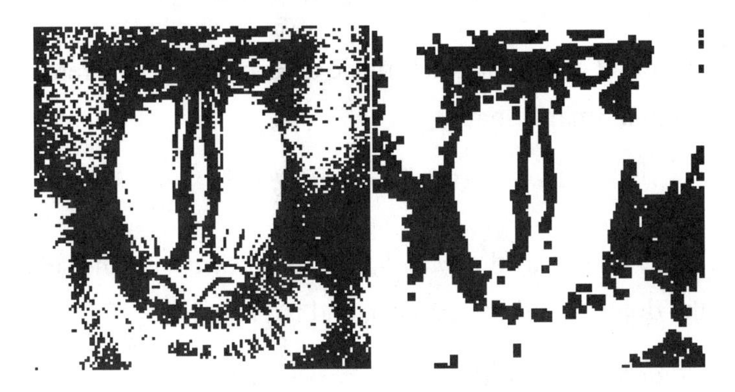

Fig. 11-5. Before and After Close(kSquare) on the Mandrill

Fig. 11-6. Before and after open(kSquare) on the Mandrill

Fig. 11-5 shows the effect of the close filter on the thresholded mandrill image. Much of the fine detail has been filled in. Fig. 11-6 shows the effect of the open filter on thresholded mandrill image.

11.4. Outlining

An *outlining* filter will find the outline of the boundary of an object. Such filters are edge detectors but are typically more noise immune than a Sobel or Roberts filter.

The outline for a set consists of an *inside contour* and an *outside contour*. The inside contour is the difference between the set and the set eroded by the structuring element:

$$insideContour(A,B) = A[/](A[-]B) \qquad (11.26).$$

The outside contour is found by dilating the image and finding the difference between the dilation and the original image, that is

$$outsideContour(A,B) = (A \oplus B)[/]A \qquad (11.27).$$

The middle contour (also known as the *morphological gradient)* is the set theoretic difference between the dilation and the erosion, and is given by

$$middleContour(A,B) = (A \oplus B)[/](A[-]B) \qquad (11.28)$$

To implement (11.26), (11.27) and (11.28), we reuse the green channel and place the results in the red channel as follows:

```
public void insideContour(float k[][]) {
 r = subtract(g, erode(g,k));
 copyRedToGreenAndBlue();
 short2Image();
}
public void outsideContour(float k[][]) {
 r = subtract( dilate(g,k),g);
 copyRedToGreenAndBlue();
 short2Image();
```

```
}
public void middleContour(float k[][]) {
    r = subtract(
      dilate(g,k),
      erode(g,k));
    copyRedToGreenAndBlue();
    short2Image();
}
```

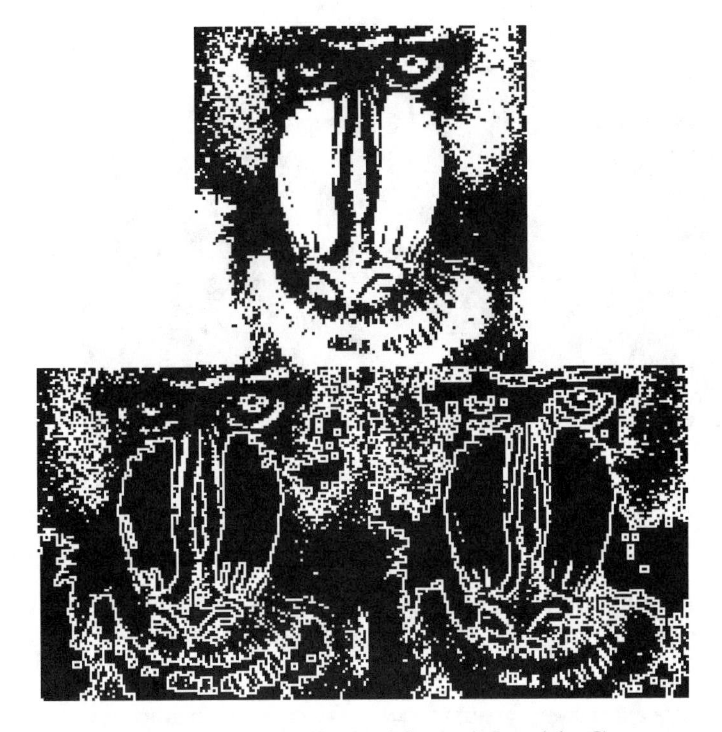

Fig. 11-7. Mandrill - Inside and Outside Contours

Fig. 11-7 shows the difference between the inside and outside contours as applied to the thresholded mandrill image.

Fig. 11-8. Icons - Inside and Outside Contours

Fig. 11-8 uses some icons and a few characters to better illustrate the difference between the inside and outside contours. These contours are created by invoking

```
insideContour(kSquare);
outsideContour(kSquare);
```

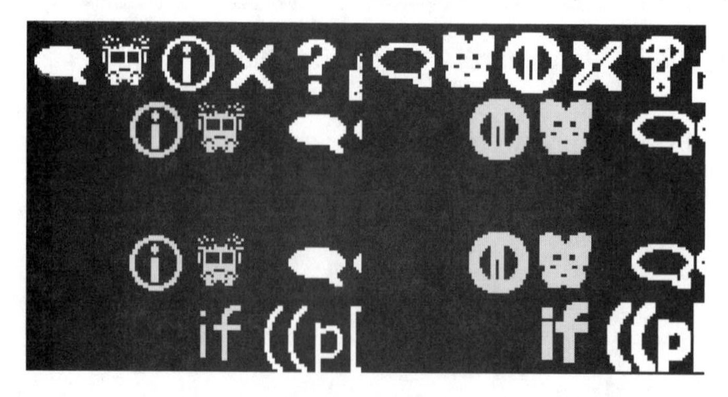

Fig. 11-9. Before and After Application of Middle Contour

Fig. 11-9 shows the icon image before and after the application of

```
middleContour(kSquare);
```

Color plate 11-6 shows the effect of the *outsideContour* on a thresholded color closed image. The boundaries appear to give us a large continuous edge showing regions of constant color. Time and space do not permit the development and exposition of a full-color contouring.

Opening and closing are sometimes combined with resampling of an image to create a pyramid [Wright]. The basic idea is that you use the same kernel to perform an open and then a close. The resulting image is resampled in both height and width. The new image has half the number of pixels. This is called a *morphological pyramid*. This is implemented in color (for the first time, as far as is known) using

```
public void colorPyramid(float k[][]) {
  r = dilategs(erodegs(r,k),k);
  g = dilategs(erodegs(g,k),k);
  b = dilategs(erodegs(b,k),k);
  r = erodegs(dilategs(r,k),k);
  g = erodegs(dilategs(g,k),k);
  b = erodegs(dilategs(b,k),k);
  resample(2);
  child.short2Image();
}
public void resample(int ratio) {
child = new MorphFrame("child");
child.width = width/2;
child.height = height/2;
child.r = resampleArray(r,2);
child.g = resampleArray(g,2);
child.b = resampleArray(b,2);
}

public short [][]
         resampleArray(short a[][], int ratio) {
int w = a.length;
int h = a[0].length;
int nw = w / ratio;
int nh = h / ratio;
short c[][] = new short[nw][nh];
for (int i=0; i < w; i++)
  for (int j=0; j < h; j++)
    c[i/ratio][j/ratio] = a[i][j];
   return c;
  }
```

Morphological pyramid level 0 is defined as the original image. Level 1 is 2:1 sub-sampled and each sub-level is, in turn, 2:1 sub-sampled. Plate 11-7 shows the *girl* image with color morphological pyramid levels 1, 2 and 3. In addition to reduced computational complexity (due to the reduced number of pixels), such techniques are useful for performing multi-resolution image processing and offer a *scale space representation*. Plate 11-8 shows the cross kernel outline contour of a thresholded version of Plate 11-7 (i.e., levels 1, 2 and 3). Plate 11-8 shows that such scale space techniques also avoid over-segmentation of traditional fixed scale algorithms for producing edge maps that have a desired resolution. Further, the technique appears to have good accuracy and is simpler to understand, easier to explain, computationally faster and easier to implement than the watershed pyramid for edge detection described by Wright and Acton [Wright]. The technique presented here also inputs color images, whereas the Wright and Acton algorithm does not. Subjective evaluation (by the biased author) indicates that the results look good. Lacking a criterion of optimality, the question of which edge detection technique is better remains open.

11.5. Thinning and Skeletonization

The *medial axis* of a thresholded region is the trace of the centers of maximal disks that are entirely contained within a region. The *medial axis transform* (MAT) produces a locus of points that lie on the medial axis.

Another term for the medial axis is the *skeleton* of the image. Thus, the *skeletonization* filter will find the medial axis transform of an object.

The skeleton represents elongated shapes formed with a single pixel-width line segments. The skeleton is useful for character recognition, road network detection in geographic information systems, path planning, etc. The medial axis forms the boundary of a *Voronoi* diagram. A Voronoi diagram is a convex network that partitions a space into cells that contain all the proximity information for a given set of points. Every vertex of the diagram has three edges and for any point in the set, the nearest neighbor of that point defines an edge. A detailed description of Voronoi diagrams may be found in the classic text, *Computational Geometry* [Preparata]. The dual of the Voronoi diagram is called the Delaunay triangulation. See <http://www.ics.uci.edu/~eppstein/gina/medial.html> for a list of applications of the Voronoi diagram and the Delaunay triangulation.

One way to perform skeletonization is by repeatedly *thinning* an image. An image is thinned by taking the set theoretic difference between the eroded image and the opening of the eroded image:

$$thin(A) = (A[-]B)[/]((A[-]B) \circ B) \qquad (11.27).$$

Direct implementation of (11.27) is computationally costly. Another approach, as given in Zhang and Suen [Zhang], uses an algorithm based on two passes through the image for each iteration. In each pass, a 3x3 window extracts pixels from the input image.

The basic idea is that in the first pass we scan for the conditions that will allow for the deletion of a pixel (e.g., setting of a pixel from 255 to 0). Our implementation uses the red color plane as the input and the green color plan as temporary storage. If a pixel in the red array is found to meet the criterion for deletion, we mark it by placing a “1” in the green array. After scanning the red array, we scan the green array. Wherever the green array is not a zero we delete the pixel from the red array.

This is implemented in the method *deleteFlagedPoints*:

```
private void deleteFlagedPoints() {
     for(int x = 1; x < width-1; x++)
      for(int y = 1; y < height-1; y++)
           if (g[x][y] != 0)
                r[x][y] = 0;
    }
```

After each pass through the image, we check to see if we are finished. In the *skeleton* method, we invoke a pass of the Zhang-Suen thinning algorithm. The pass returns *true* if a pixel is altered; otherwise the pass returns *false*. If either pass returns *false,* then we are done. To determine if the pass is the first pass or second pass, we use a boolean as an argument. If the boolean is *true,* then we are on the first pass. The conditions for thinning a pixel change slightly on the second pass. This is implemented in the *skeleton* method:

```
public void skeleton() {
      while (
        skeletonRedPassSuen(true)&&
        skeletonRedPassSuen(false)) {
       }
       copyRedToGreenAndBlue();
      short2Image();
  }
```

The answer is left in the red array, and this is copied to the green and blue arrays at the end of the method.

Fig. 11-10 shows the assignment of array elements to the location of the pixels in a 3x3 window.

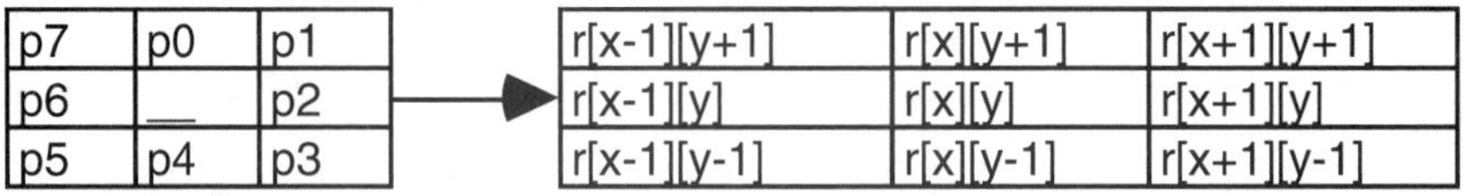

p7	p0	p1
p6	__	p2
p5	p4	p3

→

r[x-1][y+1]	r[x][y+1]	r[x+1][y+1]
r[x-1][y]	r[x][y]	r[x+1][y]
r[x-1][y-1]	r[x][y-1]	r[x+1][y-1]

Fig. 11-10. The Array Assignment for the 3x3 Window.

An eight-element boolean array is used to keep track of the non-zero pixels in the 3x3 window. If the origin of the window (*r[x][y]*) is zero, then we proceed to the next pixel by using the Java *continue* statement. During this pass, we make sure to zero out the marker array (we use the green plane for this). During the first pass we skip to the next pixel if any of the following conditions occur:

```
1. The origin, r[x][y] == 0.
2. The number of neighbors is less than two.
3. The number of neighbors is greater than six.
4. The number of zero-one transitions is not equal to one
```

```
5a. if (p[0] && p[2] && p[4]) is true
6b. if (p[2] && p[4] && p[6]) is true
```

During the second pass, we replace 5a and 6a with

```
5a. if (p[0] && p[2] && p[6]) is true
6b. if (p[0] && p[4] && p[6]) is true
```

Rule one ensures that the pixel is worth processing. Rule two ensures that isolated points and the tips of skeletons are not thinned. Rule three makes sure that r[x][y] is on the boundary of the 3x3 window, shown in Fig. 11-10. Rule four ensures that the skeleton does not become fragmented. Rules 5 and 6 keep connected the lines that are two pixels thick.

When all 6 of these tests are false, we mark the pixel for deletion and increment a counter indicating how many marks were made. To meet condition 1, we use

```
if (r[x][y] == 0) continue;
```

where the *continue* statement permits a pixel to be skipped. If this trivial condition is passed, then a boolean array is set to *true* when the local pixels, shown in Fig. 11-11, are non-zero. Thus, for each pixel in the image, the 3x3 window is mapped into a 1x8 array:

```
p[0] = r[x][y+1]   != 0;
p[1] = r[x+1][y+1]!= 0;
p[2] = r[x+1][y]   != 0;
p[3] = r[x+1][y-1]!= 0;
p[4] = r[x][y-1]   != 0;
p[5] = r[x-1][y-1]!= 0;
p[6] = r[x-1][y]   != 0;
p[7] = r[x-1][y+1]!= 0;
```

We know from conditions 2 and 3, that the number of neighbors is between 2 and 6, inclusive. Thus, if the number of neighbors is less than 2 or greater than 6, we continue to the next pixel:

```
int n = numberOfNeighbors(p);
if ((n < 2) || (n > 6)) continue;
```

Condition 4 requires that we count the number of zero-one transitions:

```
private int numberOf01Transitions(boolean p[]) {
```

```
        int n=0;
        if ((!p[0]) && p[1]) n++;
        if ((!p[1]) && p[2]) n++;
        if ((!p[2]) && p[3]) n++;
        if ((!p[3]) && p[4]) n++;
        if ((!p[4]) && p[5]) n++;
        if ((!p[5]) && p[6]) n++;
        if ((!p[6]) && p[7]) n++;
        if ((!p[7]) && p[0]) n++;
        return n;
    }
```

If the number of zero-one transitions is not equal to one, we continue to the next pixel:

```
                if (numberOf01Transitions(p) != 1) continue;
```

Conditions 5a, 6a and 5b, 6b are implemented, using

```
                if (firstPass) {
                        if ((p[0] && p[2] && p[4]) )
                                continue;
                        if ((p[2] && p[4] && p[6]))
                                continue;
                        g[x][y] = 255;
                        c++;
                        }
                else {
                    if ((p[0] && p[2] && p[6]))
                                continue;
                        if ((p[0] && p[4] && p[6]))
                                continue;
                        g[x][y] = 255;
                        c++;

                }
```

The basic difference between the implementation presented from that of Zhang and Suen is that we have inverted all their conditions so that the continue statement can move the flow of control directly to the next pixel. This is more efficient than testing all the conditions and then deciding if the pixel should be altered. The complete listing follows:

```
    public boolean skeletonRedPassSuen(
          boolean firstPass) {
        boolean p[]=new boolean[8];
```

```
    short c = 0;
    for(int x = 1; x < width-1; x++) {
      for(int y = 1; y < height-1; y++) {
          g[x][y] = 0; // use g for 1st pass
           if (r[x][y] == 0) continue;
          p[0] = r[x][y+1]   != 0;
          p[1] = r[x+1][y+1]!= 0;
          p[2] = r[x+1][y]   != 0;
          p[3] = r[x+1][y-1]!= 0;
          p[4] = r[x][y-1]   != 0;
          p[5] = r[x-1][y-1]!= 0;
          p[6] = r[x-1][y]   != 0;
          p[7] = r[x-1][y+1]!= 0;
          int n = numberOfNeighbors(p);
          if ((n < 2) || (n > 6)) continue;
          if (numberOf01Transitions(p) != 1) continue;

          if (firstPass) {
                if ((p[0] && p[2] && p[4]) )
                     continue;
                if ((p[2] && p[4] && p[6]))
                     continue;
                g[x][y] = 255;
                c++;
                }
           else {
               if ((p[0] && p[2] && p[6]))
                     continue;
                if ((p[0] && p[4] && p[6]))
                     continue;
                g[x][y] = 255;
                c++;

          }
      }
    }
    //System.out.println("c="+c);
    if (c == 0) return false;
    deleteFlagedPoints();
    return true;
  }
```

To invoke the two-pass algorithm, we alternate the arguments between *true* and *false*. If either pass returns false, then no pixels were altered, and the processing is complete.

```
public void skeleton() {
      Timer t = new Timer();
      t.start();
      while (
        skeletonRedPassSuen(true)&&
        skeletonRedPassSuen(false)) {
       }
       copyRedToGreenAndBlue();
      short2Image();
      t.print("skeletonTime=");
 }
```

Fig. 11-12 shows the Zhang and Suen algorithm applied to several icons. In each case, the number of pixels is greatly reduced. Note that the bubble has been reduced to an S-curve. In this particular case, information is lost. We also note that the left-to-right cross is missing in the X icon.

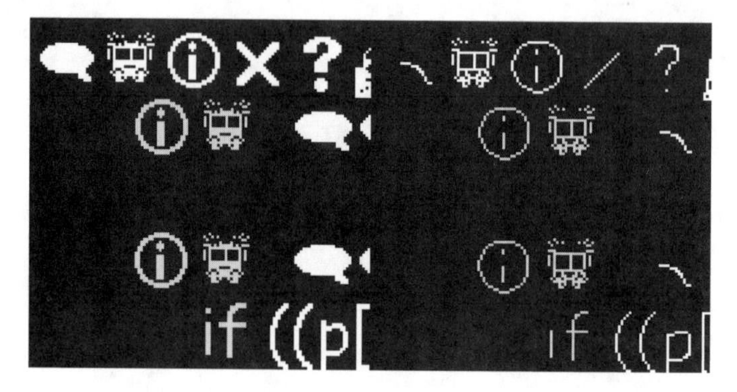

Fig. 11-11. Before and After Skeletonization

Fig. 11-11 shows that the serifs appear to be removed from some text. This greatly simplifies the geometry used to represent the characters (and may be a good OCR pre-process).

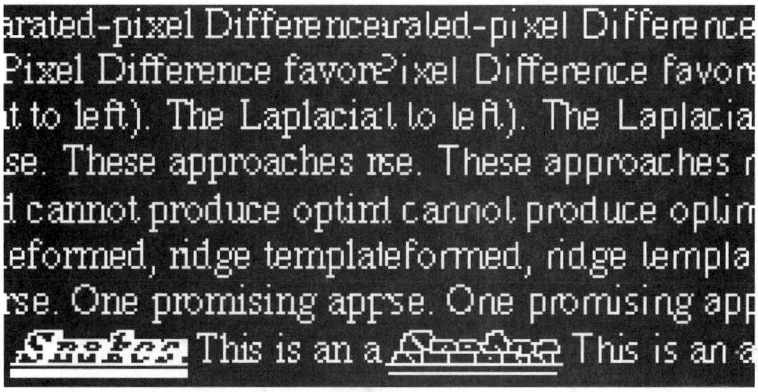

Fig. 11-12. Before and After Skeletonization

Fig. 11-13 shows that if the serifs are blown up to a resolution where they can survive the skeletization process, the serifs are preserved. Thus, skeletonizations results are not scale independent (both Fig. 11-12 and Fig. 11-13 are using *Times* font).

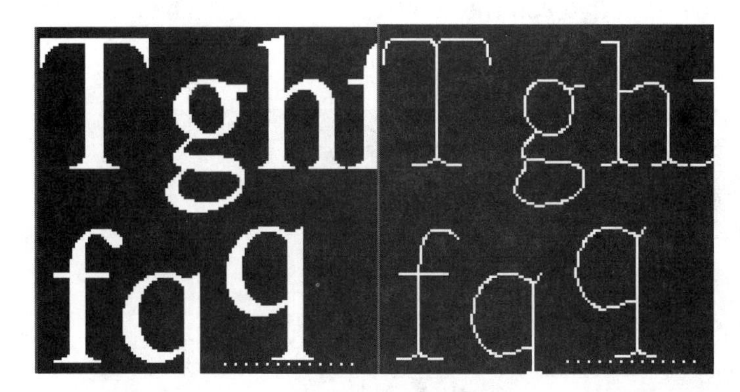

Fig. 11-13. Before and After Skeletonization

Recall that the dual is given by $[\Psi(A)]^D = [\Psi(A^c)]^c$. Thus, by negating the image, performing skeletonization and then negating the image again, we can obtain the dual to of the skeleton. Fig. 11-14 shows the dual of Fig. 11-13.

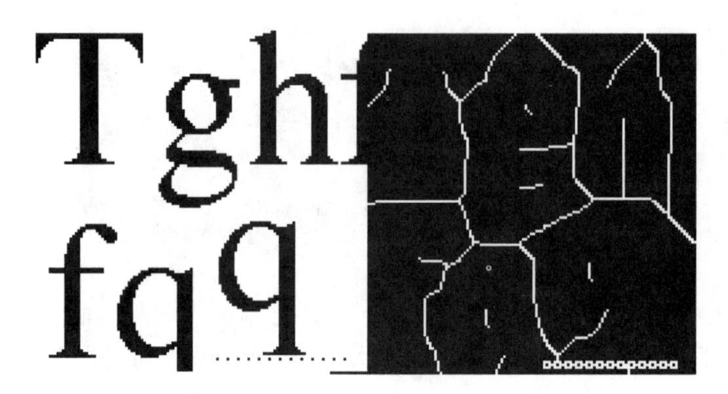

Fig. 11-14. The Dual of the Skeleton

Fig. 11-15 shows the skeletonization of Edge Man. We note that we have a single-pixel-width edge that lies at the center of each of the boundaries. Skeletonization assumes that noise has already been filtered from the image. The type of noise filter used will depend on the composition of the noise.

Fig. 11-15. Before and After Skeletonization

Consider the obstacles imposed for an autonomous land vehicle, shown in Fig. 11-16 and 11-17.

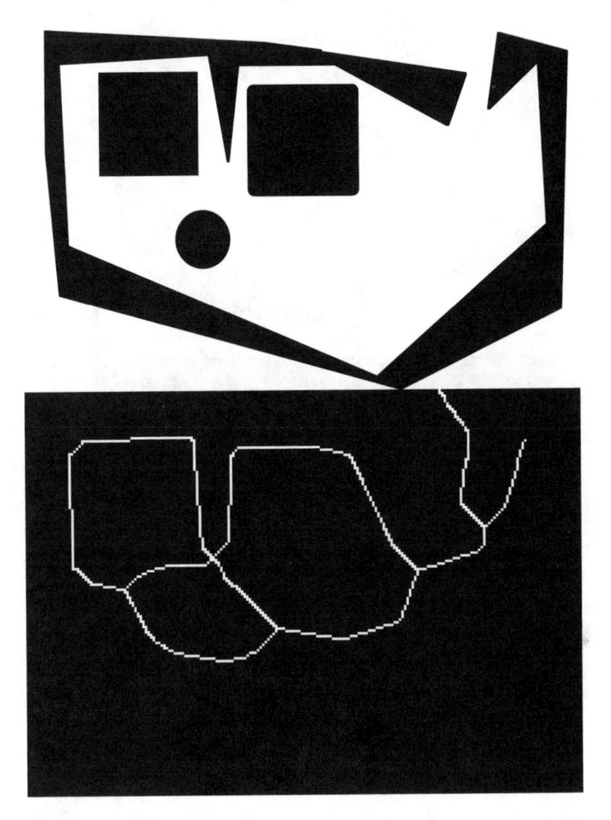

Fig. 11-16. A Sample Obstacle Course with Skeleton Showing Voronoi Diagram

Fig. 11-16 shows the safe path through the regions by providing an edge that divides the distance between the point sets. Thus, the skeleton of the open space is a Voronoi diagram.

Fig. 11-17 shows an erosion of the obstacle course in order to prevent some of the narrower paths from showing up in the Voronoi diagram.

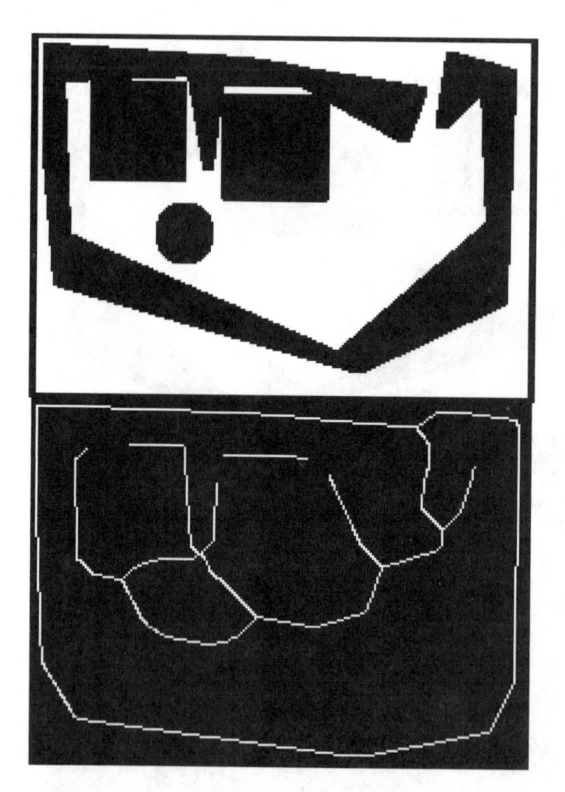

Fig. 11-17. Erosion Alters the Voronoi Diagram to Eliminate Narrow Paths.

When the structuring element does not overlap the image, the morphological image processing operations (dilation, erosion, etc.) are undefined. This causes the extra border around the original image, with the resulting extra edge on the Voronoi diagram. Skeletization took 4 seconds for the 218x156 obstacle image of Fig. 11-17 (using a PowerMac 8100/100 running the Apple MRJ VM version 2.0).

11.6. Summary

This chapter showed how set theory can be used to perform binary, gray scale and, finally, color morphological image processing. The application to binary and gray scale images is not new. The application to color images is new and without a strong theoretical frame work. Thus, the results are experimental and should be treated as such.

Erosion and dilation proved useful, both at elimination of some kinds of noise and as a technique to find safe paths in the plane (though higher-dimensioned operations could be performed).

Outlining turned out to be a useful way to obtain the edge of a boundary. One new result included the application of color morphological pyramids to obtain a good edge detection algorithm that is space-scalable. This, too, is a preliminary result and should be taken *cum grano de saline* (with a grain of salt).

The Zhang and Suen algorithm for thinning was implemented generally as presented in [Zhang]. The primary difference was in the use of the continue statement for a small constant-time speed up. Another difference is that the implementation is in Java (which is, after all, useful!).

The skeleton of a region is the locus of points that are equidistant from at least two points on a boundary. As a result, the skeleton is also called the medial axis. The skeleton suggests the shape of the region. As the Voronoi diagram, it enables the solution to the *circular range search problem*, i.e., a disk of given diameter should be able to capture all points that are within the disk diameter of each other. This can be very useful in boundary processing, as we shall see in Chapter 12. See the Projects section for future work.

11.7. Projects

1. Soft operators make use of *soft erosion* and *soft dilation*. Soft operators use a threshold that is designed to help them perform better in a noisy environment. A fast algorithm for this is presented in [Zmuda]. Get the Zmuda paper and implement the algorithm in the Kahindu program. Show the results of soft operators vs. crisp operators in your report.

2. One way to speed binary morphological operations is to convert the image into a sequence of encoded integers. The morphological operations are then performed using bitwise logical operators that are generally implemented in a single CPU cycle on a RISC machine. Many morphological operations are reported to be up to 100 times faster as a result. The details of this are found in [Boomgaard]. Implement the Boomgaard algorithm in the Kahindu program. Show the difference in speed for various operations, as well as sample images used for the testing.

3. One way to speed gray scale morphological operations is to use recursion. Read the [Déforges] paper and implement the algorithm in Kahindu. Write the methods *erodegs 2* and *dilategs2.*

4. Visit <http://www.ph.tn.tudelft.nl/Courses/FIP/frames/fip-Morpholo.html> and read about morphological smoothing, gradient and Laplacian. Implement these filters for color images, using the techniques shown in this chapter.

11.8. The *MorphFrame*

The *MorphFrame* class resides in the *gui* package and extends the *EdgeFrame* class. A *MorphFrame* instance provides the services of morphological image processing. The *MorphFrame* implements the algorithms described in this chapter.

11.8.1. Class Summary

```
package gui;
import java.awt.*;

public class MorphFrame extends EdgeFrame  {
    public MorphFrame(String title)
    public void colorPyramid(float k[][])
    public void resample(int ratio)
    public short [][]
                resampleArray(short a[][], int ratio)
    public void colorOpen(float k[][])
    public void colorClose(float k[][])
    public void open(float k[][])
```

```
    public void close(float k[][])
    public short [][]
            intersect(short a[][], short b[][])
    public short [][] complement(short s[][])
    public void dilate(float k[][])
    public void erode(float k[][])
    public void colorDilateErode(float k[][])
    public void colorDilate(float k[][])
    public void colorErode(float k[][])
    public void insideContour(float k[][])
    public void outsideContour(float k[][])
    public void middleContour(float k[][])
    public void
            clip(short a[][], short min, short max )
    public short[][]
            subtract(short a[][], short b[][])
    public void skeleton()
    public boolean skeletonRedPassSuen(
        boolean firstPass)
    public void thin()
    public short[][] erode(short f[][], float k[][])
    public short[][] dilate(short f[][], float k[][])
    public short[][] erodegs(short f[][], float k[][])
    public short[][] dilategs(short f[][], float k[][])
}

    public static final
      float kv[][] = {
          {0,1,0},
          {0,1,0},
          {0,1,0}
      };
    public static final
      float kh[][] = {
          {0,0,0},
          {1,1,1},
          {0,0,0}
      };
    public static final
      float kCross[][] = {
          {0,1,0},
          {1,1,1},
          {0,1,0}
      };
    public static final
      float kSquare[][] = {
          {1,1,1},
          {1,1,1},
          {1,1,1}
```

```
    };
    public static final
      float kThinTop[][] = {
          {0,1,0},
          {1,1,1},
          {0,0,0}
      };
    public static final
      float kThinBottom[][] = {
          {0,0,0},
          {1,1,1},
          {0,1,0}
      };
    public static final
      float kOutline[][] = {
          {0,1,1,1,0},
          {1,1,1,1,1},
          {1,1,1,1,1},
          {1,1,1,1,1},
          {0,1,1,1,0}
      };
```

11.8.2. Class Usage

To make an instance of a *MorphFrame*, use

```
String title = "my morph frame";
MorphFrame mf = new MorphFrame(title);
```

To spawn a child frame whose contents is the next level of a morphological color pyramid, use

```
mf.colorPyramid(mf.kSquare);
```

To resample the image in a frame using an integer ratio (for example, 2:1 sub-sampling), use

```
int ratio = 2;
mf.resample(ratio);
```

To resample an array of *short* by a ratio, use

```
short s[][] = mf.resampleArray(s, 2);
```

To perform the color morphological open, given a structuring element, use

```
mf.colorOpen(mf.ksquare);
```

To perform the color morphological closed, given a structuring element, use

```
mf.colorClose(mf.ksquare);
```

To perform the binary open, and closed use

```
mf.open(mf.ksquare);
mf.close(mf.ksquare);
```

To perform the set theoretic intersection between two gray scale arrays

```
short a[][], b[][];
c = mf.intersect(a,b);
```

To form the complement of a set:

```
c = mf.complement(a);
```

To dilate or erode a binary set by a structuring element, use

```
mf.dilate(mf.ksquare);
mf.erode(mf.ksquare);
```

To color dilate or color erode by a binary structuring element, use

```
mf.colorDilate(mf.ksquare);
mf.colorErode(mf.ksquare);
```

To perform 5 color dilations and 5 color erosions by a binary structuring element:

```
mf.colorDilateErode(mf.ksquare);
```

To find the inside, outside and middle contours for a binary set, given a binary structuring element, use

```
mf.insideContour(mf.ksquare);
mf.outsideContour(mf.ksquare);
mf.middleContour(mf.ksquare);
```

To perform a hard clip (that is, to limit the range of the values) to between *min* and *max*, use

```
short min =0;
short max = 0;
short a[][];
clip(a); // warning, clip alters its argument
```

To compute c = a - b, using arrays of short, with a new array being returned, use

```
short a[][];
short b[][];
short c[][];
c = subtract (a, b);
```

To run the Zhang and Suen algorithm for thinning, until a skeleton forms, use

```
mf.skeleton();
```

To run a first and second pass on the Zhang and Suen thinning algorithm, use

```
if (!mf.skeletonRedPassSuen(true)) {return};
if (!mf.skeletonRedPassSuen(false)) {return};
```

The skeletonRedPassSuen operates on the pixels in the red color plane. Typically, the pixels are set to the values of 0 and something that is non-zero. The non-zero pixel is treated like a pixel valued as on. The answer appears in the red color plane. The return value is false if no pixels are changed.

To perform a single pass of thinning, using Zhang and Suen, with a display at the end, write

```
mf.thin();
```

To perform erosion and dilation of a binary short array (0 or 255) with a binary structuring element (0 or 1), use

```
short f[][];
f = mf.erode(f, kSquare);
f = mf.dilate(f, kSquare);
```

If the input short array is gray scale, then use the gray scale versions:

```
f = mf.erodegs(f, kSquare);
f = mf.dilategs(f, kSquare);
```

12. Boundary Processing

Research is what I do
when I do not know
what I am doing.

– Wernher von Braun

Chapter 11 showed a way to obtain good multi-resolution edges using morphological image processing (outlining). In this chapter, we are concerned with transforming the images produced by edge detectors into a boundary. The primary focus is on the creation of vectors that may be used to represent the *important* edges. The important edges are identified with a combination of user input and a search through the pixel space. Such searches are known to be computationally expensive so we use both the A^* search and a heuristic depth-first search to speed the processing. We also show that such search techniques have path-planning applications, using a morphological skeleton as a local cost in the formulation of a heuristic evaluation function.

12.1. Hough Transform

The Hough transform (pronounced "huff") is used to cover a class of algorithms that detect shapes. One of the most frequent formulations of the Hough transform is for straight-line detection.

The basic idea of the Hough transform is that points in an image array are mapped into curves in a *parameter space* (also called the Hough space). The parameter space is quantized into locations called *bins*. Each bin in the parameter space (called the *accumulator-array*) corresponds to several points in the input image. As points are traced into the parameter space, they increment the bins (this is a kind of vote for a bin in the Hough space). After all the interesting points in the input image are transformed into curves in the Hough space (i.e., after all the votes are cast), a sorting process takes place. The sorting process selects those bins that have the most votes. Thus, the Hough transform is a voting algorithm [Dyer].

The Hough transform may be used to detect straight lines. Straight-line Hough transforms are classed by the kind of parameterization used to formulate the Hough space. There are two common parameterizations. The most popular parameterization is given by

$$\rho = x\cos\theta + y\sin\theta \tag{12.1}$$

where (12.1) uses the *normal* parameters, given by [Merlin]:

$\rho =$ the distance from the line to the origin

and

$\theta =$ the angle between the line and the abscissa

In summary, equation (12.1) maps a point in the image space into a curve in the Hough space. After the Hough transform is performed, the Hough space is searched for a number of maxima points. Each of these points corresponds to a straight line in the image space. The output image is created by forming the logical AND between the input image and the straight lines obtained from the maxima in the Hough space.

Another, less popular, parameterization, uses a slope-intercept parameterization for the line. This is given by

$$c = -xm + y \tag{12.2}$$

where

$$x = \text{ the slope}$$
$$y = \text{ the intercept}$$

See [Guil] for an example of the use of the slope-intercept parameterization in (12.2) to create a fast Hough transform.

Most straight-line Hough transforms may be classified as using the parameterization in (12.1) or (12.2). The Hough transform is not limited to 2D straight lines, however. Subramanian and Naylor use

$$\rho = x\sin\theta\sin\phi + y\cos\phi + z\cos\theta\sin\phi \qquad (12.3).$$

to detect planes and create partitioning trees in 3D [Subramanian].

Our primary focus is on the use of (12.1) to perform straight line detection in 2D.

The following method, in the *BoundaryFrame* class, inputs an array of shorts in the red-plane. The shorts represent an image that has been edge-detected and thresholded (i.e., a bi-level image). All non-zero pixels contribute curves to the Hough space, which is stored in the *s* array:

```
public short [][] hough() {
        int thetaMax = 360;
        int radiusMax= (int)Math.sqrt(width*width +
height*height);
        short s[][] = new short[radiusMax][thetaMax];
        for (int x=0; x < r.length; x++) {
             for (int y=0; y < r[0].length; y++) {
                  if (r[x][y] == 0) continue;
                  drawHoughLine(x,y,s);
             }
        }
        return s;
    }
```

Note that the *hough* method creates a Hough space that has a one-degree increment on θ. It also has one pixel increments in the radius parameter. This points out a fundamental drawback of Hough transforms. The drawback is that quantization in the parameter space can cause curves that are not exactly colinear to map into the same bins [Aghajan].

The method, *drawHoughLine,* traces out a curve in the Hough space, *s*. The curve is traced by incrementing the bins in the accumulator array that correspond to points computed, using the parameterization of (12.1):

```
public void drawHoughLine(int x, int y,
        short s[][]) {
        for (int theta=0; theta < s[0].length; theta++) {
                int rho = (int)(x * cos(theta) + y*sin(theta));
                if (rho >= s.length) continue;
                if (rho < 0) continue;
                s[rho][theta] ++;
        }
}
```

There is no need for the for-loop in the *drawHoughLine* to check the range on θ, since that is always well constrained. However, the various values for θ could, after computation with (12.1) yield an invalid ρ.

To display the result of the Hough transform, we copy the Hough space into the various color channels and do some housekeeping:

```
public void displayHoughOfRed() {
        r = hough();
        width = r.length;
        height = r[0].length;
        setSize(width,height);
        copyRedToGreenAndBlue();
        short2Image();
}
```

What is displayed is typically dim, so a uniform-nonadaptive histogram equalization (UNAHE) is performed, as shown in Fig. 12-1.

Fig. 12-1. A Box and Its Hough Transform

After computing the Hough transform, the Hough space is searched for the maxima. The search of the Hough space represents a computational drawback of the Hough transform. Worse, if the number of lines to be fitted is not given a priori, then local maxima may cause the search to yield incorrect results.

The reader should note that the search is performed in an unequalized Hough space (the UNAHE will alter the number and location of the maxima). To search through the image space for a max value, we use

```
public Point identifyLargestPoint() {
      int max = -1;
      Point p = null;
      for (int x=0; x < width; x++)
            for (int y=0; y < height; y++) {
                  if (g[x][y] > max) {
                        max = g[x][y];
                        p = new Point(x,y);
                  }
            }
      return p;
}
```

The *identifyLargestPoint* method returns the location of the largest bin. To get several large points from the image, we perform a search that deletes the point from the image once it is found. A point is deleted by setting the pixel to zero, hence:

```
public Point[] getTheLargestPoints(int n) {
```

```
        Point points[] = new Point[n];
        for (int i=0; i < n; i++) {
              Point p = identifyLargestPoint();
              points[i] = p;
              if (p == null) break;
              g[p.x][p.y] =0;
        }
        return points;
    }
```

Once we obtain the maxima in the Hough space, we solve (12.1) to obtain a family of points that correspond to the bin, that is, solve

$$\rho = x\cos\theta + y\sin\theta \tag{12.1}$$

for y to obtain

$$y = \frac{\rho - x\cos\theta}{\sin\theta} \quad \theta \neq 0 \tag{12.4}$$

For the case when $\theta = 0$, we use (12.1) to obtain $\rho = x$. To implement (12.4) we use

```
   public void drawHoughLines(Point points[]) {
      ...
      for (int i=0; i < points.length; i++) {
            Point p = points[i];
            int rho = p.x;
            int theta = p.y;
            int x1 = 0;
            int x2 = width;
            int y1 = (int)((rho - cos(theta) * x1) / sin(theta));
            int y2 = (int)((rho - cos(theta) * x2) / sin(theta));

            if (theta == 0) {
                  x1 = rho;
                  x2 = rho;
                  y1 = 0;
                  y2 = height;
            }
            drawLineRed(x1,y1,x2,y2);
      ...
```

The *drawLineRed* method is a Bresenham line drawing method that sets the pixels that are on the line to 255. The Java implementation (inspired by [Newman] and [Heckbert]) is beyond the scope of an image processing book.

Fig. 12-2 show the maximum points in Hough space (circled), along with the corresponding straight lines that would be drawn in the image space.

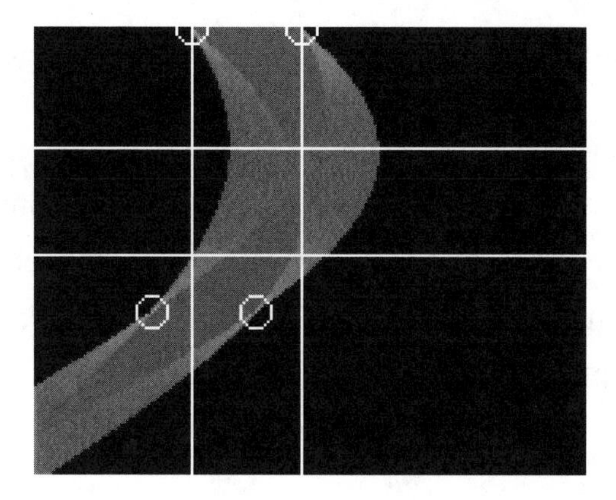

Fig. 12-2. Hough Transform of a Square

Fig. 12-2 is obtained by brightening the Hough transform and then invoking

```
public void drawSomeBigPoints() {
        Point points[] = getTheLargestPoints(4);
        drawThePoints(points);
        drawHoughLines(points);
}
```

The method, *drawSomeBigPoints* has a priori data for the number of maxima of interest in the Hough space (not something that is typically known). Generally, the Hough space is mapped into the image space, using an AND. We use an OR (i.e., we draw right into the image space) in order to illustrate the Hough lines and their relationship to the image lines.

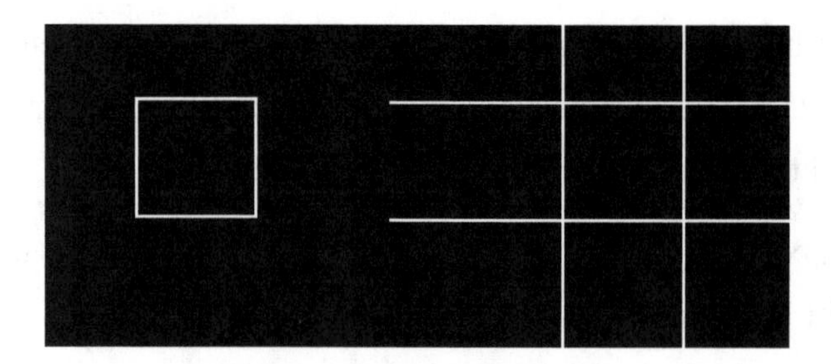

Fig. 12-3. Before and After the Hough Lines are Drawn in the Image Space

Suppose that the number of lines is not known a priori (as is typically the case). Fig. 12-4 shows what happens when 40 lines are drawn from Hough space. This emphasizes the need for the AND operation, but also shows how changing bins in the Hough space can cause a change in line orientation in the image space. This illustrates the importance of fine quantization in the Hough space.

Fig. 12-4. Drawing 40 Lines from Hough Space

Fig. 12-5 shows a tilted box and 40 lines from Hough Space.

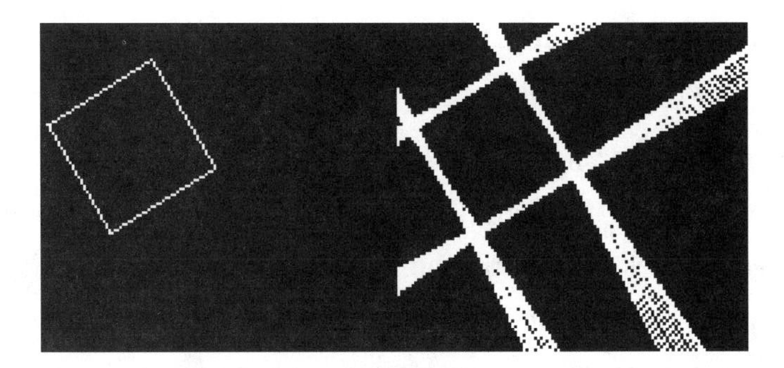

Fig. 12-5. A Tilted Box and 40 Lines from Hough Space

Fig. 12-6 shows the reconstruction of the tilted box using the Hough straight lines of Fig. 12-5 and ANDing with the original image. On the left is the tilt box ANDed with 40 lines; on the right is the tilt box ANDed with 200 lines.

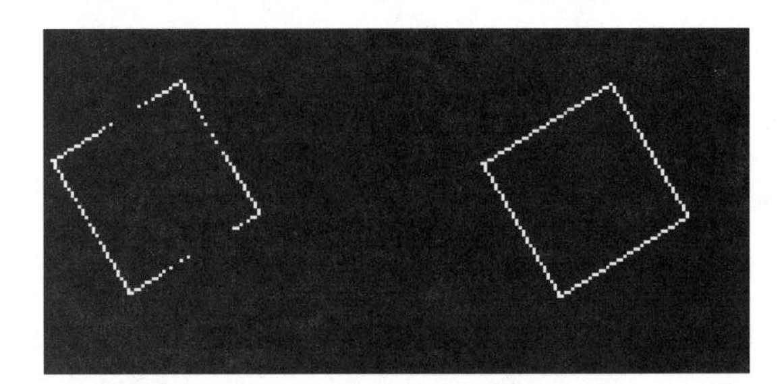

Fig. 12-6. Forty Point and 200 Point Hough Detection

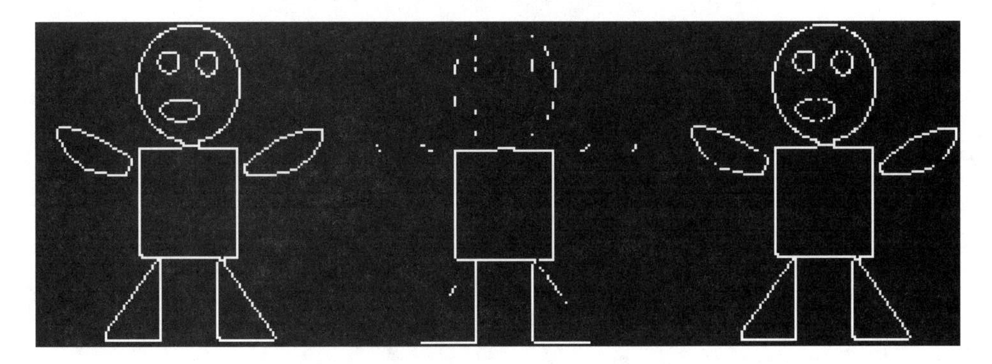

Fig. 12-7. Edge Man with 20 and 200 Point Hough Detection

Fig. 12-7 shows a basic drawback of the straight-line Hough detection: It is good only at detecting straight lines! The first 20 Hough maxima have no trouble finding the straight horizontal and vertical lines in the image. They do have a problem with the angled lines and the curves.

In order to expand the Hough transform to detect other shapes, different parameterizations are used. See [Ballard] for parameterizations for the Hough transform that recognize lines, circles, parabolas, ellipses and arbitrary shapes. Space and time do not permit an exploration into these parameterizations. Suffice it to say that the detection of an ellipse requires a parameterization such as

$$ax^2 + bxy + cy^2 = 1 \tag{12.5}$$

This requires a 3D Hough space. Another variation of the Hough transform uses the gradient magnitude and direction; our example uses only a thresholded magnitude. This is typically done by computing the Sobel gradient. The Sobel gradient yields the derivative with respect to *x* and *y* (a first-order approximation). This is used to compute the value for θ. By taking the approach of computing the gradient, we are able to start with non-edge detected images.

The basic technique is to use the red color plane to compute the gradient in the *x* and *y* directions, storing these in the green and blue color planes. The method for this computation follows:

```
// abc
// hid
// gfe
public void computeGradient(int x, int y) {
   int a1, b1, c1, d1, e1, f1, h1, i1, g1;
   a1 = r[x-1][y+1];
   b1 = r[x][y+1];
   c1 = r[x+1][y+1];
   d1 = r[x+1][y];
   e1 = r[x+1][y-1];
   f1 = r[x][y-1];
   g1 = r[x-1][y-1];
   h1 = r[x-1][y];
   i1 = r[x][y];

   g[x][y] = (short)((c1-g1) + (e1-a1) + 2 * (d1-h1));
   b[x][y] = (short)((c1-g1) + (a1-e1) + 2 * (b1-f1));
}
```

Note that the *computeGradient* destroys the original image. The green plane holds the derivative with respect to x and the blue plane holds the derivative with respect to y. This allows for the recording of a single point in Hough space for strong edges. Another variation for the straight-line Hough transform is to threshold in Hough space before taking the inversion. Then the search for maxima is avoided and the inverse Hough transform is taken for all non-zero points.

See http://iris.usc.edu/Vision-Notes/bibliography/text/contents.html for a list of citations related to the Hough transform.

12.2. Simple Edge Tracing

One very simple way to organize the pixels in an image is to trace out the connected pixels and join them into a list of points. The *java.awt* package has a built-in class called *Polygon* which can be drawn using a *drawPolygon* invocation in the *Graphics* class.

The basic idea is that we are given an edge detected image and that we return a list of *Polygon* instances. When the polygons are drawn, they trace out the edge detected image. The polygons represent a reorganization of the edge points in the image. They do not reduce the amount of data in the image; that can be done as a post-process.

We start by scanning the edge-detected image look for points that are interesting (i.e., non-zero). For each interesting point, we cast lines in one of several directions. For each direction, we check to see if the line lies on an interesting point. For each point found, a new point is added to a polygon. As points are added, they are turned off (i.e., set to zero) so that they will not appear duplicated on different polygons.

The entire procedure may be written in only a few lines of code, by using recursion:

```
public void bugWalk() {
    int n = 0;
    System.out.println("number of points = "+pointCount());
     for (int x=1; x < width-1; x++)
        for (int y=1; y < height-1; y++)
```

```
            if (r[x][y] !=0)
                buildPolygonList(x,y);
        System.out.println("polys="+polyList.size());
}
```

The *bugWalk* method starts and ends one pixel and one row short of a full image. The *pointCount* counts the number of non-zero points in the image. The following method, *buildPolygonList,* inputs a starting location and returns a new polygon. The algorithm then searches in three directions; east, southeast and south:

```
public void buildPolygonList(int x,int y) {
    Polygon p = new Polygon();
    p.addPoint(x,y);
    r[x][y]=0;
    if (r[x+1][y] != 0) {
        buildPolygonList(x+1,y,p);
        polyList.addElement(p);
        return;
    }
    if (r[x+1][y+1] != 0) {
        buildPolygonList(x+1,y+1,p);
        polyList.addElement(p);
        return;
    }
    if (r[x][y+1] != 0) {
        buildPolygonList(x,y+1,p);
        polyList.addElement(p);
        return;
    }

}
```

The recursive part of the algorithm builds the list of points into the given polygon. We are assured of continuity, but not colinearity:

```
public void buildPolygonList(int x,int y,Polygon p) {
    p.addPoint(x,y);
    r[x][y]=0;
    try {
        if (r[x+1][y] != 0) {
            buildPolygonList(x+1,y,p);
            return;
        }
        if (r[x+1][y+1] != 0) {
            buildPolygonList(x+1,y+1,p);
            return;
        }
```

```
            if (r[x][y+1] != 0) {
                  buildPolygonList(x,y+1,p);
                  return;
            }
      }
      catch (Exception e) {
      };
   }
```

Using a 233 Mhz G3 processor, 3,537 points created 675 polygons in 0.053 seconds. Fig. 12-8 shows the raster version of the image next to the vector version of the image. The vector version was created using an invocation of

```
   public void printPolys() {
      Toolkit tk = Toolkit.getDefaultToolkit();
      PrintJob printJob =
            tk.getPrintJob(
                  new Frame(),
                "print me!",
                null);
      Graphics g = printJob.getGraphics();
      for (int i=0; i < polyList.size(); i++)
            g.drawPolygon((Polygon)polyList.elementAt(i));
                  printJob.end();
   }
```

Fig. 12-8. Before and After the *bugWalk*.

The number of points in each polygon may be thinned as a post-process. The implementation for polygon thinning is a straightforward matter consisting of scanning the points on the polygons and removing them if they are colinear.

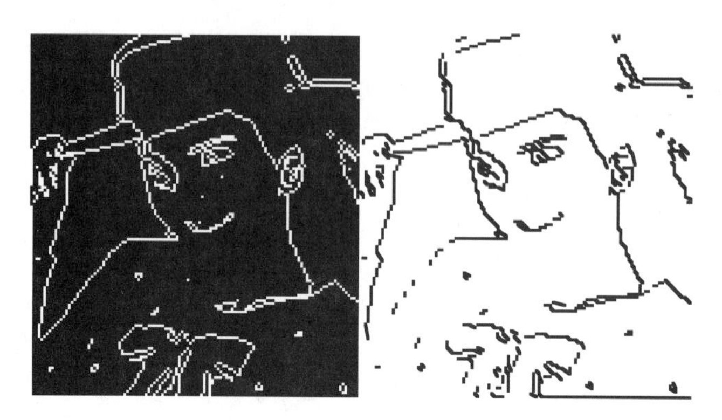

Fig. 12-9. Before and After Polygon Thinning

Fig. 12-9 shows the girl image after edge detection. The image, consisting of 1,348 points, was converted into a series of 199 polygons in 0.034 seconds on a 215 Mhz G3 processor. Thinning occurred in 0.118 seconds and yielded 649 points. It should be possible to reduce the number of points further, using a tolerance for the deviation from straight (these examples used a zero-tolerance system). The drawback of such an approach is an increase in linear distortion. The problem of performing raster to vector conversion can also be integrated with the thinning problem (at greater computational cost). See [Lyon 85] for an approach that uses search to raster to vector conversion with a vector ordering post process.

12.3. Edge Detection via Heuristic Search

One way to find a good edge is to use a search [Martelli 76]. The basic idea is that each pixel in an image may be treated like a node in a graph. The connectivity between a pixel and its neighbors is represented by arcs in the graph. The weight assigned to each arc is computed as a local function of the intensities of the surrounding pixels.

Once the graph is formulated, one must perform an efficient search for a good edge. Often the search is designed to minimize the total weight of the path used to describe the edge. There are many techniques available to search for a shortest path. An exhaustive search between a destination and all possible start nodes is typically prohibitive in computational cost. To see why this is true, we must examine the run-time of exhaustive search algorithms. One algorithm, called Dijkstra's algorithm, finds the shortest paths from one vertex to all others in $O(N^2)$. If the region of interest in the image is unconstrained, then N is the number of pixels in the image. If the region of interest is small, however, Dijkstra may be the right choice when searching for a good edge. In this section, we assume that an expert inputs a series of points that lie on an *edge of interest*. We use this data to help guide the search and obtain a series of edges that connect the points.

There are several different search strategies available, broadly classed into exhaustive and non-exhaustive. The exhaustive searches, which yield globally shortest paths are depth-first and breadth-first. The depth-first search works with the last node added and is linear in the amount of space needed. The breadth-first search works with the first node added and is exponential in the space required for the search. Both searches take exponential time and so are not further explored. The primary focus of this section is on reducing the search time for a good edge.

To reduce the search time, while still yielding a good edge, it is typical to employ a *heuristic* to help guide the search. The heuristic function is designed to use task-dependent information to help estimate the cost of a line of search. Each line of search may be considered a frontier of exploration. The heuristic identifies the most promising frontier.

The technique used in this section is to formulate a real-valued function for the heuristic, called the *evaluation function*. Evaluation functions may use the difference between the present state and the goal state (e.g., greediness). For an image, they can also use a local cost function (e.g., the difference between adjacent intensities, which is a derivative). Great care is needed to formulate good heuristics. In fact, a bad heuristic can be worse than no heuristic at all.

When a search is performed without heuristic guidance, then the search is said to be *uninformed.* There are two types of uninformed search: depth-first and breadth-first. Both types are exhaustive and are therefore computationally unsuitable for searching for edges in large images.

One way to avoid exploring a large search space is to be given a start node and an end node. Once the end node is reached, the search stops. If there is no end node, the search will continue until some other termination criterion is met. If the stopping criterion is never met, an exhaustive search ensues that will eventually fail. To guide the search toward better solutions, a fixed minimum cost may be used to force backtracking [Clocksin]. Another stopping criterion is the elapsed time (e.g., search for 10 seconds, then return the best edge you could find). Thus the stopping criterion becomes an important part of the search. When the stopping criterion is well-formulated (e.g., find a path to this node, then stop), the search will generally be sub-optimal, but very fast, compared with searches that have poor stopping criterion.

Sometimes it is more important to minimize the search effort than to minimize the solution cost. The heuristic is an encoding of a kind of knowledge called a *rule of thumb*. The rule of thumb is also called a *strategy* and is used to limit the search. Sometimes the use of heuristics actually prevents the finding of a solution, even though one may exist. A heuristic search is therefore useful for finding a solution that is adequate, most of the time [Barr].

One heuristic search used in this section is called the **A*** (say "A-star") search [Nilsson]. There are three operations that use heuristics to optimize the search:

1. Selecting the most promising frontier (e.g., expanding the node).

2. Discarding of the lines of search that appear unpromising (e.g., pruning).

Let s = the start node and let *open* be a mark associated with nodes that are queued for expansion. The method *minOpenNode* returns a minimal cost open node (in the case of ties, it does not matter which is returned). Let *nodes* be a list of all the nodes in the system. Every node has a cost associated with it. When a node is created, the default cost is zero. The method *terminateSearch* can terminate for any of several reasons. Typically, if the current node represents a goal state, the algorithm terminates. Termination may also occur due to resource limits such as time or memory. The method *expand* expands all the possible nodes available from the *lowestCostNode* and places it on a list called *expanded.* When no nodes may be expanded, we continue to the next minimum cost open node. For the purpose of edge detection, each of the nodes in our graph is actually an instance of a class called an *Edgel* which stands for *edge element*. The following Java snippet for the **A*** algorithm is a simplified version of the code contained in the method *searchFromPoint*:

```
s = new Edgel();
s.setType(OPEN);
nodes.addElement(s);
while ((lowestCostNode = minOpenNode()) != null) {
   if (terminateSearch(lowestCostNode)) break;
         expand(lowestCostNode);
         if (expanded.size() == 0) continue;
               lowestCostNode.setType(Edgel.CLOSED);
         processExpandedNodes();
         drawTracks(lowestCostNode);
   }
}
```

The *drawTracks* method allows us to view the progress of the search dynamically. This permits a visualization of the search and thus gives critical feedback regarding the quality of a heuristic. The result of the search is not nearly as helpful in formulating the heuristic as the search process. The following shows the code for the *processExpandedNodes* method:

```
private void processExpandedNodes() {
   for (int i=0; i < expanded.size();i++){
         Edgel e = (Edgel) expanded.elementAt(i);
         Edgel MarkedNode=getMarked(e);
         if (MarkedNode==null) {
               nodes.addElement(e);
               continue;
         }
```

```
            if (e.getCost()<MarkedNode.getCost())
                MarkedNode = e;
        }
    }
```

The *processExpandedNodes* method checks the successor node to see if it has already been visited. Costs for the successor node are computed and compared to existing nodes. If the successor node has a lower-cost path than an existing node, it replaces the existing node. The successor node has a reference to its parent, so the act of replacing a node with a lower cost node replaces the path to that node.

If the costs for the nodes always return the depth of the node in the search, then the algorithm becomes known as *Algorithm* **A**, a breadth-first search that returns a minimal cost path and coincides with the Dijkstra algorithm [Martelli 72]. *Algorithm* **A** becomes *Algorithm* **A*** when the heuristic evaluation function for a node, $h(n)$, has a lower bound. Typically, the lower bound is zero [Nilsson].

Plate 12-1 shows a non-optimal solution obtained by search with heuristics that were attained through experiment. An edge element consists of two horizontally adjacent pixels. The *Edgel* class contains the locations of the two pixels and the cost of the path that leads to the *Edgel* instance. The heuristic function used for producing Plate 12-1 uses the distance from the current node to the destination node. This is computed by the method called *distance*. It also uses the cost obtained from the parent node, which is accessed by the *getCost* method. The cost of a path increases as the path becomes longer. The *ply* of a search is the number of nodes between the current node and the destination node. The ply at a node is returned by the method *getPly*. To favor the longer paths, we divide the parent cost by the ply (otherwise, we expand too many nodes). The local cost consists of the difference in the intensities between the left and right pixels. By adding distance into the evaluation function, we bias the search toward minimizing the distance between the current node and the goal.

The formulation for the evaluation function follows:

```
private int C(Edgel e) {
```

```
        int pc = 0;
        Edgel p = e.getParent();
        int d = distance(e,endPoint);
        if (p != null) pc = p.getCost()/e.getPly();
        return
                clip(pc+d +
                        MaxI-(
                                g[e.p1.x][e.p1.y]-
                                g[e.p2.x][e.p2.y])
                                );
    }
```

The *clip* function ensures that the evaluation function is always greater than or equal to zero. The global, *MaxI,* is the absolute value for the maximum difference between two adjacent points in the image. The evaluation function, *C*, was attained via experiment. The horizontal derivative used in the *C* method is Martelli's [Martelli 72]. The problem with derivatives is that they typically enhance noise. We perform a morphological resampling of the image to reduce the run time and improve the edges. Plate 12-2 shows the 5 points selected by the user and a heuristic edge that passes through them. The image is 60x60 pixels and the run-time was 3 seconds on a 215 Mhz G3.

By adding points to an image, the user has made the search *more informed.* In fact, the heuristic can be made less informed, giving slower and very different results.

Plate 12-3 shows what happens when distance is removed from the evaluation function. The search tends to wander, finding boundaries that may or may not be of interest to the user. Edges continue to pass through the user points (since this is a stopping criterion), but the search otherwise ignores their existence. Run time has slowed to 48 seconds on the 60x60 image.

Plate 12-4 shows the effect of removing the ply information from the evaluation function. This leads to a single lower-cost path and is consistent with the Martelli formulation [Martelli 72]. Run time for Plate 12-4 was 118 seconds.

Plate 12-5 uses the evaluation function given in the original *C* method. The image is 119x119 (after 2 morphological resamples) and the edge is found in 3 seconds (all times are for a 215 Mhz G3). The left hand image in Plate 12-5 is inverted so that the cat tail and user selected point are more visible. The edge detection was run on the non-inverted image.

Plate 12-6 shows an application of heuristic search which values distance more highly than the finding of good edges. Because the expert is better able to identify an interesting line segment, than the search engine, we modify the evaluation function to dispense with the collective costs and focus only on locally reaching the user input destination point. As a result, the points must be input more frequently. The image may be found on the CD-ROM as *par aerial image 5*. The morphologically resampled image is 122x122 and the edge was found in 0.573 seconds. The high speed is a direct result of the focused search. Distance was weighted (a bit too much) to give the quick results:

```
d*MaxI-(
            g[e.p1.x][e.p1.y]-
            g[e.p2.x][e.p2.y])
```

Plate 12-7 introduces the experts points in a more gentle way:

```
d+MaxI-(
            g[e.p1.x][e.p1.y]-
            g[e.p2.x][e.p2.y])
```

The expert added only 5 points, shown more clearly by image inversion at the top of Plate 12-7.

The Roberts Edge-Linking Method uses a differentiated image obtained from the Roberts edge operator [Sharai]. Typically, this is thresholded to yield only the most promising candidates. Our variation uses Roberts to alter the evaluation function for the heuristic. Plate 12-8 shows results that took 46 seconds on a 215 Mhz G3 to produce. The costs were computed locally (i.e., without considering ply or parent cost!). An edge detected image makes a good heuristic function for finding edges!

Plate 12-9 shows the result of performing a heuristic search for a good edge using morphological outlining for the local cost. The search took only 0.65 seconds.

Plate 12-10 shows a color image outlined by a search through a morphological outline. The 28 points were carefully selected to fall on a good edge. The search ran in 8 seconds on a 215 Mhz G3. In order to ease the experimentation with various local costs, we modify the *C* method to assume that local costs are stored in the child image:

```
private int C(Edgel e) {
    int pc = 0;
    Edgel p = e.getParent();
    int d = distance(e,endPoint);
    if (p != null) pc = p.getCost()-e.getPly();
    return
          clip(pc+
                d+child.g[e.p1.x][e.p1.y]);
}
```

The procedure is to copy the image of interest into the child frame, then process the image so that edges are in regions with low cost. For example, Fig. 12-10 shows the girl image after processing with the Roberts operator. The magnitude is used to render a gray level, which is then negated. Points on curves of interest are selected by the user and the computer traces a path that passes through the points.

Fig. 12-10. The Inverse Roberts Search Space

Recall that the evaluation function is minimized; this explains why edges should be dark in the local cost image.

The heuristic search can also be used to find a path through a skeleton.

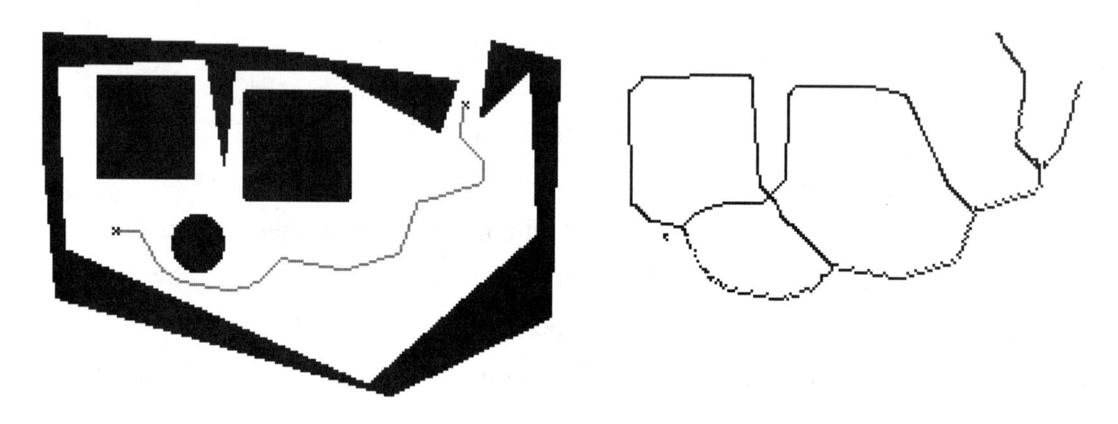

Fig. 12-11. Search for a Path Through the Skeleton

Fig. 12-11 shows that, given a start point and goal, the skeleton of an image can make a good cost function for finding a path. When the point was started a bit off the skeleton, we see the search makes a straight line toward the skeleton, then follows the skeleton until it reaches the goal. The child image is animated with an invocation to *drawTracks*. This enables the programmer to visualize the progress being made in the search. The search took 2 seconds on a 215 Mhz G3. There was backtracking on one branch before the search terminated. Getting a real robot to track such a path presents a set of problems in control and sensor fusion. For example, a passenger in a car would have some trouble with the sudden changes in direction. The derivative of speed is called acceleration and the derivative of acceleration is called *jerk*. Following a curve that has a discontinuity in the acceleration will cause an infinite jerk (passengers become spam in a can). Additionally, cars have a finite turning radius and so curvature constraints are placed on the paths for a solution to be feasible. For more information on the issues of sensor fusion and control of cars, as well as feasible trajectory formulations, see [Lyon] and [Lyon 90].

12.4. Summary

In this chapter we examined two techniques for finding edges in images. One, the Hough transform, used a transform space to perform a kind of optimized voting. Each pixel selected in the Hough space was projected back (as a straight line) into the image space. Hough transforms are useful for other shapes as well (see the following section for suggested projects).

The other technique that we examined was the heuristic search technique. The basic idea is that Dijkstra's algorithm runs in $O(N^2)$, which is too slow. As a result, we used user input to help guide the search. The guidance was provided by an evaluation function that enabled a fast answer. The trade-off is that the heuristic search is non-optimal and has a non-deterministic run-time. The experts who guide such a search for a good edge are expert at photo-interpretation (i.e., good at spotting an edge) as well as experts in the use of the heuristic search program (i.e., good at selecting the right spots to optimize the search).

The **A*** search uses a search space that is exponential in path length. If we eliminate the summation of the parent node into the cost, we have a heuristic depth-first search. This is locally minimal in the evaluation function and has a linear search path length.

The heuristic depth-first search is a much faster search than **A*** and is excellent for finding obvious paths in low-noise images [Poole].

12.5. Projects

1. Use equation (12.5) to implement the Hough transform for detecting an ellipse. Try your results on Edge Man. Also try your transform with the mandrill. See [Ballard] for more details.

2. Section 12.2 described a simple tracing algorithm that created a series of polygons. A post-processor called *thinPoly* inputs a list of polygons and removes the points that are on the same line. Write another post-processor that inputs the list of polygons and arranges them in order. The ordering of vectors should be performed so that a pen plotter would plot the vectors in minimum time (i.e., minimize the amount of time the pen is up). This type of problem is called the Chinese Postman problem and is known to be NP-Complete [Roberts].

3. Klette and Zamperoni give an algorithm for contour following on pp. 332 of their book [Klette]. Implement this algorithm in the Kahindu system and compare run-times, image quality and polygon statistics.

4. Heuristic search for an edge was shown on a two-dimensional image. Extend this algorithm to output edges across a sequence of images. How will you formulate the heuristics?

5. One problem with using the derivative to compute the heuristic for edge detection is that the heuristic becomes sensitive to noise. Improve the heuristic by filtering the search space. Which filters work best? Show several experiments on several images.

6. In the example for computing the heuristics, we showed that heuristics were computed every time a pixel was visited. Use the green plane to pre compute the local costs. Does this speed the search? By how much?

7. The heuristic search in the *MartelliFrame* looks for an open node with a minimal cost. The search for a minimal cost node runs in a time that is $O(N)$. One way to speed the search for a minimal cost node is to use a priority queue. Such a system can delete the minimum node in $O(\log N)$ time. Re-implement the search for the minimum cost node using a priority queue. Measure the speed-up, if any. How large is the speed up? Java code for a priority queue may be found in [Standish].

8. The Hough transform is implemented to work only with the red channel. Modify the Hough transform to work with color images. Keep the input channels and output channels from mixing (i.e., do not do a simple average of the red, green and blue channels). The output should be a color image.

9. The Hough transform may be inverted without ANDing into the original image. When this occurs, a series of lines are drawn into the image space. What happens if you use this image space as a search space for finding good edges? Are the edges better? Draw the lines by adding their intensities together in the image space and use as the local cost function. To perform this operation properly, the drawing of lines should have a cumulative effect (like the voting in Hough space). Thus, when a pixel is on a line, do not set its value to 255, simply increment it. Will this make for a better result when performing a search? How long does it take? Try this on several example images and keep track of execution time, image size and your heuristic formulation.

10. The*buildPolygonList* assures us of continuity for various points in a polygon, but not colinearity. Write *buildLineList* which returns a list of line segments that are formed from an image. How many lines did you make? How many points did you start with? What was the running time of the algorithm? Repeat the experiment and introduce a slope tolerance to reduce the number of vectors (while introducing distortion).

12.6. The *BoundaryFrame*

The *BoundaryFrame* extends the *MorphFrame*, resides in the *gui* package and provides basic vector processing services. These services include support for the straight-line Hough transform. They also include support for the *bugWalk* method, described in Section 12.2.

The *BoundaryFrame* keeps a list of polygons that may be drawn as an overlay on the image. The drawing of the polygons does not alter the image data and makes use of the *java.awt drawPolygon* method built into the *Graphics* class.

12.6.1. Class Summary

```
package gui;
public class BoundaryFrame extends MorphFrame
  public boolean onLine(int x1, int y1, int x2, int y2,
int x3, int y3)
  public BoundaryFrame(String title)
  public Point[] getTheLargestPoints(int n)
  public Polygon thinPoly(Polygon p)
  public short [][] hough()
  public short [][] houghGray()
  public short [][] trim(int dx, int dy, short s[][])
  public static void main(String args[])
  public static void PrintContainer(Container c)
  public Vector getPolyList()
  public void andWithChild()
  public void bugWalk()
  public void computeHoughAndDraw()
  public void copyToChildFrame()
  public void displayHoughOfRed()
  public void drawHoughLine(int x, int y,
  public void drawHoughLines(Point points[])
  public void drawLineRed (int x , int y , int x , int y )
  public void drawPoly(Polygon p)
  public void drawPolys()
  public void drawThePoints(Point points[])
  public void filterPolys()
  public void setPolyList(Vector v)
  public void grabChild()
  public void grayPyramid(float k[][])
  public void houghDetect()
  public void houghEdge()
  public void inverseHough()
```

```
    public void inverseHoughToRed()
    public void listPolys(Vector v)
    public void polyStats()
    public void printPolys()
}
```

12.6.2. Class Usage

Suppose that a new instance of a *BoundaryFrame* is defined as follows:

```
String title = "Boundary Frame";
BoundaryFrame bf = new BoundaryFrame(title);
```

Then, to determine if three points are colinear, use

```
int x1, y1;
int x2, y2;
int x3, y3;
boolean colinear = bf.online(x1,y1,x2,y2,x3,y3);
```

To get the 10 largest points from a Hough transform (i.e., largest bins), use

```
int n = 10;
Point[] points = bf.getTheLargestPoints(n);
```

To remove colinear-adjacent points from a *Polygon* instance, use

```
Polygon p = bf.thinPoly(p);
```

To take the straight-line Hough transform and return the Hough space, use

```
short s[][] = bf.hough();
```

To trim the edges from an array and return a new array, use

```
int dx = 10; // takes off dx/2 from left and right
int dy = 10; // takes off dy/2 from top and bottom
short s[][] = bf.trim(dx,dy,s);
```

There is a main method in the *BoundaryFrame* and it may be invoked directly:

```
String args[] = {" "};
BoundaryFrame.main(args);
```

To print a *Container*, use

```
Container c = bf;
BoundaryFrame.PrintContainer(c);
```

To perform an AND of a *BoundaryFrame* with its child, use
```
bf.andWithChild();
```

The result destroys the image in *bf* by over-writing. The AND works by taking the minimum of the intensities on a pixel-by-pixel basis.

To perform the bugWalk scan conversion discussed in Section 12.2, use
```
bg.bugWalk();
```

The bug walk creates a list of polygons in a vector. To get and set this vector, use
```
Vector v = bf.getPolyList();
bf.setPolyList(v);
```

To compute the Hough transform and display it both in Hough and image space (without the AND operation), use
```
bf.computeHoughAndDraw();
```

To make a copy of the *BoundaryFrame* into the child frame, use
```
bf.copyToChildFrame();
```

To take the Hough transform of the red color channel and display the Hough space, use
```
bf.displayHoughOfRed();
```

To draw a line in Hough space, given a location in image space, use
```
short s[][];
int x = x1;
int y = y1;
bf.drawHoughLine(x,y,s);
```

To draw several lines in image space, given points in Hough space, use
```
Point points[];
bf.drawHoughLines(points);
```

To draw a line in the red channel, using Bresenham's line drawing algorithm (see [Newman] and [Heckbert]), use
```
// public void drawLineRed (int x , int y , int x , int y )
        drawLineRed(0,0,width,height);
        drawLineRed(width,0,0,height);
```

```
        drawLineRed(width/2,height/2,0,height/2);
        copyRedToGreenAndBlue();
        short2Image();
```

To draw a series of points, stored in a polygon, not connecting the first point to the last, use

```
// public void drawPoly(Polygon p)
bf.drawPoly(p);
```

To draw the polygon list kept internally by the *BoundaryFrame,* use

```
//public void drawPolys()
bf.drawPolys();
```

To draw a series of circles about a list of points, use

```
//public void drawThePoints(Point points[])
bf.drawThePoints(points);
```

To filter the internal polygon list so that colinear points are removed, use

```
//public void filterPolys()
bf.filterPolys();
```

To replace the current image with the image in the child frame, use

```
//public void grabChild()
bf.grabChild();
```

To perform a gray-scale pyramid that trims the border and resamples the image, use

```
//public void grayPyramid(float k[][])
grayPyramid(bf.kSquare);
```

To perform a Hough transform and use it to perform edge detection, use

```
//public void houghDetect()
bf.houghDetect();
```

To make a copy of the parent frame, histogram equalize the child, threshold and use for Hough transform based edge detection, use

```
//public void houghEdge()
bf.houghEdge();
```

To perform the inverse Hough transform from 100 of the most popular Hough bins (stored in the child frame) into the image space, use

```
//public void inverseHough()
bf.inverseHough();
```

To take the inverse Hough transform and draw the lines in image space (without ANDing with the image), use

```
//public void inverseHoughToRed()
bf.inverseHoughToRed();
```

To list a series of polygons, stored in a vector, *v*, to the console, use

```
//public void listPolys(Vector v)
bf.listPolys(v);
```

To print the statistics for the polygon that have been created by the *bugWalk* method, use

```
//public void polyStats()
bf.polyStats();
```

To send the polygons out to the printer (or a postscript file), use

```
//public void printPolys()
bf.printPolys();
```

12.7. The *MartelliFrame*

The *MartelliFrame* resides in the *gui* package and provides heuristic search services for the purpose of edge detection.

12.7.1. Class Summary

```
package gui;
public class MartelliFrame extends PaintFrame
   public MartelliFrame(String title)
   public static void main(String args[])
   public Polygon getPath(Edgel pos)
   public void drawPath(Edgel pos)
   public Edgel searchFromPoint(Point startPoint)
}
```

12.7.2. Class Usage

To make an instance of the *MartelliFrame,* use

```
//      public MartelliFrame(String title)
MartelliFrame mf = new MartelliFrame("MartelliFrame");
```

To run the *MartelliFrame,* use

```
// public static void main(String args[])
String args[] = {" "};
MartelliFrame.main(args);
```

To get an instance of an *Edgel* class (an edge element) that represents a terminal node in a path, use

```
    //public Edgel searchFromPoint(Point startPoint)
Edgel e = mf.searchFromPoint(startPoint);
```

To turn an edge element into a list of points in an instance of a *Polygon*, use

```
    //public Polygon getPath(Edgel pos)
Polygon p = mf.getPath(
         mf.searchFromPoint(startPoint));
```

To draw the path in the overlay plane, use

```
    //public void drawPath(Edgel pos)
mf.drawPath(e);
```

An interface was designed for the *MartelliFrame* that allows the user to experiment with heuristics. The basic procedure for use is to:

1. Open an image
2. Make a copy of the image into the child frame (see Fig. 12-12).

SpatialFilter Xform Help 9:03
LowPass
Median
Hi-pass
Edge
Morph
Boundary
Paint
[E-p]grayPyramid
[E-H]houghDetect
houghDetectGray
[E-C]copyToChildFrame
l Bar

Fig. 12-12. Copy the Original Image to the Child Frame

3. Process the child image into a reasonable cost function. Fig. 12-13 shows an image that is a 13x13 Gaussian low-pass filtered image on the left, with the magnitude of a negated 2x2 Roberts operator on the right.

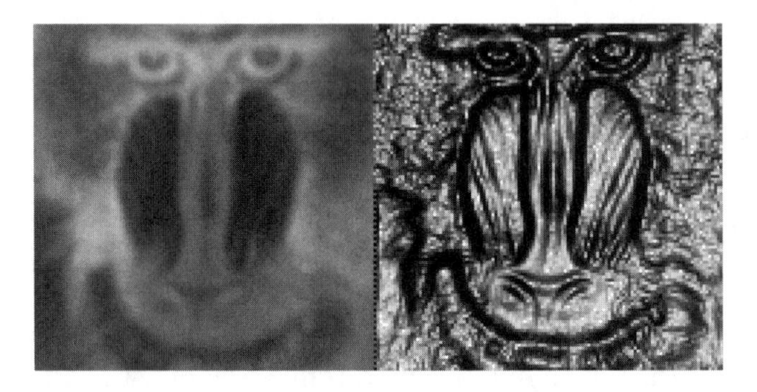

Fig. 12-13. Make a Good Cost Function

4. Select points in the parent frame that are interesting and on an edge (just click the mouse in the parent frame). Fig. 12-14 shows the *X* marker icon that must be selected from the Kahindu toolbar in order to enter the points into the parent image.

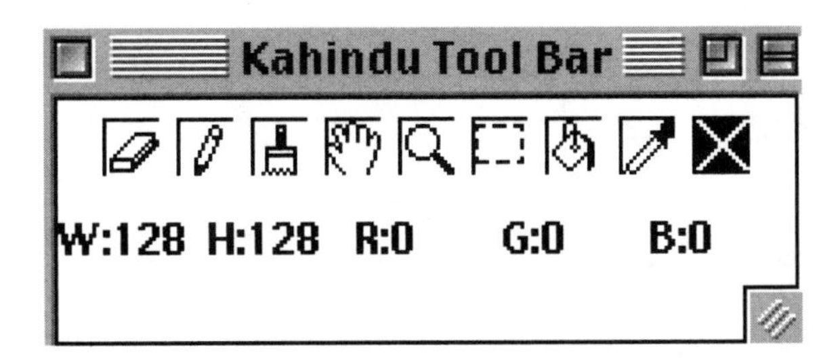

Fig. 12-14. Select the Marker Icon to Enter the Edge Points

5. Execute the search for a good path (use the *process User Points* from the *Heuristic Edge* menu).

13. Chromatic Light

Better to light one candle
than curse the darkness

– Chinese proverb

13.1. Introduction to Color

Light is a kind of electromagnetic radiation that propagates via electromagnetic laws. The energy in light is distributed in its spectra. Objects that are heated to the point of glowing have a radiancy that is a function of their temperature, shape and material properties. Some materials are used as standards against which radiancy is measured. When a material is shaped like a block of metal with a small hole drilled in it, it is said to be a *cavity radiator*.

Cavity radiators exhibit a radiancy that is dependent on temperature and is independent of their material. The spectral density of radiant energy per unit wavelength of a blackbody radiator is characterized by *Planck's Formula:*

$$u_{e\lambda} = \frac{8\pi hc}{\lambda^5 (e^{hc/kT^{\lambda}} - 1)} \tag{13.1}$$

where

$T =$ degrees Kelvin $= 273.15 +$ degrees Celsius

$c =$ speed of light $= 2.997925 \times 10^{10}$ cm/s

$h =$ Plank's constant $= 6.624940 \times 10^{-27}$ erg s

$k =$ Boltzmann's constant $= 1.380410 \times 10^{-16}$ erg/(deg K)

and

$\lambda =$ wavelength of light

When materials used in the construction of displays (such as the phosphors in a TV tube) are compared in their spectral radiance with cavity radiators, they are said to have a *color temperature* (that is, a spectral radiance that is equivalent to a cavity radiator of a given temperature). The intensity of the radiance is given in power over a unit area (i.e., watts per square meter). When a blackbody radiator is cooled to room temperature, it is black. At 800 K, it is red hot. At 3,000 K it is yellow, white at 5,000 K, pale blue at 8,000 K and brilliant blue at 60,000 K. A device resembling a small electric furnace is used for testing [Sturrock].

Color is the human perception of spectral radiance. To measure spectral radiance, we use a spectrometer. The spectral response of the human eye varies according to age and from person to person. Therefore, the human eye is not a good way to measure spectral radiance. The perception of intensity is called *brightness*.

There are different theories about how humans see color [Teevan]. In the *tristimulus* theory of color perception, the cones of the eye are divided into three types, R, G and B [Hunt]. These cones have a peak spectral response around 444, 526 and 645 nm [Cowen et al].

The three spectral response peaks in the eye form the basis for a three-dimensional color specification. The eye has a logarithmic response curve and so uniform quantization of color in the color space will often lead to an exponential response in the display.

Recall the response curve of (4.14):

$$f(v_{ij}) = 255\left(\frac{v_{ij}}{255}\right)^{pow} \tag{4.14}$$

which led to our discussion of *gamma* in a color monitor:

$$f(v) = v^{\gamma} \tag{5.3}$$

The sensitivity of the human visual system to light varies as a function of frequency. The standards body in charge of determining the sensitivity of a *standard observer* to various wavelengths of light is the *Commission Internationale de L'Éclairage* (CIE). The standard observer is a statistical profile for the average human viewer with normal color vision.

13.2. Tristimulus Transforms

The typical linear tristimulus transformation inputs a 3D point in a color space, performs a 3x3 matrix multiplication, and returns a 3D point in a new color space [Pratt]. This amounts to a homogeneous coordinate transform of the form:

$$C' = AC \tag{13.2}$$

where

$$C = \begin{bmatrix} R \\ G \\ B \end{bmatrix} = \text{primary colors in the old coordinate system} \tag{13.3}$$

$$C' = \begin{bmatrix} R' \\ G' \\ B' \end{bmatrix} = \text{primary colors in the new coordinate system} \tag{13.4}$$

and

$$A = \begin{bmatrix} a_{00} & a_{01} & a_{02} \\ a_{10} & a_{11} & a_{12} \\ a_{20} & a_{21} & a_{22} \end{bmatrix}$$

The primary vectors are linearly independent so that:

$$a_{00}R + a_{01}B + a_{02}G = 0 \quad (13.5)$$

iff (if and only if):

$$a_{00} = a_{01} = a_{02} = 0 \quad (13.6)$$

Further, the transform is invertable, iff the inverse of A exists, that is

$$C = A^{-1}C' \quad (13.7)$$

if the determinant of $A \neq 0$ [Wyszecki].

Homogeneous color space transforms can lead to points in the color space that have negative coordinates.

In 1931, the CIE defined the *CIE 1931 XYZ* standard color space. The XYZ space can represent the colors of normal human vision without negative coordinates. The CIE 1931 XYZ color space is a device-independent color space that is based upon normal human color vision (i.e., the standard observer). The Y component of the XYZ space represents the luminance. Luminance corresponds to the perceptual quality of light known as brightness. It is photo metrically measured in units of candelas per meter squared [Travis].

Up to this point, we have spoken about the RGB color space without making reference to the *type* of RGB color space. There are several types of RGB color spaces. They are characterized by the placement of the RGB components in the XYZ space, their use of a *reference white* and their gamma correction (which can be more complicated that equation 5.3). The reference white refers to the color temperature of the color created when equal and maximal amounts of red, green and blue are mixed.

Each RGB color image created for digital distribution must make reference to the type of RGB color space for which it was designed. Without knowing the basis for the vectors describing the set of points in a color space, it is impossible to formulate a transform from one space to another with any assurance of accuracy. Typically, assumptions are made regarding the type of RGB image. These assumptions are based upon an estimate of the origin of the image. Sometimes these assumptions are unstated. For example, the early scans of such test images as Lena were made without any statement regarding the gamma correction used in the scanner. As a result, the Lena image, which is so common in image processing literature, is uncalibrated and often has a reddish appearance!

Once the type of RGB color space is identified, it can be transformed into the CIE 1931 XYZ color space using equations of the form of (13.1) and (5.3) [Kasson]. It is typical to gamma correct an RGB color space before transformation into the XYZ color space.

Most video cameras will gamma correct before transmission of the color primaries [Poynton]. A video camera's gamma correction creates a more perceptually uniform space that compensates for the non-linearity in most display systems. Gamma is typically applied to the linear color space before a monitor displays the image. Modern display systems permit the user to control the gamma of the display. Some systems (e.g., Apple Mac and Silicon Graphics) use multiple look-up tables (both before and after the frame buffer) in order to perform gamma correction [Poynton 96a]. Other systems (e.g., video systems) gamma correct in the camera before display on the monitor. In such systems, the monitors have built-in gamma correction for display [Poynton 96b].

Gamma corrected video is generally designed for the display on a specific device. In comparison, the XYZ color specification is display independent.

When performing gamma correction on a color space that has negative primaries, gamma correction must be defined to yield a result that is not imaginary. Thus, we rewrite (5.3):

$$f(v) = v^{\gamma} \tag{5.3}$$

as

$$f(v) = sign(v)|v|^{\gamma} \tag{13.8}$$

The inverse gamma correction then becomes

$$v = sign(f(v))|f(v)|^{1/\gamma} \tag{13.9}$$

If the primaries are stored in floating point arrays that have been normalized to vary from zero to one, then we implement (13.8) with

```
public void powArray(float a[][], float gamma) {
   for (int x=0; x < width; x++)
         for (int y=0; y < height; y++)
               if (a[x][y] < 0)
                     a[x][y] = (float)(-Math.pow(-
a[x][y],gamma));
               else
                     a[x][y] =(float) Math.pow(a[x][y],gamma);
}
```

The procedure for gamma correction, with a change in color coordinates, is to gamma correct first, then perform the transform. After processing is complete, transform back using the inverse of the A matrix, and then perform the inverse gamma correction (13.9).

13.2.1. RGB and XYZ for Illuminant D65

When calibrating a color monitor, it is important to have a white on the monitor that corresponds to the standard established for the monitor. Without proper calibration, a monitor cannot accurately reproduce color.

The Society of Motion Picture and Television Engineers (SMPTE) has adopted the CIE Illuminant D_{6500} as the white reference for color television studio monitors [RP 37-1969]. Illuminant D_{6500} has a spectral distribution that is equal to daylight plus some ultraviolet light. A blackbody radiator with a color temperature of 6504 K has a spectral distribution that is close to D_{6500}. Illuminant D_{6500} approximates the spectrum of sunlight on a cloudy day [Glassner].

D_{6500} is the accepted broadcast standard for white in the NTSC (National Television Systems Committee), PAL (Phase Alternating Lines), PAL-M, and the SECAM (Sequentiel Couleur avec Memoire) broadcast standards. NTSC is used in North America, Japan, Taiwan, Korea and Central America. SECAM is used in France, the former USSR and Eastern Europe. PAL is used in Europe (except France), Singapore and Hong Kong. PAL-M is used in Brazil.

In Japan, NTSC is used with a standard white called Illuminant D_{9300}. Illuminant D_{9300} is slightly off the 9300 Kelvin blackbody radiator curve. This is a singular exception to the usage of Illuminant D_{6500} with NTSC.

The SMPTE-C RGB, primaries represent a gamma-corrected color space that uses the D_{6500} reference white. To convert from SMPTE-C RGB to XYZ(D65), we gamma correct the SMPTE-C RGB, to make linear RGB components. Then we apply a homogeneous transform to the XYZ D65 coordinates:

```
1.      package gui;
2.
3.      public class Xyzd65 extends FloatPlane {
4.
```

Line 3 mentions a class called the *FloatPlane* class. The *FloatPlane* class consists of 3 floating point two-dimensional arrays, with associated methods for processing them. Floating point format image data is high-precision, but it also represents a high cost in memory usage. The 32 bit floating point operations are getting cheaper on modern machines (most RISC systems perform them as fast an fixed point operations). They do have a memory usage requirement, but RAM is getting cheaper all the time...so perhaps this is also an acceptable cost. The coverage of the *FloatPlane* is left to Appendix B.

Lines 5-9 set the A matrix in (13.4), that is

$$\begin{bmatrix} X_{65} \\ Y_{65} \\ Z_{65} \end{bmatrix} = \begin{bmatrix} 0.3935 & 0.3635 & 0.1916 \\ 0.2124 & 0.7011 & 0.0865 \\ 0.0187 & 0.1119 & 0.9582 \end{bmatrix} \begin{bmatrix} sign(R_s)|R_s|^{2.2} \\ sign(G_s)|G_s|^{2.2} \\ sign(B_s)|B_s|^{2.2} \end{bmatrix} \quad (13.10).$$

where $[X_{65} \quad Y_{65} \quad Z_{65}]^t$ represents the XYZ D65 vector and $[R_s \quad G_s \quad B_s]^t$ represents the SMPTE-C RGB primaries. Equation (13.10) assumes that the primaries are normalized to vary from zero to one; otherwise, scale factors must be applied. In order to keep the XYZ D65 vector from exceeding one, we divide the RGB values by 255 (their maximum). Before conversion back, we multiply by 255.

```
5.    double A[][] = {
6.      { 0.3935, 0.3653,  0.1916},
7.      { 0.2124, 0.7011,  0.0865},
8.      { 0.0187, 0.1119,  0.9582}
9.      };
10.
11.    float gamma = 2.2f;
12.    float oneOnGamma = 1/gamma;
13.
14.    Mat3  rgb2xyzMat = new Mat3(A);
15.    Mat3  xyz2rgbMat = rgb2xyzMat.invert();
16.
17.    public Xyzd65(ColorFrame _cf){
18.     super(_cf);
19.    }
20.
21.    public void fromRgb() {
```

Line 22 is used because we want to normalize the range on the RGB input. Lines 22 and 23 perform the operation of gamma correcting the RGB input signal to make a linear RGB input signal:

```
22.      scale(1/255f);
23.      pow(gamma);
```

Line 24 performs a floating point matrix multiplication that performs the linear transform into the XYZ D65 color space.

```
24.        convertSpace(rgb2xyzMat);
25.        rgb2xyzMat.print();
26.       }
27.
28.       public void toRgb() {
```

The matrix multiplication of line 26 yields a linear RGB color space. The inverse gamma correction is applied in line 30. Line 31 scales the colors back to the original values.

```
29.        convertSpace(xyz2rgbMat);
30.        pow(oneOnGamma);
31.        scale(255f);
32.        xyz2rgbMat.print();
33.       }
34.
35.       }
```

The gist of the work in the color space conversion is performed in the *FloatPlane* and *Mat3* classes. They are summarized in Appendix B.

13.2.2. Sub-Sampling in Chroma

One color space held in common use is a *Y* (luminance), C_R (R - *Y*) and C_B (B - *Y*) color space. This is known as color difference component and comes from Recommendation *CCIR* 601-2 for digital television studios, so it has some applicability to digital image processing [CCIR-601-2]. CCIR stands for Comite Consultant International des Radio communications (International Radio Consultative Committee). The CCIR is part of the ITU (International Telecommunication Union). The CCIR acts as a coordinator for international radio communication services [Graf]. This is the same color space as the one used by professional quality video equipment.

The color difference component space represents all the chroma information in the orthogonal C_R and C_B vectors. Due to the narrow band allocated for the transmission of color, the CCIR 601-2 presents several methods for sub sampling the C_R-C_B components of the signal.

CCIR-601 specification uses an *A:B:C* notation to represent sub-sampling. For example, 4:4:4 means no sub-sampling, with 24 bits per pixel (8 bits on each component). CCIR-601 4:2:2 means 2:1 sub sampling in chroma. This means that for every two *Y* pixels, there will be one value for C_R and C_B. If there are 8 bits per sample, then there are 8+4+4 = 16 bits for each 2:1 sub sampled CCIR-601 pixel.

Another sub-sampling technique is called CCIR-601 4:1:1. If there are 8 bits per sample, then there are 8+2+2 = 12 bits per 4:1 sub sampled CCIR-601 pixel. There are those who say that 4:1:1 should refer to sub-sampling in a scanline only. That is held in comparison with 4:2:0, which means 2:1 sub-sampling in the horizontal and vertical directions. The CCIR 601-2 4:2:2 is supposed to refer to the multiple of 3.375 Mhz the component is sampled at. For example *Y* is sampled at 4*3.375 Mhz = 13.5 Mhz. The chroma components are sampled at 2*3.375 Mhz = 6.75 Mhz. Sorry to say, the *A:B:C* notation is held as internally inconsistent (see <http://www.crs4.it/~luigi/MPEG/mpeggloss-a.html#a:b:c%20notation>).

The first step in performing the conversion from RGB to YC_RC_B is to address the gamma correction. CCIR-601 assumes a standard gamma of 2.2, while SMPTE has a standardized gamma of (1/0.45) with a linear segment for intensities less than 0.0228 [Kasson]. Our approach is to assume that the RGB quantities are already gamma corrected and that we can transform the data directly into the YC_RC_B color space. This assumption is probably wrong for images that are obtained from the network. Such images may not have a known gamma associated with them. Since we cannot know what gamma correction has been applied to an image, we shall assume that gamma correction was already performed.

The CCIR 601-2 $C_R C_B$ conversion assumes that we start with SMPTE-C RGB primaries using a D65 reference white and CCIR-709 (HDTV) non linearity. The transform is defined by

$$\begin{aligned} Y &= 0.299R + 0.587B + 0.114G \\ C_R &= R - Y \\ C_B &= B - Y \end{aligned} \tag{13.11}$$

We can recast (13.11) in the homogeneous form:

$$\begin{bmatrix} Y \\ C_R \\ C_B \end{bmatrix} = \begin{bmatrix} 0.299 & 0.587 & 0.114 \\ 0.701 & -0.587 & -0.114 \\ -0.299 & 0.587 & 0.886 \end{bmatrix} \begin{bmatrix} R \\ G \\ B \end{bmatrix} \tag{13.12}$$

The assumption is that the RGB signals are already gamma corrected and that the pixels have an 8 bit range from 0 to 255, inclusive. When the input ranges from 0 to 255 and the output ranges from 0 to 255, round-off errors and range problems lead to a scaling that results in an integer compromise:

$$\begin{bmatrix} Y \\ C_R \\ C_B \end{bmatrix} = \frac{1}{256} \begin{bmatrix} 77 & 150 & 29 \\ 131 & -110 & -21 \\ -44 & -87 & 131 \end{bmatrix} \begin{bmatrix} R \\ G \\ B \end{bmatrix} + \begin{bmatrix} 0 \\ 128 \\ 128 \end{bmatrix} \tag{13.13}$$

Because (13.13) and (13.12) both appear in the specification, the numbers reported for these color space transforms have appeared in several forms in literature [Mattison]. We take (13.12) to be the correct form when computing floating point images and so perform the transform, using

```
public class Ccir601_2cbcr extends FloatPlane {

// See CCIR 601-2
// available at
// http://www.igd.fhg.de/icib/tv/ccir/rec_601-2/scan.html
double A[][] = {
```

```
        { 0.299, 0.587,  0.114},
        { 0.701,-0.587, -0.114},
        { 0.299,-0.587,  0.886}
        };
    ...
```

Class *Ccir601_2cbcr* is identical with class *Xyzd65* , except that no gamma correction is performed and the transform matrix is different. Conversion into the XYZ D65 color space requires reliable information on the gamma used to create the RGB input, which is often not available (particularly in network sourced images).

In order to provide for the 2:1 sub sampling in chroma, we include a method in the *FloatFrame* class for resampling the *g* and *b* arrays (which are assumed to contain the chroma component):

```
   public float[][] oneDSubsampleTwoTo1(float f[][]) {
        int width = f.length;
        int height = f[0].length;
      float h[][]=new float[width][height];
      double sum = 0;

      for(int y = 0; y < height; y++) {
        for(int x = 0; x < width-2; x=x+2) {
            float a = (float)((f[x][y] + f[x+1][y]) / 2.0);
              h[x][y] = a;
              h[x+1][y] = a;
        }
      }
      return h;
   }
```

The average value between two adjacent pixels is computed as *a* and then duplicated along the row. A bi-directional chroma sub sampling, as suggested by [Mattison], is useful for creating the CCIR-601 4:2:0 sub sampling. Bi-directional chroma sub sampling involves a 2:1 sub sampling in chroma in both the horizontal and vertical directions. This is a simple modification of the doubly-nested for-loop:

```
        for(int y = 0; y < height-2; y=y+2) {
          for(int x = 0; x < width-2; x=x+2) {
             float a =
   (f[x][y]+f[x+1][y]+f[x+1][y+1]+f[x][y+1])/4f;
             h[x][y] = a;
             h[x+1][y] = a;
```

```
            h[x][y+1]  = a;
            h[x+1][y+1]  = a;
        }
    }
```

Plate 13-1 shows the *girl* image vs. 2:1 Sub sampling in chroma. Plate 13-2 shows the *girl* image vs 4:1 sub Sampling in chroma. To create a CCIR-601 4:1:1 image, see Section 13.5.

Sub sampling images in chroma represents a classic lossy perceptual coding scheme. The basic idea is that large areas in an image should have higher bandwidth allocated for edge information than for color information. In the following section, we show how NTSC takes this idea one step further.

13.2.2. YIQ and NTSC Encoders

The NTSC (National Television System Committee) color broadcast system was designed around a pre-existing monochrome analog broadcast. As a result, there were bandwidth and compatibility constraints on the addition of color to the existing market.

A modulation scheme was devised that created a basis set for the color space that could be band limited in a non-uniform fashion. The basis set is based upon human perception of color. A linear transform relates the gamma-corrected NTSC RGB colors to a new color space called *YIQ*:

$$\begin{bmatrix} Y \\ I \\ Q \end{bmatrix} = \begin{bmatrix} 0.2989 & 0.5866 & 0.1144 \\ 0.5959 & -0.2741 & -0.3218 \\ 0.2113 & -0.5227 & 0.3113 \end{bmatrix} \begin{bmatrix} R_n \\ G_n \\ B_n \end{bmatrix} \quad (13.14).$$

where Y is the luminance, I stands for In phase and Q stands for Quadrature. The in-phase signal refers to the modulation scheme that performs a *balanced modulation* of the in-phase signal against a clocking signal known as *color sub-carrier*.

Balanced modulation is a technique that creates two side-bands. They are at the frequency + carrier and at the frequency - carrier. This is also known as *double-sideband modulation*.

Sub-carrier is a crystal controlled clock that runs at a frequency of 3.579545 Mhz. The idea is that the quadrature signal results from delaying the sub-carrier by 90 degrees in phase. Then the quadrature signal is balanced modulated against the quadrature shifted sub-carrier (hence the name, *quadrature*). The sub-carrier frequency is derived from the relationship of the number of active horizontal lines divided by two (due to interlace) times the line rate. That is 455/ 2 * 15734.26374 = 3579545. The line frequency comes from the total number of lines in the image, both active and inactive, times the field rate. That is 525 * field rate = 525 * 29.97002627 Hz = 15734.26374.

The YIQ color space is not a perceptually uniform space and so the *Y* signal is allocated more bandwidth than the *I* signal. Further, the *I* signal is allocated more bandwidth than the *Q* signal. These band limits are created with low-pass filters that are applied before modulation. In fact, the *Y* signal is given about 4.2 Mhz of bandwidth, the *I* signal 1.5 Mhz, and the *Q* signal only 0.5 Mhz. Thus, NTSC color television uses a lossy analog perceptual coding technique. If 8 bits are used for the *Y* signal, then 2.8 bits are used for the *I* signal, and 0.95 bits are used for the *Q* signal. This totals to 11.75 bits per pixel. This allocation is about the same bit budget as that for CCIR 601-2 4:1:1. However, bandwidth limits do not stop at the transmitter.

Video recorders are broadly classified into two types, *direct color* and *color-under*. The direct color machine interleaves the color signal with the luminance signal for both NTSC and PAL type encoders. The color-under machine heterodynes (i.e., balance modulates) the chroma signal to a frequency in the 650-750 kHz band (with a 1.4 Mhz bandwidth). This is a bandwidth reduction and frequency shift of the chroma band to one which is below that of the luminance signal (hence the term, color-under). The luminance component on an 8 mm VTR (Video Tape Recorder), for example, is band limited to 2.5 Mhz. Thus the total record bandwidth is 3.9 MHz.

Given the initial sampling for CCIR 601-2, the 8 mm camcorder is providing only 4.76 bits for *Y,* 1.96 bits for *I,* and .665 bits for *Q,* leaving a total of 7.4 bits per pixel [Inglis].

Finally, such band limiting filters are devised to work unidirectionally. That is, they can operate only on a scanline-at-a-time. Thus, the bi-directional sub-sampling, as shown in Section 13.3 (e.g., CCIR-601 4:2:0), cannot be used to band limit an NTSC transmission. Direct color machines are typically, of professional quality and are too expensive to be generally affordable (e.g., $10k to $100k). Such machines include BetaCam, MII, 2 inch Quad and most 1 inch machines. Color-under (also known as *hetrodyne*) machines are more common in the home. They include U-Matic, U-Matic-SP, VHS, S-VHS, BetaMax, LaserDisc, and 8 mm. See the projects section at the end of this chapter for an experiment in sampling in YIQ color space.

13.3. Linear Cut Color Reduction

One subject that the last two sections touched upon is the compression of an image by bandlimiting color space. In the YIQ and CCIR-601 YC_RC_B color spaces, bandwidth was reduced by sub-sampling in chroma. In this section, we address the bandwidth reduction by reducing the dynamic range of the pixel element. This is done by cutting off bits from the pixel, least significant bits first. This is called the *linear cut* algorithm.

Consider the integral RGB color space. Each component is constrained to range from 0 to 255 and resides in a 16 bit *short* array. To perform the linear cut algorithm on such an array, we need to mask the low order bits that we want to "cut" out of the pixel. The *ColorFrame* has a method designed for this task:

```
	public void linearCut(short a[][], int numberOfBitsToCut)
   {
		int mask = 255 << numberOfBitsToCut;
		for (int x=0; x < width; x++)
			for (int y=0; y < height; y++)
```

```
                    a[x][y] = (short)(a[x][y] & mask);
    }
```

The *mask* in the *linearCut* method is computed, assuming that there are only 8 bits per color. If the programmer performs a linear cut of the last two bits, for example, we shift 255 over 2 bits (e.g., 11111111 << 2 becomes 1111111100). The mask will thus remove the least significant two bits from the *short* type number stored in *a[x][y]*. Plate 13-3 shows a linear cut in RGB space of the girl image to 5 bits per color. Note the modeling artifacts.

It is possible to combine the spatial resampling of the previous section with the linear cut algorithm. Naturally, this leads to further degradation in the image. See the Projects section (Section 13.5) at the back of this chapter for further suggestions.

13.4. The Median Cut Algorithm and Computing SNR

There are several techniques available for reducing the number of colors in an image (or image sequence). One of the more popular algorithms is the *median cut* algorithm [Heckbert 82]. The basic idea is that a color histogram of the image is computed in RGB color space. The histogram is then clustered into a *k* group. Once the clustering is performed, the pixels are mapped to the centroids of the clusters in order to minimize the color error in the image.

The goal is to have each color represent approximately the same number of pixels. This is accomplished by creating a tightly fitting color cube and then cutting it at the median of the longest axis (hence the name, median cut algorithm). The median cut procedure is applied to subcubes until there are *k* cubes. The centers of the cubes are used for the *k* output colors in the color map. Given a color map, each pixel is mapped from its original color to its nearest color neighbor. This mapping is done to minimize a color error metric. The color error metric is central to the proper functioning of such algorithms, so we spend a little time addressing it.

One way to implement the median cut algorithm is to use a queue to enable a breadth-first cutting of the sub-cubes. The idea is that an instance of the *Cube* class has a static member variable that is used to keep track of the total number of Cube instances. The pseudo-Java follows:

```
int k = 256 // number of colors
CubeQueue cq = new CubeQueue();
//get a cube that fits tightly around the list of colors.
Cube c = new Cube(colorList);
cq.enqueue(c);

while (c.ncubes < k) {
    c = cq.dequeue();
    Cube childArray[] c.split();
    cq.enqueue(childArray[0]);
    cq.enqueue(childArray[1]);
}
```

The use of the *CubeQueue* is for illustration only. The implementation in the *MedianCut* class does not make use of explicit recursion, due, in part, to the computational expense. In fact, given the a priori knowledge of the number of cubes needed, we allocate an array that is exactly *k* cubes in length. Then we simulate the queue, using an array index. Once the list of sub-cubes is known, it is a matter of assigning the colors to the centroids of the cubes that minimize the *color error*.

As initially formulated by Heckbert [Heckbert 80], the color error is measured as the Euclidean distance from the pixel's original color to the remapped color. The sum of all the errors in the remapping is the objective function whose minima is sought. In fact, this is a standard metric in *clustering*.

In the clustering problem we are given a set of points in a Euclidean space and are asked to group them into *k* partitions to minimize a distortion function.

The importance of the distortion measure choice cannot be understated. The distortion measure used by the median cut algorithm is the mean-square distortion, *D*. The mean-square distortion is computed by taking the expectation of the square of the difference between the quantized value and the actual value, then multiplying by the probability of the value. In the continuous one-dimensional domain, we write

$$D = \int_{-\infty}^{\infty} [Q(x) - x]^2 p(x) dx \tag{13.15}$$

where

$$D = \text{mean - square distortion measure}$$
$$p(x) = \text{probability of value } x \text{ and}$$
$$Q(x) = \text{quantized value for } x$$

Typically, the quantizer's performance is measured using the signal-to-noise ratio (SNR), which is given in dB as

$$SNR_{dB} = 10 \log_{10}(\sigma^2 / D) \tag{13.16}$$

where

$$\sigma^2 = \text{variance of the input}$$
$$= E(x^2) - [E(x)]^2$$

Recall that $E(x)$ is the expected value for x.

Unfortunately, distortion measures, such as the SNR, are not necessarily reflective of any physiological metric for improving the subjective appearance of an image. Hence their use is open to question. For example, a histogram equalization has been shown to improve an image's appearance; however, according to (13.16), such a process will lower the SNR. The reason that the appearance is improved may actually have to do with the improved contrast ratio of the image. Such a subjective improvement is not taken into account with (13.16).

using a discrete point set, the Heckbert quantizer modifies (13.15) to reflect the distortion function by summing the Euclidean distances between the color of each pixel and its map. This is expressed in

$$D_{tqe} = \sum_{y=0}^{height-1} \sum_{x=0}^{width-1} e_{x,y}^2 \qquad (13.17)$$

where

$$e_{x,y} = Q(C_{x,y}) - C_{x,y}$$
$$D_{tqe} = \text{ total quantization error}$$
$$C_{x,y} = \text{ color at location } x, y$$
$$Q(C_{x,y}) = \text{ quantized color at location } x, y$$

In fact, we could obtain the mean-square distortion measure from the total quantization error by dividing it by the total number of pixels, that is;

$$D = \frac{D_{tqe}}{width * height} \qquad (13.18).$$

This is computed by subtracting the original image from the quantized image, squaring the resulting error pixels, summing their color components, then dividing by the total number of pixels. The mean square error (MSE) represented by (13.18) is a widely used measure of distortion and is also called the *coding noise power*[Netravali].

Another measure of the coding is the bit rate. This is computed by the number of bits needed per pixel. One method for computing bit rate is to measure the image file size, in bits, then divide by the total number of pixels.

Perceptual difficulties aside, it is useful to have an objective fidelity citerion, such as the SNR, to use when evaluating a lossy coding scheme. In addition, SNR is one of the most used fidelity criteria. One way to compute the SNR in dB is

$$SNR_{dB} = 10 \log_{10} \left[\frac{1}{D_{tqe}} \sum_{y=0}^{height-1} \sum_{x=0}^{width-1} Q(C_{x,y}) \right] \qquad (13.19).$$

The SNR defined in (13.19) is consistent with [Weeks] and can also be used on image compression algorithms (where the quantized image is replaced with the compressed image). In the Kahindu program, we assume that the original image is contained in the child frame and that the processed image is contained in the parent frame. Thus, to compute (13.19), we use a method in the *ColorFrame* called *getSNRinDb*.

For example, the girl image shown in Plate 13-4 was median cut to 16 colors and saved as a GIF image. The resulting SNR was 17 dB using 1.6 bpp (bits per pixel). This represents a 15:1 compression ratio, since the full-color version of the girl image was a 24 bit image. Keep in mind that the GIF compression employs an lossless Lempel-Ziv compression, but it may also have some file system over-head, so the compression ratio is a conservative estimate.

Now that we have completed the preliminary discussion of distortion measurement, we can focus on the implementation of the median cut algorithm. The implementation of the median cut algorithm is optimized to handle just one kind of data: 24 bit RGB image data. As a result, modifying the implementation to work with other data-types is not trivial, but quite worth while! See the Projects section at the end of this chapter (Section 13.4).

The first step in the algorithm is to sample the original image to obtain an estimate for the PMF (Probability Mass Function) for the colors in the image. In theory, such a computation may be performed with a three-dimensional array. The 3-D array is used for creating a color histogram. Each dimension in the array should be able to span the available values in the intensity range for each color component. Such an array would, for a 24 bit image, be about 4 MB of data. For example:

```
byte colorHist[][][] = new byte[256][256][256];
```

Then for each occurrence of a color in an image, we increment that cell in the array. For example:

```
colorHist[255][0][0]++;
```

records the discovery of a red pixel. Such a solution is feasible, but only if large memory stores are available. The *MedianCut* class, which implements the median-cut algorithm, performs a linear cut on the image so that there are only 15 bits per pixel (i.e., 5 bits per color). The pixels are then packed into a single integer. The integer is used to index into an array, using

```
for (int i=0; i<width*height; i++) {
        color16 = rgb(pixels32[i]);
        hist[color16]++;
}
```

The *rgb* method strips the 32 bit pixel and turns it into a 15 bit color.

The second step is to compute the color map using the *hist* array. This is done by creating a list of tightly fitting cubes that span the color space. The cube's longest axis is cut so that the number of colors on the left side of the cut is about the same as the number of colors on the right side of the cut. The two new cubes are placed on the list of cubes and are queued for processing. This procedure repeats until there are k cubes.

The third step is to map the color of each pixel in the original image into the centroid of one of the k color cubes to minimize the distance between the original color and the quantized color. This is performed in the method, *makeInverseTable.* The *makeInverseTable* computes the centroid of each cube on the list. These are used to create three look-up tables which serve to map a color index into a 24 bit color. Once these are computed, the color for a pixel in the input image can be quickly remapped into the quantized color (this is step 4).

Computing the inverse color map can be performed optimally using an exhaustive search. Thus, the best color for a given pixel is obtained by searching through all k cubes. Such a search is done in the *makeInverseTable* method. This is a very slow, though optimal, approach. Because of the speed, *makeInverseTable* is never invoked. Instead, a fast remapping is used, as described by [Kruger]. The fast remapping is a sub-optimal remapping that allows the centroid of the color cube to represent all the colors contained in the cube. Thus the error metric used is not really the Euclidean distance, but it is close. This approximation is performed in the *makeInverseMap* method.

13.4. Summary

This chapter described a few basic color space transforms. They involved a matrix multiplication and a gamma correction. Space and time constraints do not permit a discussion of many of the important color spaces (e.g., L*u*v*, L*a*b*, HSV, HLS, HLS, etc.). For more information about color science, one of the most often cited books is [Wyszecki]. For suggestions for a color-space type project, see the Project section.

This chapter also disclosed two ways to perform color quantization: the linear-cut and median-cut algorithms. Space and time do not permit us to describe all the algorithms available for this task. Color quantization has applications in color image compression, segmentation and video indexing. Experiments for video indexing are described in the following section.

13.5. Projects

1. Modify the Kahindu program to perform the CCIR-601 4:1:1 sub-sampling in chroma. How does this compare with the 4:2:0 sub-sampling? Suppose you wanted one color for every 8 pixels. Write another version of your code to resample chroma 8:1.

2. Modify the Kahindu program to simulate color-under video resampling in the YIQ color space.

3. Write a routine that thresholds all regions in an image that contains “hot video” colors. Hot colors are defined as all colors that have a chroma magnitude greater than

a limit, i.e., $\sqrt{I^2 + Q^2} > \text{limit}$. Is this a good way to perform color based edge detection? Why or why not? See [Martindale] for more information.

4. Implement the statistical color quantization technique of Wu in the Kahindu package. Compare it with the median cut and linear quantization algorithms. Which is better? See [Wu] for details. For sample code, in C, <http://www.uni-kiel.de:8080/Logik/persons/mc/quant.c> might be of help.

5. Implement the octree color quantization technique of Thomas [Thomas]. Compare it with at least two other color quantization techniques.

6. Reimplement the median-cut algorithm to work with an error metric formulated in the YUV color space. Quantize several test images. Compare the SNR's of the quantized images using both the YUV and RGB error metrics. Which color space leads to better SNR? Which error metric leads to better looking images?

7. In Xiaolin Wu's paper on image quantization [Wu 97], he suggests a diagonal resampling of an image, using

$$\mu_{i,j} = \left\lfloor \frac{I_{2i,2j} + I_{2i+1,2j+1}}{2} \right\rfloor$$

for

$$0 \le i \le \frac{W}{2}, 0 \le j < \frac{H}{2}$$

Apply this resampling in chroma for a YUV image. Compare your results with CCIR-601 4:2:2. Which looks better? Now compare the SNR. Try several images.

8. Use color quantization to compute look-up tables for an image sequence. Compare the look-up tables for images that represent two different scenes. Compare the look-up tables for images that are contained within the same scene. Use SNR to compare the look-up tables. Can SNR between look-up tables be used as a good indicator

between scene changes? Experiment with several image sequences and justify your answer with data.

9. Use the k-d range searching techniques described in [Bentley] to speed up the creation of an inverse color map in the median cut algorithm. Compare the SNR of several test images after the "optimal" color quantization with the SNR of the test images after the fast-remapping. How much better is the SNR? How much slower is the optimal quantization? Which images look better?

10. Section 10.3 discussed the linear cut algorithm for compressing an image. What happens if you combine the linear cut algorithm with resampling? Try 4:2:0 resampling of a linear cut image. Are there more colors or fewer? Why? Now modify the 4:2:0 resampling so that no averaging occurs. Save the images to GIF files. What are the compression ratios, SNR's and run-times for your results? Show the images.

11. One technique for color quantization is the use of a Kohonen Neural Network. Such systems use learning to match the distribution of colors in an input image. Implement this technique, as described in <http://www.ozemail.com.au/~dekker/NEUQUANT.HTML>. Compare your results using several test images. What are the compression ratios, SNR's and run-times for your results? Show the images.

12. Another technique for color quantization is the popularity algorithm. The popularity algorithm selects the most used colors as the ones to be employed for color quantization. An implementation of the popularity algorithm may be found in [Watkins]. Implement the popularity algorithm. Compare your results using several test images. What are the compression ratios, SNR's and run-times for your results? Show the images.

13.6. The *ColorHash* Class

One problem that we mentioned in the median-cut algorithm was that of counting the number of colors in an image. It uses too much ram to build a 8x8x8 histogram matrix for a 24 bit image. The number of colors in an image is strictly bound by the number of pixels and is generally much smaller than the number of pixels.

Since the 8x8x8 histogram matrix would be sparse, we employ a hash algorithm for counting the number of colors in an image. This algorithm is implemented in the *ColorHash* class. The *ColorHash* class resides in the *gui* package and provides an object oriented storage and processing facility for instances of the *Color* class.

13.6.1. Class Summary

```
package gui;
public class ColorHash extends Hashtable {

   public void
         addShortArrays(
               short r[][],
               short g[][],
               short b[][])
   public int countColors()
   public void printColors()
   public Vector makeVector()
}
```

13.6.2. Class Usage

The default constructor for the *ColorHash* class takes no arguments. To create an instance of the *ColorHash* class, use

```
ColorHash ch = new ColorHash();
```

To add instances of colors to the *ch* instance, use

```
short r[][];
short g[][];
```

```
short b[][];
ch.addShortArrays(r, g, b);
```

Once the colors have been added to the *ColorHash* instance, we can count the total number of different colors, using

```
    int number of colors = ch.countColors();
```

Also, very useful, we can obtain a list of the unique colors in an image, using:

```
Vector colorList = ch.makeVector();
```

To print all the colors in an image to the console, use

```
ch.printColors();
```

13.7. The *ColorFrame* Class

The *ColorFrame* class extends the *MartelliFrame* class and resides in the *gui* package. It provides the service of color conversion and color quantization of an image. While the majority of the implementations do not reside in the *ColorFrame* class, they are all presented with a unified API from within the *ColorFrame* class. The color conversions available from the *ColorFrame* class are by no mean complete. As more are added, the number of conversion methods is likely to grow. The same may be said for the color quantizations available to the *ColorFrame*. Currently, only the linear-cut and median-cut algorithms are implemented, but there are plans to extend these features, too.

13.7.1. Class Summary

```
package gui;
import java.awt.*;
import java.awt.event.*;

public class ColorFrame extends MartelliFrame {
   ColorFrame(String title);
   public void printSNR()
   public double getSNRinDb()
   public float getSNR()
   public float getTotalSignalPower()
   public float getTotalNoisePower()
   public void copyToFloatPlane()
   public void subSampleChroma4To1()
```

```
    public void subSampleChroma2To1()
    public void rgb2Ccir601_2cbcr()
    public void ccir601_2cbcr2rgb()
    public void rgb2xyzd65()
    public void xyzd652rgb()
    public void rgb2iyq()
    public void iyq2rgb()
    public void rgb2hsb()
    public void hsb2rgb()
    public void rgb2yuv()
    public void yuv2rgb()
    public void medianCut(int k)
    public void linearCut(int sr, int sg, int sb)
    public void linearCut(
          short a[][], int numberOfBitsToKeepAtHead)
    public int [] getPels()
    public void printNumberOfColors()
    public int computeNumberOfColors()
    public void printColors()
}
```

13.7.2. Class Usage

To make an instance of a color frame, use

```
String title = "colorFrame";
ColorFrame cf = new ColorFrame(title);
```

To print the SNR, in dB, to the console (assuming that the child frame has the original image):

```
cf.printSNR();
```

To get the SNR, in dB, (assuming that the child frame has the original image):

```
double snr = cf.getSNRinDb();
```

To get the SNR, not in dB, (assuming that the child frame has the original image):

```
float snr = cf.getSNR();
```

To get the signal power, as measured by the sum of the square of the values in each color in each pixel, use

```
float totalSignalPower = cf.getTotalSignalPower();
   public float getTotalSignalPower()
```

To get the total noise power, as measured relative to the square error between the current image and the child image, use

```
float totalNoisePower = cf.getTotalNoisePower();
```

To copy the short arrays held internally in the *ColorFrame* instance to an internally held instance of the *FloatPlane* , use

```
cf.copyToFloatPlane();
```

Typically, the *short* resolution of the pixels in the frames discussed in this book is sufficient for most image processing. The *FloatPlane* is a repository for floating-point color images. It is used to enable high-precision computation (such as exact color transforms) without having to worry about round-off error.

To subsample the image using CCIR-601-2 4:2:0 (i.e., horizontally and vertically) use

```
cf.subSampleChroma4To1();
```

To subsample the image using CCIR-601-2 4:1:1 (i.e., 2:1 in chroma) use

```
cf.subSampleChroma2To1();
```

To convert the image to the internally held *FloatPlane* instance in CCIR-601-2 YCbCr format, use

```
cf.rgb2Ccir601_2cbcr();
```

To convert from the CCIR-601-2 YCbCr *FloatPlane* instance back into the *ColorFrame* instance, use

```
cf.ccir601_2cbcr2rgb();
```

To convert the *ColorFrame* image data to the internally held *FloatPlane* instance in XYZ D65 format, use

```
cf.rgb2xyzd65();
```

To convert from the XYZ D65 *FloatPlane* instance back into the *ColorFrame* instance, use

```
cf.xyzd652rgb();
```

To convert the *ColorFrame* image data (assumed to be gamma-corrected SMPTE-C RGB) to the internally held *FloatPlane* instance in SMPTE YIQ format, use

```
cf.rgb2iyq();
```

To convert from the SMPTE YIQ *FloatPlane* instance back into the *ColorFrame* instance, use

```
cf.iyq2rgb();
```

To convert the *ColorFrame* image data to the internally held *FloatPlane* instance in Hue Saturation and Brightness (HSB) format, use

```
cf.rgb2hsb();
```

To convert from the HSB *FloatPlane* instance back into the *ColorFrame* instance, use

```
cf.hsb2rgb();
```

For more information about the HSB conversion, see the *Color* class in the Java *awt* package. The same algorithm is used here. The brightness, hue and saturation are all normalized to range from 0..1. Time and space do not permit a complete disclosure of the HSB color space.

To convert the *ColorFrame* image data to the internally held *FloatPlane* instance in YUV format, use

```
cf.rgb2yuv();
```

To convert from the YUV *FloatPlane* instance back into the *ColorFrame* instance, use

```
cf.yuv2rgb();
```

See [Martindale] for more information about the YUV format. The YUV conversion assumes that the illuminant is D65 and that the RGB primaries are EBU reference primaries. This is known as PAL encoding. The YUV encoding should probably be used only when the primaries are known to be of a PAL source. Otherwise, the matrix weightings (and possibly the gamma) will have to be altered.

To quantize the number of colors to *k* values using the median cut algorithm, use

```
int k = 256;
cf.medianCut(k);
```

To keep the most significant *sr* bits in red, *sb* bits in blue and *sg* bits in green, use

```
int sr = 3;
int sg = 4;
int sb = 2;
cf.linearCut(sr,sg,sb);
```

To keep the first *b* bits in an array of *short* (and alter the array), use

```
short a[][];
cf.linearCut(a,b);
```

To get the 3 arrays of *short* RGB pixels packed into a long 32 bit array, use

```
pels = cf. getPels();
```

To print the number of colors in an RGB image, use

```
cf.printNumberOfColors();
```

To compute the number of colors in the RGB image of the *ColorFrame*, use

```
int n = cf.computeNumberOfColors();
```

To print the colors in a *ColorFrame* instance, use

```
cf.printColors()
```

14. Warping

There are 2 possible outcomes:
If the result confirms the hypothesis,
then you've made a measurement.
If the result is contrary to the hypothesis,
then you've made a discovery.

– Enrico Fermi

The process of image warping is a kind of image processing that causes pixels to change their location. Typically, a mapping function describes the relationship between the coordinates in a source and target image. In this chapter, we address homogenous transformations as applied in two-dimensions. The homogenous transforms include scaling, translation, rotation and shear which, collectively are special cases of *affine* transforms. The dictionary defines "affine" as preserving *finiteness*. That is, for the formula $y = ax + b$, where a is not equal to zero, a finite value of y will result from a finite value of x and vice versa.

The implementation seen here is done without the use of the *Java2D* or Java advanced imaging packages. This means that the code will work wherever JDK 1.1 is implemented. It also means that once JDK 1.2 becomes widely available, we can retrofit the code to take advantage of the (hopefully) higher-speed implementations available.

14.1. Translation

The simplest of transforms is that of translation. Using homogeneous coordinate transforms, we are able to treat every point in an image in the same manner (hence the name, homogeneous, which means *uniform*). The translation transform is given by

$$\begin{bmatrix} x' \\ y' \\ 1 \end{bmatrix} = \begin{bmatrix} 1 & 0 & t_x \\ 0 & 1 & t_y \\ 0 & 0 & 1 \end{bmatrix} \begin{bmatrix} x \\ y \\ 1 \end{bmatrix} \tag{14.1}$$

To see how this works, simply multiply the matrix out and note that

$$\begin{aligned} x' &= x + t_x \\ y' &= y + t_y \end{aligned} \tag{14.2}$$

The 2-D homogeneous transform is represented by a 3x3 matrix:

$$A = \begin{bmatrix} a_{00} & a_{01} & a_{02} \\ a_{10} & a_{11} & a_{12} \\ a_{20} & a_{21} & a_{22} \end{bmatrix} \tag{14.3}$$

Using (14.3) we write a method into the *Mat3* class that sets the elements of the A matrix in accordance with (14.1). This is reflected in the following snippet:

```
package gui;
public class Mat3 {
   double a[] []  = new double[3][3];

   private static final double piOn180 = Math.PI / 180.0;

public void setTranslation(double tx, double ty) {
   a[0][0] = 1;
   a[1][1] = 1;
   a[2][2] = 1;
   a[0][2] = tx;
   a[1][2] = ty;
}
```

To effect the multiplication, shown in (14.1), we employ a simple routine for multiplying the matrix.

When the A matrix, given in (14.3), is *nonsingular,* then the transform is called a nonsingular linear homogeneous transformation of the plane. Recall that a matrix is singular iff its determinant is zero [Pettofrezzo].

14.2. Scaling

With the scaling of a point, about the origin, we use a transform of the form

$$\begin{bmatrix} x' \\ y' \\ 1 \end{bmatrix} = \begin{bmatrix} s_x & 0 & 0 \\ 0 & s_y & 0 \\ 0 & 0 & 1 \end{bmatrix} \begin{bmatrix} x \\ y \\ 1 \end{bmatrix} \quad (14.4).$$

Multiplying (14.4) yields

$$\begin{aligned} x' &= s_x x \\ y' &= s_y y \end{aligned} \quad (14.5).$$

The basic idea is that the scale factors will scale relative to the origin. If the goal is to scale relative to any point, then the point must be translated to the origin, scaled, then translated back. This is done by matrix *contatenation*. We concatenate the transforms by multiplying them together. Thus, to scale about any point, use

$$\begin{bmatrix} x' \\ y' \\ 1 \end{bmatrix} = \begin{bmatrix} 1 & 0 & t_x \\ 0 & 1 & t_y \\ 0 & 0 & 1 \end{bmatrix} \begin{bmatrix} s_x & 0 & 0 \\ 0 & s_y & 0 \\ 0 & 0 & 1 \end{bmatrix} \begin{bmatrix} 1 & 0 & -t_x \\ 0 & 1 & -t_y \\ 0 & 0 & 1 \end{bmatrix} \begin{bmatrix} x \\ y \\ 1 \end{bmatrix} \quad (14.6)$$

Note that the transform is multiplied out before the individual points are transformed. This saves a great deal of computation. Combining the terms in (14.6) yields

$$\begin{bmatrix} x' \\ y' \\ 1 \end{bmatrix} = \begin{bmatrix} s_x & 0 & t_x \\ 0 & s_y & t_y \\ 0 & 0 & 1 \end{bmatrix} \begin{bmatrix} 1 & 0 & -t_x \\ 0 & 1 & -t_y \\ 0 & 0 & 1 \end{bmatrix} \begin{bmatrix} x \\ y \\ 1 \end{bmatrix} \quad (14.7)$$

which simplifies to

$$\begin{bmatrix} x' \\ y' \\ 1 \end{bmatrix} = \begin{bmatrix} s_x & 0 & t_x - s_x t_x \\ 0 & s_y & t_y - s_y t_y \\ 0 & 0 & 1 \end{bmatrix} \begin{bmatrix} x \\ y \\ 1 \end{bmatrix} \qquad (14.8).$$

Thus,

$$\begin{aligned} x' &= s_x x + t_x - s_x t_x = s_x(x - t_x) + t_x \\ y' &= s_y y + t_y - s_y t_y = s_y(y - t_y) + t_y \end{aligned} \qquad (14.9).$$

Hence, we have translated the components, scaled them, then translated them back.

In Java, we can set the scaling parameters using

```
public void setScaling(double sx, double sy) {
   a[0][0] = sx;
   a[1][1] = sy;
   a[2][2] = 1;
}
```

Note that this does not concatenate the transform, it simply sets it. As an example, consider the following test program:

```
public static void main(String args[]) {
   Mat3 tr1 = new Mat3();
   Mat3 tr2 = new Mat3();
   Mat3 sc = new Mat3();
   Mat3 at ;

   tr1.setTranslation(1,1);
   sc.setScale(2,2);
   tr2.setTranslation(-1,-1);

   at = tr1.multiply(sc);
   at = at.multiply(tr2);
   at.print();

}
```

which outputs

```
2.0 0.0 -1.0
0.0 2.0 -1.0
0.0 0.0 1.0
```

This provides a satisfying check against (14.8).

14.3. Rotation

Two-dimensional rotation about the origin is performed, using

$$\begin{bmatrix} x' \\ y' \\ 1 \end{bmatrix} = \begin{bmatrix} \cos\theta & -\sin\theta & 0 \\ \sin\theta & \cos\theta & 0 \\ 0 & 0 & 1 \end{bmatrix} \begin{bmatrix} x \\ y \\ 1 \end{bmatrix} \quad (14.10).$$

Multiplying (14.10) yields

$$\begin{aligned} x' &= x\cos\theta - y\sin\theta \\ y' &= x\sin\theta + y\cos\theta \end{aligned} \quad (14.11)$$

We can show (14.11) using the double angle formulas for sine and cosine, namely;

$$\begin{aligned} \cos(\theta+\phi) &= x\cos\theta - y\sin\theta = \cos\phi\cos\theta - \sin\phi\sin\theta \\ \sin(\theta+\phi) &= x\sin\theta + y\cos\theta = \cos\phi\sin\theta - \sin\phi\cos\theta \end{aligned} \quad (14.12)$$

where

$$\begin{aligned} x &= \cos(\phi) \\ y &= \sin(\phi) \end{aligned} \quad (14.13).$$

In Java, we write

```
public void  setRotation(double theta) {
    theta = theta  * Math.PI/180;
    double cas = Math.cos(theta);
    double sas = Math.sin(theta);
    a[0][0] =  cas;
    a[1][1] =  cas;
    a[0][1] = -sas;
    a[1][0] =  sas;
}
```

Note that we have specified the angle in degrees, then converted it into radians. Concatenation of the rotation matrix proceeds as in the scaling or translation case. That is, to rotate about any point, we first translate the point to the origin, perform the rotation, then translate back. Rotation and scaling can be concatenated into a single affine transform.

14.4. Shear

The shear transform in the x and y directions is given by

$$\begin{bmatrix} x' \\ y' \\ 1 \end{bmatrix} = \begin{bmatrix} 1 & sh_x & 0 \\ sh_y & 1 & 0 \\ 0 & 0 & 1 \end{bmatrix} \begin{bmatrix} x \\ y \\ 1 \end{bmatrix} \tag{14.14}$$

Expanding (14.14) yields

$$\begin{aligned} x' &= x + sh_x y \\ y' &= y + sh_y y \end{aligned} \tag{14.15}$$

The shear in the x direction is seen as a translation by a scale factor in the y component (and vice versa). In Java, we write

```
public  void setShear(double shx, double shy) {
    a[0][0] = 1;
    a[1][1] = 1;
    a[2][2] = 1;
    a[0][1] = shx;
    a[1][0] = shy;
}
```

The shear matrix may be concatenated just like the other transforms. That is, translation to the origin is the first step, then the affine transform is applied, and finally, the point is translated back.

14.5. The AffineFrame

There are no good interfaces pre-built in the Java AWT for specifying the parameters of the affine transform. In the past, we allowed users to type in the parameters using a dialog box [Lyon and Rao]. This turned out to be a non-intuitive interface, particularly with regard to shear and scale factors. Further, it takes so long to manipulate an image that some means of quickly previewing the result is needed to give the program an interactive feel. As a result, we have devised an interface that allows the user to exercise the different transforms. This interface resides in the

AffineFrame, a class that resides in the *gui* package. The *AffineFrame* extends the *ShortCutFrame* and provides for several menu options, which are shown in Fig. 14-1.

Transform menu He
[r]otate
[s]cale - xy
[X]scale - x
[Y]scale - y
[*]scalexy+rotate
[x]shear x
[y]shear y
[R]otate+shear
[a]pply transform

Fig. 14-1. The *AffineFrame* Menu

The user must have the ability to control which parameters are altered when moving the mouse. For example, changing rotation and shear at the same time gives the feeling of a loss of control. Fig. 14-2 shows the rotation transform as applied to the quadrilateral.

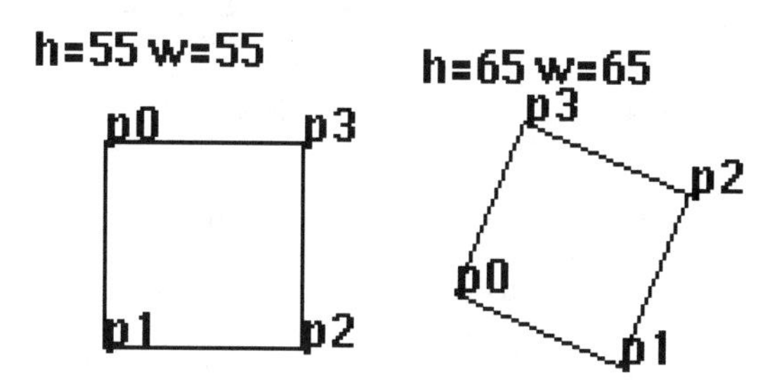

Fig. 14-2. The Quadrilateral, Before and After Rotation

Fig. 14-3 shows the uniform scaling performed on the quadrilateral. Uniform scaling will maintain the aspect ratio of an image. The corner points of the cube are labeled in accordance with the numbering of the points in the *Polygon* instance. The most tightly fitting bounding box for the polygon yields a height and width for the new image after the transform. The bounding box dimensions are shown, so the user knows how large the image will be after the transform is applied. The dimensions and corner-point labels float around the quadrilateral in a manner consistent with its position, size and orientation.

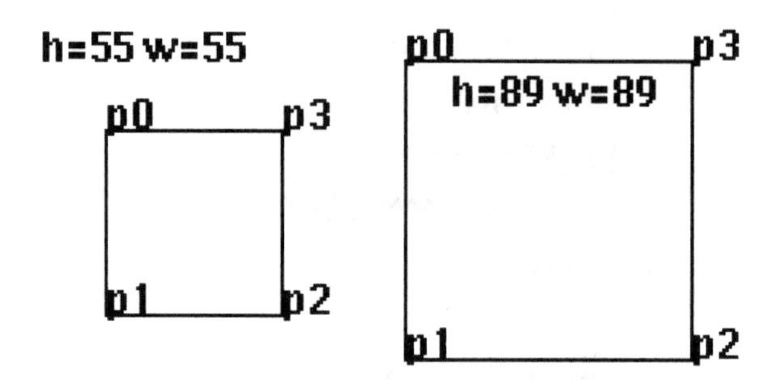

Fig. 14-3. Before and After Scaling in x and y.

Fig. 14-4 shows the non-uniform scaling in x vs. y.

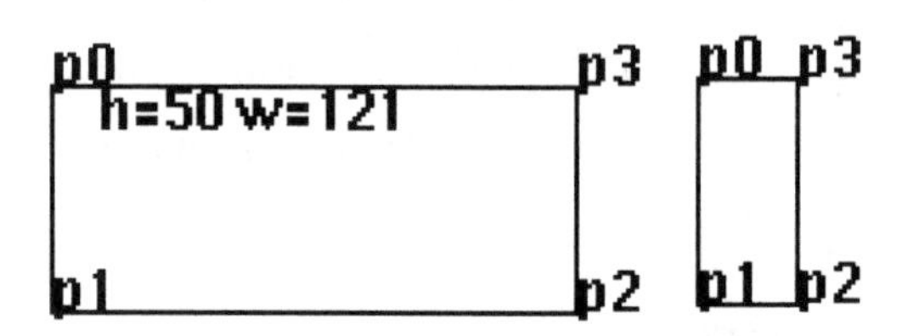

Fig. 14-4. Scale in x vs. Scale in y.

Fig. 14-5 shows the shear in x vs. y. All the affine operations are computed and redrawn with a high update rate (i.e., no flicker).

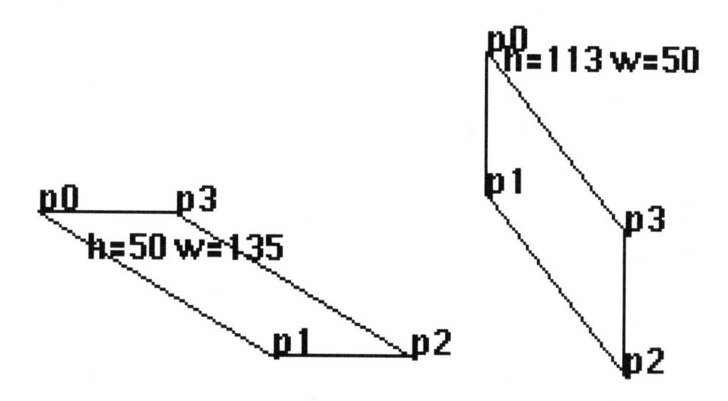

Fig. 14-5. Shear in x vs. Shear in y.

Various affine transforms may be cascaded by applying them to the underlying geometry of the quadrilateral. The *Mat3* class performs its affine transform on each point in a polygon, using the *transform* method:

```
public  Polygon transform(Polygon p ) {
   Polygon pp = new Polygon();
   int x[];
   for (int i = 0; i < p.npoints; i++) {
         x = multiply(p.xpoints[i],p.ypoints[i]);
         pp.addPoint(x[0],x[1]);
   }
   return pp;
}
```

The *AffineFrame* class has a *paint* method that transforms the polygon and makes a new instance of the polygon every time the polygon is drawn:

```
public void paint(Graphics g) {
      g.drawPolygon(at.transform(p));
}
```

where *at* is an affine transform (an instance of the *Mat3* class). This makes the application of the affine transform to all the points in the polygon a trivial matter:

```
public void apply() {
      p = at.transform(p);
}
```

Once the points on the polygon are set, we can perform an image warping.

14.6. Applying the Transforms to an Image

The last section described how basic 2D affine transforms are used to translate, rotate, scale and shear a simple 2D geometry. In this section, we are given an affine transform and must find a way to apply the transform to an image. In the last section, points in the source space were multiplied by a 3x3 matrix in order to obtain a mapping into a destination space. This is called a *forward mapping*. A forward mapping uses a transform to take points in the source space and map them into the destination space.

When forward mapping is applied to pixel coordinates in an image, it can lead to dark pixels in the image. This is caused by producing floating point coordinates for the destination pixels. The destination pixels are positioned at integer coordinates, and the act of rounding or truncating the floating point coordinates means that some pixels will be unchanged during the mapping. As a result, we use *destination scanning*. Destination scanning performs an inverse mapping from the destination image into the source image. To implement destination scanning, given a transformation matrix, it is necessary to perform a matrix inversion. The *xform* method in the *XformFrame* class shows an implementation of destination scanning:

```
public void xform(Mat3 transform) {
  int w = width;
  int h = height;
  short rn[][] = new short[w][h];
  short gn[][] = new short[w][h];
  short bn[][] = new short[w][h];
  int p[] = new int [3];
  int red, green, blue;
  int xp, yp, i, j;
  transform = transform.invert();
  for (int x = 0; x < w; x++)
    for (int y=0; y < h; y++) {
      p=transform.multiply(x,y);
      xp = (int) p[0];
      yp = (int) p[1];
      if ((xp < w) && (yp < h) && (xp >= 0) && (yp >= 0)) {
        rn[x][y] = r[xp][yp];
        gn[x][y] = g[xp][yp];
        bn[x][y] = b[xp][yp];
      }
  }
  r = rn;
  g = gn;
  b = bn;
  short2Image();
}
```

Using the *xform* method in the *XformFrame,* we can modify the *AffineFrame* to perform a forward transform on the vector geometry while invoking an inverse transform on the image:

```
public void apply() {
      p = at.transform(p);
      xf.xform(at);
}
```

Recall that *at* is the affine transform. An instance of the invoking *XformFrame* is stored in the *xf* variable. Thus, when a user invokes *apply* in the *AffineFrame*, the local geometry and the image are transformed. Fig. 14-6 shows an example of the rotation transform. Fig. 14-7 shows image scaling in *x* vs. *y*. Fig. 14-8 shows an example of shear in the *x* direction. Fig. 14-9 shows an example of shear in the *y* direction.

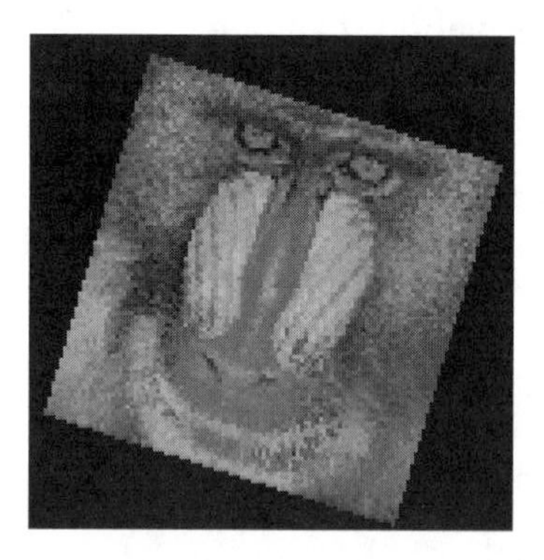

Fig. 14-6. Rotation

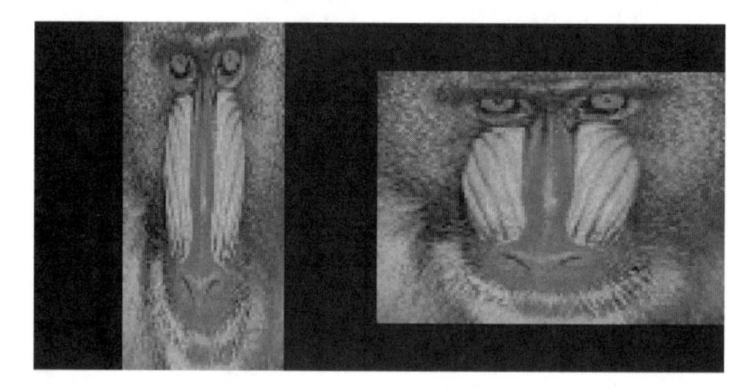

Fig. 14-7. Scale in the *x* vs. *y* Direction

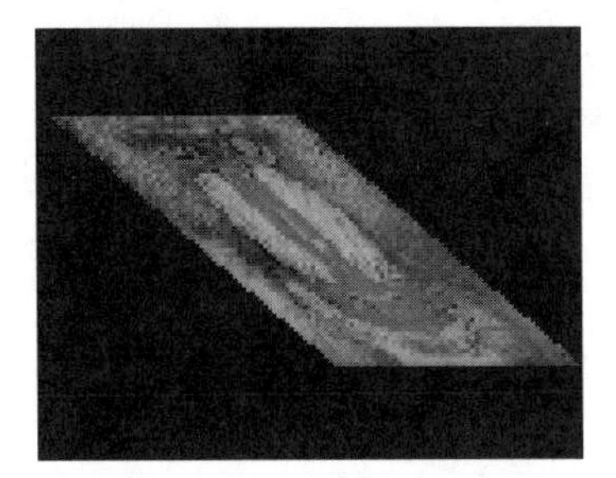

Fig. 14-8. Shear in the x Direction

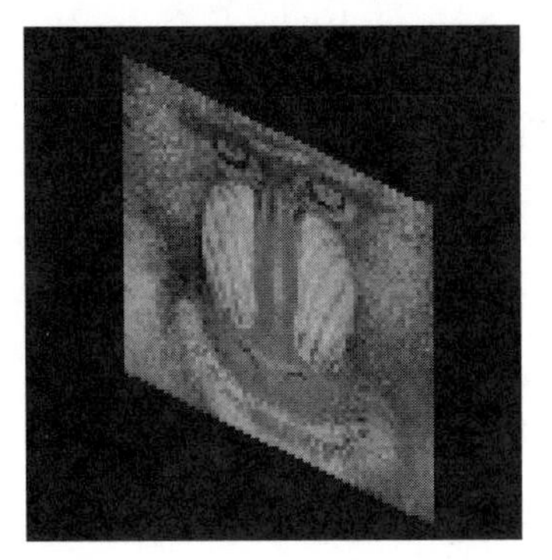

Fig. 14-9. Shear in the y Direction

In this section, we used an inverse mapping and simple point sampling to determine the intensity of a pixel. See the Project section at the back of this chapter for more advanced sampling exercises (Section 14.10).

14.7. Inferring a 3 Point Affine Mapping

In the last section we assumed that the affine transform was given. To formulate the inverse transform, we needed only to find the inverse of a 3x3 matrix. In this section we are given 3 points in the source and destination image and have to find the affine transform.

Recall that the affine transform was given by

$$P' = AP \tag{14.16}$$

where the destination point is given by

$$P' = \begin{bmatrix} x' \\ y' \\ 1 \end{bmatrix} \tag{14.17},$$

the source point is given by

$$P = \begin{bmatrix} x \\ y \\ 1 \end{bmatrix} \tag{14.18}$$

and

$$A = \begin{bmatrix} a_{00} & a_{01} & a_{02} \\ a_{10} & a_{11} & a_{12} \\ a_{20} & a_{21} & a_{22} \end{bmatrix} \tag{14.19}.$$

Suppose that A is unknown, but that we are given 3 points. In that case, we could set up a system of linear equations:

$$\begin{bmatrix} x'_0 & x'_1 & x'_2 \\ y'_0 & y'_1 & y'_2 \\ 1 & 1 & 1 \end{bmatrix} = \begin{bmatrix} a_{00} & a_{01} & a_{02} \\ a_{10} & a_{11} & a_{12} \\ a_{20} & a_{21} & a_{22} \end{bmatrix} \begin{bmatrix} x_0 & x_1 & x_2 \\ y_0 & y_1 & y_2 \\ 1 & 1 & 1 \end{bmatrix} \tag{14.20}$$

and if the A matrix is non-singular, we can solve for A inverse, using

$$\begin{bmatrix} a_{00} & a_{01} & a_{02} \\ a_{10} & a_{11} & a_{12} \\ a_{20} & a_{21} & a_{22} \end{bmatrix}^{-1} = \begin{bmatrix} x_0 & x_1 & x_2 \\ y_0 & y_1 & y_2 \\ 1 & 1 & 1 \end{bmatrix} \begin{bmatrix} x'_0 & x'_1 & x'_2 \\ y'_0 & y'_1 & y'_2 \\ 1 & 1 & 1 \end{bmatrix} \tag{14.21}.$$

Recall that A inverse is exactly what we needed to perform an image transform. Suppose that

$$D_3 = AS_3 \tag{14.22}.$$

Where the 3 points in the destination image are given by

$$D_3 = \begin{bmatrix} x'_0 & x'_1 & x'_2 \\ y'_0 & y'_1 & y'_2 \\ 1 & 1 & 1 \end{bmatrix} \quad (14.23)$$

and the 3 points in the source image are given by

$$S_3 = \begin{bmatrix} x_0 & x_1 & x_2 \\ y_0 & y_1 & y_2 \\ 1 & 1 & 1 \end{bmatrix} \quad (14.24).$$

then the forward transform is given by

$$A = D_3 S_3^{-1} \quad (14.25)$$

iff S_3 is non-singular.

In Java, we infer the *A* matrix by using a *Polygon* instance as a point storage facility. The source points come from a *Polygon* instance called *sp* (for source polygon). The destination points come from a *Polygon* instance called *dp* (for destination polygon):

```
public Mat3 infer3PointA(Polygon sp, Polygon dp) {
 // D3 is destination
 // S3 is source
 double s3 [][] ={ // (14.24)
  {sp.xpoints[0],sp.xpoints[1],sp.xpoints[2]},
  {sp.ypoints[0],sp.ypoints[1],sp.ypoints[2]},
  {1           ,            1,           1}};
 double d3 [][] ={ // (14.23)
  {dp.xpoints[0],dp.xpoints[1],dp.xpoints[2]},
  {dp.ypoints[0],dp.ypoints[1],dp.ypoints[2]},
  {1           ,            1,           1}};
 Mat3 d3Mat = new Mat3(d3);
 Mat3 s3Mat = new Mat3(s3);
 Mat3 s3MatInverse = s3Mat.invert();
 Mat3 a = d3Mat.multiply(s3MatInverse); //(14.25)
 return a;
}
```

With 3 point affine mappings, any triangular outline of a source image may be transformed into any triangular outline in the destination image. Further, a rectangle in the source image may be transformed into a parallelogram in the destination image

[Heckbert 89]. To test the 3 point mapping, we use the first 3 points of the *Polygon* instance in the *AffineFrame* as our destination points. For the source points, we use 3 points from the corners of the source image.

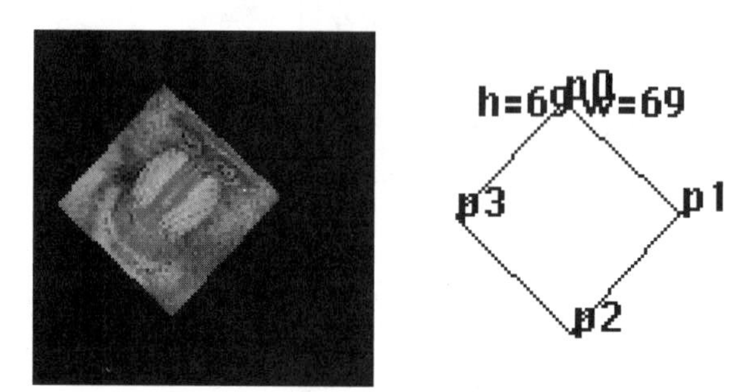

Fig. 14-10. Inferring the Transform from 3 Points

Fig. 14-10 shows the result of inferring the transform from the first 3 points in the *AffineFrame* instance. The Java snippet for this resides in the *XformFrame*:

```
public void applyAffineFrameThreePoints() {
  Polygon sourcePoly = new Polygon();
  sourcePoly.addPoint(0,0);
  sourcePoly.addPoint(width,0);
  sourcePoly.addPoint(width,height);
  xform(
        infer3PointA(sourcePoly,
              af.getPolygon()));
}
```

The *af* variable contains an instance of the *AffineFrame*. The *sourcePoly* instance consists of three *corresponding* points. Once we infer the *A* matrix, we can perform the transform, just like the last section.

For further work with 3 point affine maps, the reader is directed to the Projects section at the back of this chapter (Section 14.10).

14.8. Inferring a 4 Point Bilinear Map

In the last section, we were given 3 points in the source and destination image and we found the affine mapping that would enable triangles to be mapped into triangles. In this section, we are given 4 points in the source and destination image. Our objective is to map quadrilaterals into quadrilaterals.

Thus we are seeking a bilinear mapping from one four-point shape into another. Also, because we are performing destination scanning, we must find the inverse of the bilinear mapping. In the previous sections, we examined special cases of affine transforms.

In the last section, we saw an example for the forward affine transform mapping from P into P'. This took the form of a matrix multiplication:

$$P' = AP \quad (14.26).$$

where A was a 4x4 matrix. Now we assume that we have a bilinear model for the mapping between the source plane and the destination plane. Thus, for the bilinear model, we define a mapping function of the form;

$$\begin{bmatrix} x' \\ y' \end{bmatrix} = \begin{bmatrix} a_{00} & a_{01} & a_{02} & a_{03} \\ a_{10} & a_{11} & a_{12} & a_{13} \end{bmatrix} \begin{bmatrix} x \\ y \\ xy \\ 1 \end{bmatrix} \quad (14.27).$$

Equation (14.27) is linear in x and linear in y, or bilinear in x and y, hence the name, *bilinear transformation* [Churchill]. Equation (14.27) shows A as a 2x4 multiplied by a 4x1 P matrix of source coordinates. The resulting P' matrix is a 2x1. Equation (14.27) also shows that there are 8 degrees of freedom, so we will need 4 points in both the source and destination plane to solve for the A matrix. The system of equations is described by

$$D_4 = AS_4 \quad (14.28)$$

The four destination points are given by the 2x4:

$$D_4 = \begin{bmatrix} x'_0 & x'_1 & x'_2 & x'_3 \\ y'_0 & y'_1 & y'_2 & y'_3 \end{bmatrix} \quad (14.29)$$

and the four source points are given by the 4x4:

$$S_4 = \begin{bmatrix} x_0 & x_1 & x_2 & x_3 \\ y_0 & y_1 & y_2 & y_3 \\ x_0y_0 & x_1y_1 & x_2y_2 & x_3y_3 \\ 1 & 1 & 1 & 1 \end{bmatrix} \quad (14.30)$$

Just like the case of the simple inverse affine mapping, iff (14.30) is non-singular, then the A matrix is given by

$$A = D_4 S_4^{-1} \quad (14.31).$$

The *XformFrame* class uses the *infer4PointA* method to compute (14.31) from a source and destination polygon.

```
public double[][] infer4PointA(Polygon sp, Polygon dp) {
// D is destination
// S is source
int xd[] = dp.xpoints;
int yd[] = dp.ypoints;
int xs[] = sp.xpoints;
int ys[] = sp.ypoints;

// d4 is a 2x4 (14.29)
double d4 [][] ={
 {xd[0],xd[1],xd[2],xd[3]},
 {yd[0],yd[1],yd[2],yd[3]},
};
// s4 is a 4x4 (14.30)
double s4[][]={
 {      xs[0],       xs[1],       xs[2],       xs[3]},
 {      ys[0],       ys[1],       ys[2],       ys[3]},
 {xs[0]*ys[0], xs[1]*ys[1], xs[2]*ys[2], xs[3]*ys[3]},
 {          1,           1,           1,           1},
};
Mat4 s4Mat = new Mat4(s4);
Mat4 s4MatInverse = s4Mat.invert();
// 2x4*4x4 = 2x4 (14.31)
double [][] a = s4MatInverse.multiply2x4(d4);
return a;
}
```

Unlike the case of the simple inverse affine mapping, however, the inverse mapping of the forward transform,

$$P' = AP \qquad (14.26).$$

is not obtained by inverting the A matrix (since the A matrix is a non-square matrix and AP is bilinear). There are several ways to approach this problem. Assuming that the source or destination shapes are squares is one way to obtain a simplification (see the Projects section for more information on this). Generally, the inverse mapping of:

$$\begin{bmatrix} x' \\ y' \end{bmatrix} = \begin{bmatrix} a_{00} & a_{01} & a_{02} & a_{03} \\ a_{10} & a_{11} & a_{12} & a_{13} \end{bmatrix} \begin{bmatrix} x \\ y \\ xy \\ 1 \end{bmatrix} \qquad (14.27)$$

is described by

$$x = \frac{x' - a_{10}y - a_{03}}{a_{00} + a_{02}y} \qquad (14.32)$$

and

$$y = \frac{-B \pm \sqrt{B^2 - 4A_sC}}{2A_s} \qquad (14.33)$$

where

$$A_s = a_{12}a_{01} - a_{11}a_{02} \qquad (14.34)$$

$$B = a_{02}y' + a_{10}a_{01} - a_{00}a_{11} - a_{12}x' + a_{12}a_{03} - a_{02}a_{13} \qquad (14.35)$$

and

$$C = a_{00}y' - a_{10}x' + a_{10}a_{03} - a_{13}a_{00} \qquad (14.36)$$

The inverse transform is multi-valued and we have to be sure that $A_s \neq 0$. Such equations are solved symbolically, by hand or by using the Maple code [Chat et al.]:

```
eq1:=xp=(a[0,0]*x+a[0,1]*y+a[0,2]*x*y+a[0,3]);
eq2:=yp=(a[1,0]*x+a[1,1]*y+a[1,2]*x*y+a[1,3]);
sols:=solve({eq1,eq2},{x,y});
```

Selecting from the two roots is done by checking for an imaginary answer in (14.33). If the argument to the square root is negative, we ignore it and return the real part. If the denominator is zero, we set it to something small so that we get a large value for y. Otherwise, we select from the two roots by checking for the range. Since the roots are to yield a coordinate for y, we check to see if the root is negative or greater than the height of the source image. If no root fits the bill, we return a clipped value for y. In Java, we write

```
public double quadraticRoot(double a, double b, double c) {
  if (a == 0) a = 0.00001;
  double sqrtArg = b*b-4*a*c;
  double aa = 2*a;
  if (sqrtArg < 0) return -b/aa; // ignore imaginary part.
  double root1 = (-b + Math.sqrt(sqrtArg))/aa;
  if ((root1 >= 0) && (root1 < height)) return root1;
  double root2 = (-b - Math.sqrt(sqrtArg))/aa;
  if ((root2 >= 0) && (root2 < height)) return root2;
  if (root1 > height) return height;
  return 0; // (14.33)
}
```

To perform the inverse mapping, then, we must invoke the *quadraticRoot* method on each point to be inverted. We implement (14.32) - (14.36), using

```
public double[] inverseMap4(
            double a[][],
            double xp,double yp) {
   double as = // (14.34)
         -a[1][1]*a[0][2]
         +a[1][2]*a[0][1];
   double b = // (14.35)
          a[0][2]*yp + a[1][0]*a[0][1] - a[0][0]*a[1][1]
         -a[1][2]*xp + a[1][2]*a[0][3] - a[0][2]*a[1][3];
   double c =yp*a[0][0] // (14.36)
             - a[1][0]*xp
             + a[1][0]*a[0][3]
              -a[1][3]*a[0][0];
   double y = quadraticRoot(as,b,c); // (14.33)
   double x =  // (14.32)
         (xp-a[0][1]*y-a[0][3])/(a[0][0]+a[0][2]*y);
   double p[] = {x,y};
   return p;
  }
```

An example of the image warping is shown in Fig. 14-11. Plate 14-1 shows some examples of color image warping.

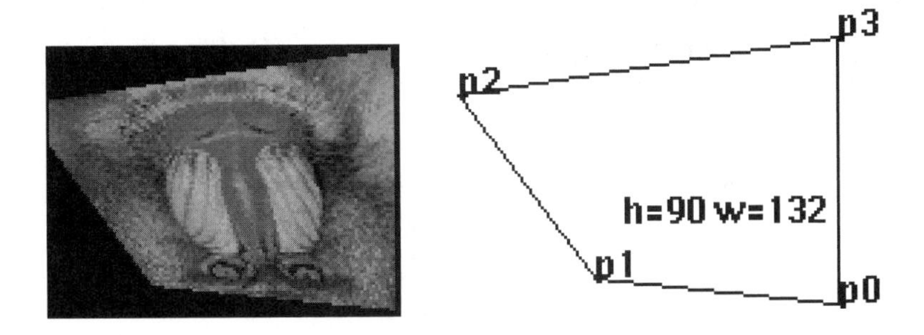

Fig. 14-11. Mapping Quadrilaterals to Quadrilaterals

14.9. Effects

All work and no play
makes Jill a dull girl.

– DL

It is possible to put a little fun into image processing. We have enough tools in place to create some interesting special effects. One idea, borrowed from [Holzmann], is to compute the polar coordinate location of each pixel, perform a polar coordinate transform, then use destination scanning to alter the location of the pixels. This is a non-linear warping which enables an effect (created by Tom Duff) called *fishEye*:

```
public void fishEye(int xc, int yc, double gamma) {
  int w = width;
  int h = height;
  short rn[][] = new short[width][height];
  short gn[][] = new short[width][height];
  short bn[][] = new short[width][height];
  double p[] = new double [2];
  int red, green, blue;
  int xp, yp, i, j;
  double R = Math.sqrt(w*w/4+h*h/4);
  for (int x = 0; x < w; x++)
    for (int y=0; y < h; y++) {
          double dx = x-xc;
          double dy = y-yc;
          double radius = Math.sqrt(dx*dx + dy*dy);
          // From [Holzmann] pp. 60
          double u = Math.pow(radius,gamma)/R;
          double a = Math.atan2(dy,dx);

    p[0] = u *Math.cos(a);
    p[1] = u *Math.sin(a);
    xp = (int) p[0]+xc;
```

```
      yp = (int) p[1]+yc;
      if ((xp < w) && (yp < h) && (xp >= 0) && (yp >= 0)) {
        rn[x][y] = r[xp][yp];
        gn[x][y] = g[xp][yp];
        bn[x][y] = b[xp][yp];
      }
    }
    r = rn;
    g = gn;
    b = bn;
    short2Image();
  }
```

Plate 14-2 shows the fish eye effect.

Another idea from Holzmann's book is that of performing a polar transform. Changing the angle, *a,* in the code above so that it is in degrees, instead of radians, we create the *polar* method in the *XformFrame*class. We then add the radius to the angle to obtain the following snippet:

```
      double radius = Math.sqrt(dx*dx + dy*dy);
      double a = (180/Math.PI)*Math.atan2(dy,dx);

       a = a + radius;

      a = a * Math.PI/180;
      p[0] = radius *Math.cos(a);
      p[1] = radius *Math.sin(a);
      xp = (int) p[0]+xc;
      yp = (int) p[1]+yc;
```

Plate 14-3 shows the effect of the polar transform.

The square root transform leaves the angle alone, but it takes the square root of the radius:

```
       a = a;
       radius = Math.sqrt(radius*R);
```

This is implemented in the *sqrt* method in the *XformFrame*class. Plate 14-4 shows the effect of the square root transform.

One effect favored by video artists is video feedback. Video feedback is obtained when a video camera points at a monitor that displays the camera's output. The optical equivalent is seen when two mirrors face each other. Affine transforms and the bilinear mapping may have feedback effects added to them.

To arrive at feedback, we need only to recurculate the buffer used to store the image.

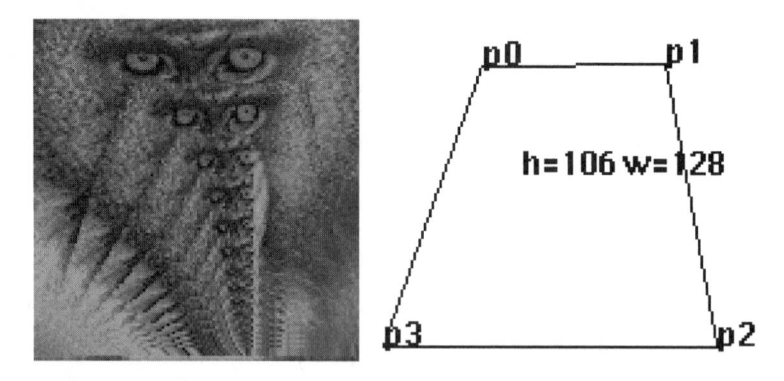

Fig. 14-12. Inferred 4 Point Bilinear Feedback

Fig. 14-12 shows a sample of inferred 4 point bilinear mapping used with a recirculating array. This type of effect is easy to reproduce in a TV studio using post-production equipment available there (e.g., ADO, Quantel, Harry, etc.). Such multi-million dollar post-production suites are typically beyond the reach of most people. It is nice to see a simple Java program that can achieve the same effect. Plate 14-5 shows an example of a color inferred 4 point bilinear feedback effect. Another effect, which comes from complex variables, is called *conformal mapping*. Conformal mapping is a mapping using an analytic function whose derivative is non-zero across the domain of the function. If w and z are complex variables, then we can define a conformal map of the form (using Euler's formula),

$$w = z^2 = \left(re^{i\theta}\right)^2 = r^2 e^{2i\theta}$$

Thus, we square the radius and double the angles. In Java, we write

```
a = a*2;
radius = radius*radius/R;
```

Where R is a scale factor that comes from the maximum value for the radius. An example of the conformal map is shown in Fig. 14-13.

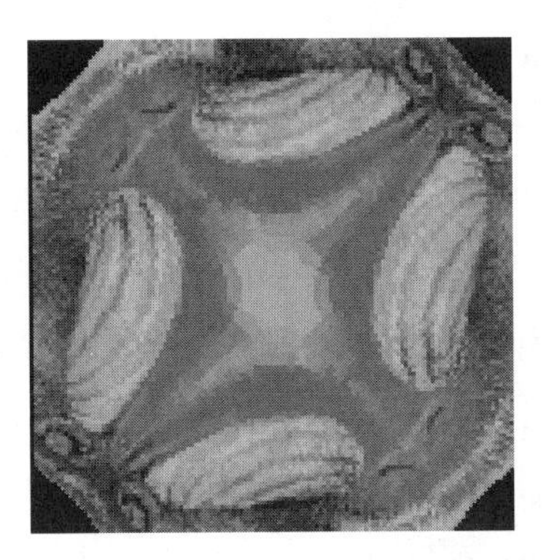

Fig. 14-13. Conformal Mapping

Plate 14-6 shows an example of iterative conformal mapping with $w = z^2$. Plate 14-7 shows an example of $w = z^2$ followed by fish eye and square root. For more ideas involving conformal mapping, see the Projects section.

14.10. Projects

1. In Section 14.5, we used a simple point sampling to get image data. This is done by using the inverse transform to obtain floating point coordinates that index into the image. Our approach was to truncate the coordinates, since pixels are indexed using integers, not floats. There are several techniques available that are generally more accurate than coordinate truncation, such as nearest neighbor, linear interpolation and bilinear interpolation. Nearest neighbor selects the value of the closest pixel to the floating point coordinate. Linear interpolation uses a weighted average for the value of the intensity. The weighting comes from the distance the pixel is away from the corner pixels. Another technique, using bilinear interpolation, is also available [Jain]. Still another technique is to filter the image using a low-pass filter (such as a Gaussian) before resampling the image.

Select two of the three techniques described and implement them in the Kahindu program. Show examples of affine transforms comparing the techniques. Which look best?

2. In Section 14.7 we disclosed a technique for determining the inverse map of an object by using three points. Create a triangulation of an image and display it as a mesh superimposed over the image. Enable points to be moved using the mouse. After the user is finished moving the points, warp each triangular section of the image in accordance with the shape of the altered mesh.

3. One way to simplify the solution to inverse bilinear mapping is to make some simplifying assumptions. Heckbert suggests mapping from a quadrilateral to a rectangle, then mapping from a rectangle to a quadrilateral [Heckbert 86]. Unfortunately, the mapping from a rectangle to a quadrilateral still requires solving the quadratic equation.

$$S_{4c} = \begin{bmatrix} x_0 & x_1 & x_2 & x_3 \\ y_0 & y_1 & y_2 & y_3 \end{bmatrix} = \begin{bmatrix} 0 & 1 & 1 & 0 \\ 0 & 0 & 1 & 1 \end{bmatrix} \quad (14.37)$$

Substituting (14.37) into (14.30):

$$S_4 = \begin{bmatrix} x_0 & x_1 & x_2 & x_3 \\ y_0 & y_1 & y_2 & y_3 \\ x_0y_0 & x_1y_1 & x_2y_2 & x_3y_3 \\ 1 & 1 & 1 & 1 \end{bmatrix} \quad (14.30)$$

results in the four point grid matrix:

$$S_{4g} = \begin{bmatrix} 0 & 1 & 1 & 0 \\ 0 & 0 & 1 & 1 \\ 0 & 0 & 1 & 0 \\ 1 & 1 & 1 & 1 \end{bmatrix} \quad (14.38).$$

The inverse of (14.38) is found, using Maple:

```
with(linalg):
 s4g:=array(
```

```
    [
     [0,1,1,0],
     [0,0,1,1],
     [0,0,1,0],
     [1,1,1,1]
    ]);
> inverse(matrix(s4g));
```

which produces

$$S_{4g}^{-1} = \begin{bmatrix} -1 & -1 & 1 & 1 \\ 1 & 0 & -1 & 0 \\ 0 & 0 & 1 & 0 \\ 0 & 1 & -1 & 0 \end{bmatrix} \tag{14.39}$$

Substituting (14.39) into:

$$A_g = D_4 S_{4g}^{-1} \qquad (14.36).$$

yields

$$A_g = \begin{bmatrix} x'_0 & x'_1 & x'_2 & x'_3 \\ y'_0 & y'_1 & y'_2 & y'_3 \end{bmatrix} \begin{bmatrix} -1 & -1 & 1 & 1 \\ 1 & 0 & -1 & 0 \\ 0 & 0 & 1 & 0 \\ 0 & 1 & -1 & 0 \end{bmatrix} \tag{14.40}$$

use (14.40) to solve the inverse mapping problem. Implement your solution in Kahindu and time the result. Use images of several different sizes. Tabulate the time it takes for the general inverse mapping solution, compared to your special case. What is the average speed up?

4. We have seen how quadrilaterals map to quadrilaterals. Devise a program that allows the user edit a vertex mesh. Anchor quadrilateral to the vertices. Then warp each sub-image in the each quadrilateral.

5. Section 14.8 showed how to use conformal mapping to warp images. Get a book on complex variables (such as [Churchill]) and try 3 or 4 conformal mapping equations for image warping. You may need to introduce scale factors to keep the

image in view. Write out the conformal mapping equations in terms of polar coordinates. Show your results.

15. Unitary Transforms

The man who claims to be the boss in his own home
will lie about other things as well.

– Amish saying

15.1. Introduction

The last chapter introduced warping transforms. Such transforms move pixels from one place to another. This chapter describes transforms that do not change a pixel's location, but instead provide an isomorphic (one-to-one) mapping of the pixel from one *space* into another *space*. For example, the Fourier transform provides a mechanism to perform an isomorphic mapping between the pixel space and the frequency space. A large class of these mappings is performed by taking an inner product between an image and a *kernel*. When the kernel has a magnitude of one, the transforms are called *unitary transforms.* In Chapter 5, we introduced the notation for the inner product, $< f|g >$:

$$< f|g > \equiv \int_{-\infty}^{\infty} f(x)g(x)dx \equiv \text{inner product} \qquad (5.7a).$$

Typically, g is called the kernel of the transform. The kernel of a transform is often an orthogonal basis that can be used to reconstruct the original waveform, f. If the inner product between two functions is zero, then the functions are said to be *orthogonal*. The discrete unitary transform is defined as

$$<f \mid g> \equiv \sum_{x=0}^{X-1} f(x)g(x) \qquad (15.1).$$

The functions f and g may be two vectors over the field of complex numbers. For example, the one-dimensional Fourier transform has a complex basis function

$$F(u) = \int_{-\infty}^{\infty} f(x)e^{-j2\pi ux}dx = <f \mid e^{-j2\pi ux}> \qquad (15.2).$$

The discrete version of (15.2) is given by

$$F(u) = <f \mid e^{-j2\pi ux}> = \sum_{x=0}^{X-1} f(x)e^{-j2\pi ux} \qquad (15.3).$$

The two-dimensional version of (15.1) is given by

$$F(u,v) = <f \mid g> \equiv \sum_{x=0}^{W-1}\sum_{y=0}^{H-1} f(x,y)g(x,y,u,v) \qquad (15.4).$$

where W is the width of the image, H is the height, and (u,v) are indices that parameterize the kernel. Typically, (u,v) are given as discrete values and so the number of evaluations needed for g is WH. This is because the input image and output image are typically WxH in size.

For example, the 2-D Discrete Fourier Transform (DFT) is given by

$$F(u,v) = <f \mid e^{-2\pi i(ux/W+vy/H)}> = \frac{1}{WH}\sum_{x=0}^{W-1}\sum_{y=0}^{H-1} f(x,y)e^{-2\pi i(ux/W+vy/H)} \qquad (15.5).$$

Thus, unitary transforms are named according to the kernel used. There are an infinite number of kernels that could be used for a transform. There are also an infinite number of two-dimensional unitary transforms of the form of (15.4). It is easy to see that sine and cosine are orthogonal under the chosen inner product of (15.1).

15.2. The Discrete Fourier Transform

There has been a great deal of literature devoted to the acceleration of (15.5),

$$F(u,v) = < f \mid e^{-2\pi i(ux/W+vy/H)} > = \frac{1}{WH} \sum_{x=0}^{W-1} \sum_{y=0}^{H-1} f(x,y) e^{-2\pi i(ux/W+vy/H)} \quad (15.5)$$

and its inverse

$$f(x,y) = < F \mid e^{2\pi i(ux/W+vy/H)} > = \sum_{u=0}^{W-1} \sum_{v=0}^{H-1} F(u,v) e^{2\pi i(ux/W+vy/H)} \quad (15.6).$$

The reason for the interest in the Fourier transform is the convolution theorem. The convolution theorem states that convolution in the spatial domain is equal to multiplication in the frequency domain (and vice versa). Another motivation for using the Fourier transform lies in the existence of fast algorithms for its evaluation. As shown in [Lyon and Rao], direct evaluation of (15.5) is an $O(N^2)$ task, where N is the number of points in the image. Further, direct evaluation of (15.5) results in a complex, floating point number.

The first step in simplification of the Fourier transform is to express (15.5) in a separable form. Using the laws of exponents, we obtain

$$F(u,v) = \frac{1}{W} \sum_{x=0}^{W-1} \left[\frac{1}{H} \sum_{y=0}^{H-1} f(x,y) e^{-2\pi ivy/H} \right] e^{-2\pi iux/W} \quad (15.7)$$

for the forward transform and

$$f(x,y) = \sum_{u=0}^{W-1} \left[\sum_{v=0}^{H-1} F(u,v) e^{2\pi ivy/H} \right] e^{2\pi iux/W} \quad (15.8)$$

for the reverse transform. There is a large class of algorithms that are designed to accelerate the one-dimensional FFTs that use the separability shown in (15.7) and (15.8). Typically, they make some assumptions about the number of samples in the input. For example, if N is an integral power of two, we may use a Radix-2 FFT (of which there are several). For background on the Radix-2 FFT, see [Lyon and Rao].

To use the Radix-2 FFT for images that have dimensions that are not integral powers of two, the images may be truncated or padded to the next integral power of two number. Image truncation may be acceptable in some applications. Padding is performed by adding a black pixel border to the image.

The Radix-2 FFT uses a *divide-and-conquer* algorithm to accelerate its computations. The unaccelerated discrete Fourier transform (DFT) is an $O(N^2)$ algorithm. Using the divide-and-conquer technique, we recursively split N until there are only two samples left. The DFT is taken on all sets of two samples and then is recombined. This is called the Radix-2 FFT, and it has an $O(N \log N)$ execution time. The splitting of a sequence is justified using the *Danielson-Lanczos lemma*. The Danielson-Lanczos lemma enables the sample set to be split into its odd and even components. The sub-sets are redivided until there are only two samples left. Such an algorithm may be implemented in a recursive fashion [Wolberg] or a non-recursive fashion [Lyon and Rao]. A non-recursive implementation gives a 6x speed-up over the recursive implementation, at a cost of greater implementation complexity. An implementation of the non-recursive Cooley-Tukey Radix-2 FFT is contained in the *vsFFT1D* class which resides in the *vs* package. The details of the derivations and implementations are given in [Lyon and Rao].

The reader is now apt to ask, "O.K., when should I use an FFT?" This is a difficult question to answer. Typically, the difficulty arises because the size of the image is not known. Under the most favorable conditions, the dimensions of the image are an integral power of two. Let us assume that the image to be processed is a 128x128 image. The Kahindu implementation of the Cooley-Tukey Radix-2 FFT (using a 233 Mhz G3) computes and displays the result in 1.282 seconds. To make the test fair, an inverse FFT with a complex-number domain multiplication must be used to complete the process. We have not tried to optimize the FFT further by removing displays of the intermediate results; however, an expectation of 2.4 seconds is not unreasonable. Table 15-1 shows the convolution times for two RGB images and three different kernel sizes.

NxN Kernel	128x128	512x512
3x3	0.25 secs	3.47 secs
7x7	0.81 secs	13.3 secs
15x15	3.91 secs	60 secs

Table 15-1. Convolution Times for two RGB Images

For a 512x512 image, the Radix-2 FFT implementation took 18.489 seconds. So, estimate 36 seconds for an FFT based filtering operation, using a 512x512 image.

In summary, for a practical range of image sizes (up to 512x512) and kernel sizes (7x7), the convolution appears to out-pace the FFT, even under the favorable circumstances of using a square image that is an integral power of two in size. Results will be worse in pathological cases. For example, suppose that the advisary postulates a worst-case scenario image that is 513x513. Such an image must be padded to 1024x1024. This is a four fold increase in the number of pixels. See the Projects section (Section 15.8) for suggestions on how to further optimize the FFT computation.

Fig. 15-1 shows a picture of *EdgeMan* and the log of the power spectral density (i.e., log of the magnitude) of his FFT.

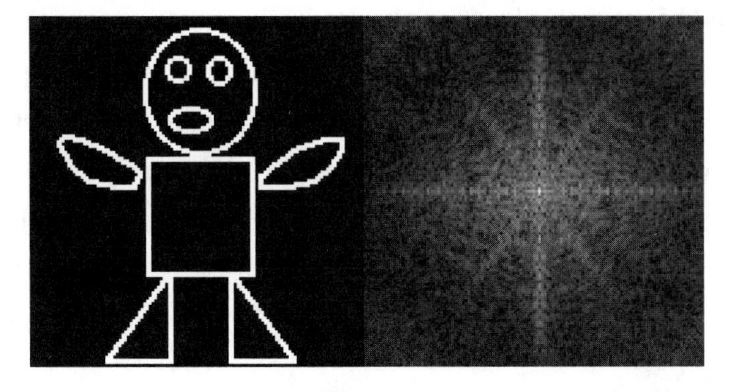

Fig. 15-1. Edge Man and His FFT

Sometimes the FFT is the preferred way to design filters and experiment with various algorithms. On the other hand, small convolutions on small images can be very fast. It really depends on the application.

To take the Radix-2 FFT, use the *FFTFrame*. Make sure that an image is displayed. For example:

```
FFTFrame fftf= new FFTFrame("FFTFrame");
fftf.fftR2(); // takes the FFT
fftf.ifftR2(); // takes the inverse FFT.
```

Both of these operations may be performed by using menu item selections in the *Xform* menu, shown in Fig. 15-2.

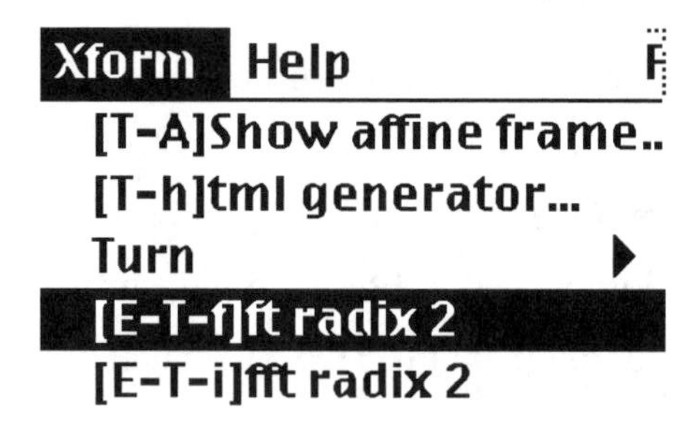

Fig. 15-2. The Xform Menu

Once the FFT is performed, it is possible to do filtering in the frequency domain. This is done by computing a filter and multiplying (using a complex multiplication) in the frequency domain. Very large filters are possible using this technique. The easiest way to generate and display such filters is to create images that are saved to disk. These images are read into the *Xform* frame and then multiplied in the frequency domain. For the purpose of display, the image color component data is restricted to vary from 0 to 255. The multiplication is normalized by dividing by 255, so that an overflow does not result. Complex multiplication, by a real number, may be represented by

$$rz = rx + iry \tag{15.9}$$

That is to say, two multplications are required, for each pixel.

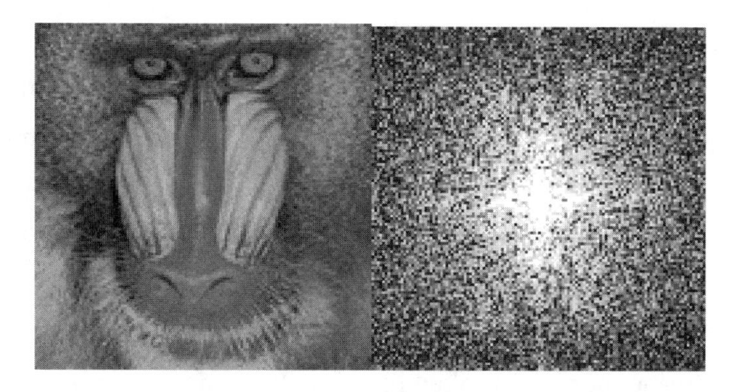

Fig. 15-3. The Mandrill and the PSD of the FFT

Fig. 15-4. A Notch Filter and the Inverse FFT

Fig. 15-3 shows an image of the mandrill and the log of the PSD of the FFT. Fig. 15-4 shows a notch filter that removes some of the high-frequency components, along with the inverse of the FFT. See the Projects section (Section 15.8) for suggestions related to FFT-based filtering.

In an effort to speed the FFT (particularly for non-square images that are not a power of two), it is worthwhile exploring another of the many FFT algorithms.

Another possible approach to the FFT is to use the Radix-4 Cooley-Tukey FFT. In this approach, we read the one-dimensional array of numbers into a two-dimensional array. The DFT is performed on the columns of the two-dimensional array, the results are rearranged, then the DFT is performed on the rows. Data is read in row wise, but read out column wise. For example, a 20-sample one-dimensional array may be read into a 4x5 element two-dimensional array. The columns are transformed with 5 DFTs. The data is then rearranged, then the rows are transformed using 4 DFT's. This is called the Radix-4 Cooley-Tukey FFT. The DFT is an order $O(N^2)$ algorithm, so that, for $N = 20$, the DFT run-time is order 400. The Radix-4 Cooley-Tukey FFT would take order $5(4^2)+4(5^2)=180$. However, optimizations do not end there. Once a long sequence is broken into small, often used, sub-sequences, there is a chance for optimization. Such optimizations may be performed using a hybrid-symbolic-numeric computation for automatically generating optimal DFT computations for given size sequences. The automatic nature of such approaches permits the computation to go beyond the sizes that are currently practical to optimize by hand.

For example, a Maple procedure for generating optimized DFT's of length *N:*

```
readlib(optimize):
f := proc(n)
local yr,yi,a,k,cosTheta,theta,sinTheta;
yr := array(0..n-1);yi := array(0..n-1);
for k from 0 by 1 while k < n do
   theta := 2*Pi*k*m/n;
   cosTheta := cos(theta):
   sinTheta := sin(theta):

   yr[k] := sum(aRe[m]*cosTheta+aIm[m]*sinTheta, m=0..(n-1));
   yi[k] := sum(-aRe[m]*sinTheta+aIm[m]*cosTheta, m=0..(n-1));
od:
C([optimize(yr),optimize(yi)],optimized):
end:f(2);
```

When function *f* is invoked with an argument of $N = 2$, the following is emitted:

```
f(2);
      t1 = aRe[0];
      t2 = aRe[1];
      yr[0] = t1+t2;
      yr[1] = t1-t2;
      t3 = aIm[0];
      t4 = aIm[1];
```

```
        yi[0] = t3+t4;
        yi[1] = t3-t4;
```

The real and imaginary parts of the output are stored in *yr* and *yi*. The temporary variables are typically declared using a *static* modifier, so that they are not re-allocated during each method invocation. To save memory, the temporary variables are reused by other DFT methods. Program length and memory requirement increase rapidly as the DFT length increases. One common optimization results when the forward FFT is assumed to be a function of only real-valued samples (as is the case with an image). When this occurs, the imaginary part is set to zero and the result is seen as symmetric about the sample midpoint. The modified Maple procedure follows:

readlib(C):

```
readlib(optimize):
realFFT := proc(n)
local y,a,k,cosTheta,theta,sinTheta;
y := array(0..1,0..n-1);
for k from 0 by 1 while k < n do
   theta := 2*Pi*k*m/n;
   cosTheta := cos(theta):
   sinTheta := sin(theta):
   y[0,k] := sum(a[m]*cosTheta, m=0..(n-1));
   y[1,k] := sum(-a[m]*sinTheta, m=0..(n-1));
od:
C(y,optimized):
end:
```

The result is now being stored in a two-dimensional array. The zero row is storing the real-valued part, the one row is storing the imaginary part. This makes the symmetry easy to see in the following output:

```
realFFT (8);
      t2 = sqrt(2.0);
      t3 = a[1]*t2;
      t4 = a[3]*t2;
      t5 = a[5]*t2;
      t6 = a[7]*t2;
      t7 = a[0]+t3/2-t4/2-a[4]-t5/2+t6/2;
      t8 = a[0]-a[2]+a[4]-a[6];
```

```
t9 = a[0]-t3/2+t4/2-a[4]+t5/2-t6/2;
y[0][0] = a[0]+a[1]+a[2]+a[3]+a[4]+a[5]+a[6]+a[7];
y[0][1] = t7;
y[0][2] = t8;
y[0][3] = t9;
y[0][4] = a[0]-a[1]+a[2]-a[3]+a[4]-a[5]+a[6]-a[7];
y[0][5] = t9;
y[0][6] = t8;
y[0][7] = t7;
y[1][0] = 0.0;
y[1][1] = -t3/2-a[2]-t4/2+t5/2+a[6]+t6/2;
y[1][2] = -a[1]+a[3]-a[5]+a[7];
y[1][3] = -t3/2+a[2]-t4/2+t5/2-a[6]+t6/2;
y[1][4] = 0.0;
y[1][5] = t3/2-a[2]+t4/2-t5/2+a[6]-t6/2;
y[1][6] = a[1]-a[3]+a[5]-a[7];
y[1][7] = t3/2+a[2]+t4/2-t5/2-a[6]-t6/2;
```

It is also easy to see the sign alternation in the imaginary part of the output sequence. Note also that the symmetry is not being exploited by the optimizer. A more optimal solution is:

```
y[1][5] =  - y[1][3]
y[1][6] =   -y[1][2];
y[1][7] =   -y[1][1];
```

This only goes to show that we must check the output of symbolic manipulators carefully. The suboptimal code comes from a bug in Maple VR4 and is fixed in Maple VR5 [Maple 98]. See the Projects section (Section 15.8) for suggestions on exploring this topic further.

15.3. The Wavelet Transform

The Fourier transform uses a basis that consists of sine and cosine waves. These functions start at time minus infinity and end at time infinity. When we take a DFT, the signal is treated as if it were periodic. That is, the sample window that is used on the signal is assumed to be on a single period of that signal. The periodic waveform assumption is invalid for most real waveforms.

To address the aperiodic nature of most waveforms, a technique was used to resample the data using divide-and-conquer techniques (just like the FFT). Recall that the FFT placed the odd samples in one set, the even samples in another. In contrast, the wavelet transform places the high frequencies into one sample set and the low frequencies into another. Both sets are then scaled to one-half their original size.

Wavelets and FFT's differ in other ways, too. The wavelet uses a more complicated basis function than the sine and cosine waveforms of the FFT. The basis function is called the *analyzing* (or mother) wavelet. The FFT can use only sine and cosine basis function for the kernel. The wavelet transform has access to an infinite set of available basis functions.

The term *wavelet* suggests that this is a small (i.e., non-infinite and therefore localized) waveform. The feature of localization means that wavelets are non-zero over a finite range and zero elsewhere. This means that wavelets have *compact support.* Support is the domain where the function is non-zero. Compact means that the interval of support is bounded. Thus, compact support means that the function is non-zero over a bounded interval.

15.4. The Haar Basis

The simplest of all wavelet basis functions is the Haar basis function. The basic idea is that the low-pass filter is computed with a simple average and the high-pass filter is computed with a difference. These filters are expressed as

$$H_0 : f(n) = (x_n + x_{n+1})/2 \tag{15.10}$$

and

$$H_1 : f(n) = (x_n - x_{n+1})/2 \tag{15.11}$$

It is easy to see that (15.10) represent a moving average and (15.11) represents a moving difference. When both (15.10) and (15.11) are transmitted, the original sample sequence can be reconstructed. After the filtering is performed, the output sequence is 2:1 sub-sampled. To reconstruct the original waveform, after the 2:1 sub-sampling, a procedure called 1:2*up-sampling* must be performed. The 1:2 up-sampling duplicates the input sample, doubling the data rate.

The one-dimensional Haar transform may be performed by taking the one-dimensional transform on all the rows, then taking the result and performing the one-dimensional transform on the columns. For simplicity, we show an example where a one-dimensional Haar transform is performed on each row, but not on the columns.

Few things are harder to put up with
than the annoyance of a good example.

– Mark Twain, (1835-1910)

For example, suppose you are given 16 numbers to send:

$$\begin{bmatrix} 9 & 7 & 5 & 3 \\ 3 & 5 & 7 & 9 \\ 2 & 4 & 6 & 8 \\ 4 & 6 & 8 & 10 \end{bmatrix} \tag{15.12}$$

Proceed along the rows and note that if the numbers did not work out evenly, we truncate. As this is a carefully contrived example, no truncation occurs:

$$\begin{bmatrix} 8 & 4 \\ 4 & 8 \\ 3 & 7 \\ 5 & 9 \end{bmatrix} \tag{15.13}$$

The detail coefficients are given by

$$\begin{bmatrix} 1 & 1 \\ -1 & -1 \\ -1 & -1 \\ -1 & -1 \end{bmatrix} \tag{15.14}$$

Note that (15.13) and (15.14) actually fit into a matrix that has the same number of dimensions as (15.12),

$$\left[\begin{array}{cc|cc} 8 & 4 & 1 & 1 \\ 4 & 8 & -1 & -1 \\ 3 & 7 & -1 & -1 \\ 5 & 9 & -1 & -1 \end{array}\right] \tag{15.15}$$

Recall that the matrix shown in (15.12) was initially positive. Image data is typically stored as an 8 bit unsigned quantity. Having negative values for image data, as shown in (15.15), requires that we store signed quantities. In fact, the dynamic range for 8 bit data, after the transform, is 9 bits. Now treat the averages in (15.15) as the input for the next computation, to obtain:

$$\begin{bmatrix} 6 & 2 & 1 & 1 \\ 6 & -2 & -1 & -1 \\ 5 & -2 & -1 & -1 \\ 7 & -2 & -1 & -1 \end{bmatrix} \qquad (15.16).$$

When we apply the Haar transform in two dimensions, we are performing a high-pass and low-pass filter sequence. The image's low-frequency data is focused into the upper left-hand corner of wavelet transformed output image. Fig. 15-5 shows a diagram depicting the layout of the sub-bands in the image.

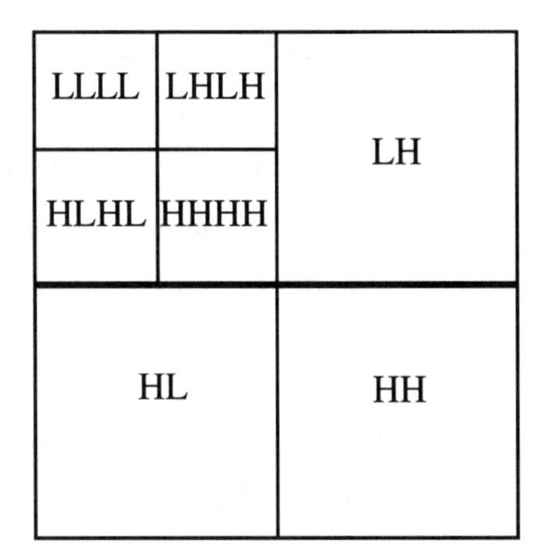

Fig. 15-5. A Depiction of the Sub-Band Layout

The sub-bands are constructed by cascading a series of filters with a down-sampling. Fig. 15-6 shows a diagram depicting the wavelet transform operation, as carried out by a sub-sampling filter bank.

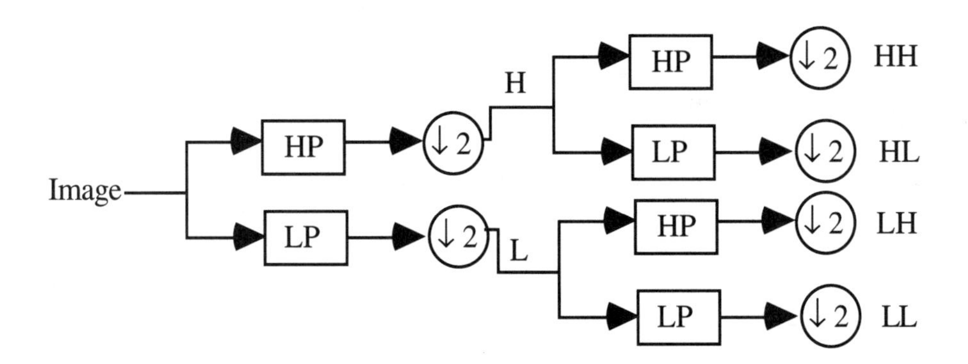

Fig. 15-6. A Sub-Sampling Filter-Bank.

In order to compute all the sub-bands shown in Fig. 15-5, the filter-bank of Fig. 15-6 would have to be performed once on the image, then again on sub-band *LL*. Note that at each stage in the filter bank, the number of samples is cut in half. As a result, the operation runs in $O(N \log N)$ time. One problem with the filter is that it produces a floating point result. The beginner in this field is tempted to truncate the floating point operation and then to proceed. This causes an imperfect reconstruction after the inverse wavelet transform is performed. For example, on the Mandrill image, the imperfect reconstruction has an SNR of 37 dB with respect to the original image. To illustrate, consider the following example:

To implement the simple, 1-D Haar wavelet transform with truncation (for demonstration purposes) there is a class called *Wavelet* in the *gui* package. By running the *main* in the *Wavelet* class, the following is emitted:

```
9  7     5     3
3  5     7     9
2  4     6     8
4  6     8     10
-------------------
Running Forward Haar Wavelet on the matrix:
8  4     , 1   1
4  8     , -1  -1
3  7     , -1  -1
5  9     , -1  -1
-------------------
6  , 2
6  , -2
5  , -2
7  , -2
-------------------
Running Inverse Haar Wavelet on the matrix:
8  4
4  8
3  7
5  9
-------------------
9  7     5     3
3  5     7     9
2  4     6     8
4  6     8     10
-------------------
```

Note that this carefully crafted example has a perfect reconstruction. Suppose the matrix were not so carefully selected:

```
1  2     3     4
4  3     2     1
```

```
1  2     3     4
4  3     2     1
-------------------
Running Forward Haar Wavelet on the matrix:
1  3     , 0   0
3  1     , 0   0
1  3     , 0   0
3  1     , 0   0
-------------------
2  , -1
2  , 1
2  , -1
2  , 1
-------------------
Running Inverse Haar Wavelet on the matrix:
1  3
3  1
1  3
3  1
-------------------
1  1     3     3
3  3     1     1
1  1     3     3
3  3     1     1
-------------------
```

We can see that the effect of rounding has caused an imperfect reconstruction of the original sequence. One approach to solving this problem is to use floating point numbers to represent the wavelet transform. The trouble is that the image started as an integer and that floating point numbers take more space. Also, truncation of the floats causes an imperfect reconstruction. One way around this problem is to employ a technique called a *lifting scheme* [Sweldens].

The simplest of lifting schemes is the integer version of the Haar transform [Uytterhoeven]. It consists of a forward and inverse transform. The forward transform is given by a multi-step algorithm. The first step is to split the input sequence into its odd and even components. A snippet follows:

```
for (int i=0; i < s.length; i++) {
   s[i]=x[2*i];    // The even part
   d[i]=x[2*i+1]; // The odd part
}
```

Then we perform the dual lifting step:

```
for (int i=0; i < s.length; i++)
    d[i]=d[i]-s[i];
```

Finally, we perform the primal lifting step:

```
for (int i=0; i < s.length; i++)
    s[i]=s[i]+round(d[i]/2.0);
```

Lifting computes the difference between the true coefficient and its prediction [Calderbank].

For the inverse transform, we perform inverse primal lifting:

```
for (int i=0; i < s.length; i++)
    s[i]=s[i]-round(d[i]/2.0);
```

This is followed by the inverse dual lifting:

```
for (int i=0; i < s.length; i++)
    d[i]=d[i]+s[i];
```

And finally, merging into a single sequence:

```
for (int i=0; i < s.length; i++) {
    x[2*i]=s[i];    // The even part
    x[2*i+1]=d[i]; // The odd part
}
```

This description leaves out many details that are required for an implementation. For example, the input must be checked to make sure that it is an integral power of two:

```
public static int log2(int n) {
 return (int) (Math.log(n)/Math.log(2));
}
public boolean powerOf2(int n) {
 return (1<<log2(n) == n);
}
```

Note that *log2* is taking the log of *n* to the base 2. The *powerOf2* check returns true if *n* is an integral power of 2. As a check, we devise an example that shows a one-dimensional forward lifting Haar transform and its inverse:

```
1  2      3      4      2      4      3      1      6      7      10
   13     17     12     13     14
Lifting Forward Haar Wavelet:
2  4      3      3      7      12     15     14
```

```
1  1      2     -2     1     3     -5     1
Lifting Inverse Haar Wavelet:
1  2      3      4     2     4      3     1      6      7      10
   13     17     12    13    14
```

Note that there is a perfect reconstruction. All that is left is the creation of a recursion mechanism to perform the lifting on the averages. The recursive implementation outputs:

```
1  2      3      4     2     4      3     1      6      7      10
   13     17     12    13    14
Lifting Forward Haar Wavelet:
averages
2  4      3      3     7     12     15    14
3  3      10     15
3  13
8
details
1  1      2     -2     1     3     -5     1
2  0      5     -1
0  5
10
```

Transmission of the wavelet transform consists of the last number from the averages, plus all the details, given from bottom to top. For example:

```
transmission
8  10     0      5     2     0      5    -1      1      1      2      -
2  1      3     -5     1
```

Thus, the number of elements to be transmitted remains the same as the input to the transform, and the transform is integer to integer. As a result, an in-place formulation may occur with perfect reconstruction.

In implementation, there are a few tricks that are used to perform a binary subdivision of the transmission sequence. In the following example, we see the binary sequence is invoked with exactly half of its elements. This continues until there are only two elements left in the sequence. When the recursion unwinds, we print the intermediate results:

```
    private void backwardHaarPhaseOne(short in[]) {
          int n = in.length;
          if (n < 2) return;
```

```
        int nOn2 = n / 2;
        short s[] = new short[nOn2];
        for (int i = 0; i < nOn2; i++)
            s[i] = in[i];
        backwardHaar(s);
        print(s);
    }
```

In *backwardHaarPhaseOne* we see the beginning of the inverse lifting transform. The output appears below:

```
8
8  10
8  10  0  5
8  10  0  5  2  0  5  -1
8  10  0  5  2  0  5  -1  1  1  2  -
2  1   3  -5 1
```

Thus, we are dividing the sequence into powers of two groups.

To complete the inverse lifting Haar transform, we write:

```
    public static void backwardHaar(short in[]) {
        int n = in.length;
        if (n < 2) return;
        int nOn2 = n / 2;
        short s[] = new short[nOn2];
        for (int i = 0; i < nOn2; i++)
            s[i] = in[i];
        backwardHaar(s);
        print(in);
        nOn2 = in.length/2;
        short d[] = new short[nOn2];
        for (int i=0; i < d.length;i++)
            d[i] = in[i+nOn2];
        for (int i=0; i < s.length; i++)
            s[i]=(short)(s[i]-round(d[i]/2.0));
        for (int i=0; i < s.length; i++)
            d[i]=(short)(d[i]+s[i]);
        for (int i=0; i < s.length; i++) {
            in[2*i]=s[i];    // The even part
            in[2*i+1]=d[i]; // The odd part
        }
    }
```

Note that the internal representation is *short*. Thus, this is a pure, integer-to-integer transform. The output follows:

```
Lifting Inverse Haar Wavelet:
8  10
8  10    0    5
8  10    0    5    2    0    5    -1
8  10    0    5    2    0    5    -1    1    1    2    -
2  1     3    -5   1
1  2     3    4    2    4    3    1     6    7    10
   13    17   12   13   14
```

This represent a perfect reconstruction. We still have a problem with the increased dynamic range in the detail coefficients. Recall that the image data was initially just byte (i.e., 8-bit) data. Now we require 9-bit data.

See problem 7 in the Projects section (Section 15.8) for more information on this.

15.5. Implementing the Two-Dimensional Haar Lifting Transform

To implement the two-dimensional Haar-lifting transform, we transform all the rows, transpose the image, then transform the columns, and finally, we transpose the image back. The *Lifting* class in the *gui* package performs all these functions using two static methods. The two-dimensional integer forward Haar transform (also known as the *standard decomposition*) is given by:

```
public static void forwardHaar(short in[][]) {
        for (int x=0; x < in.length; x++)
                forwardHaar(in[x],in.length);
        in = transpose(in);
        for (int x=0; x < in.length; x++)
                forwardHaar(in[x],in.length);
        in = transpose(in);
}
```

Similarly, the two-dimensional integer backward Haar transform (also known as the *standard reconstruction*) is given by:

```
public static void backwardHaar(short in[][]) {
        for (int x=0; x < in.length; x++)
                backwardHaar(in[x]);
        in = transpose(in);
        for (int x=0; x < in.length; x++)
                backwardHaar(in[x]);
        in = transpose(in);
}
```

The integer Haar transform is able to perform a perfect reconstruction. Fig. 15-7 shows the effect of the integer Haar transform. The arrangement of sub-bands into quadrants, as shown in Section 15.5, is called a non-standard decomposition. With the standard decomposition, you can transmit the image in scan-line order (from top to bottom) and the resolution will gradually improve.

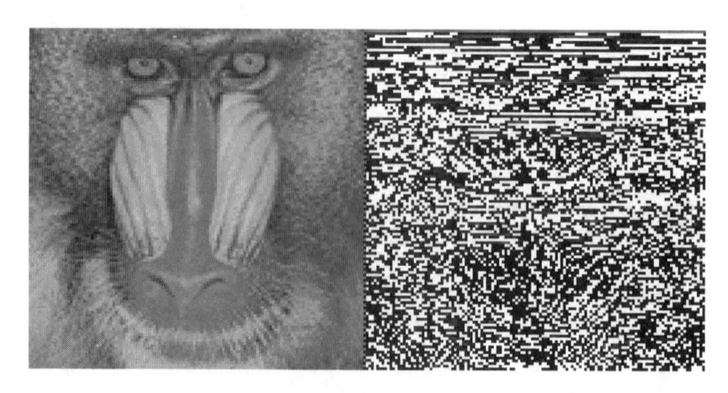

Fig. 15-7. Before and After the Integer Haar Transform

Fig. 15-8 shows the effect of eliminating 50% of the detail from the transformed image.

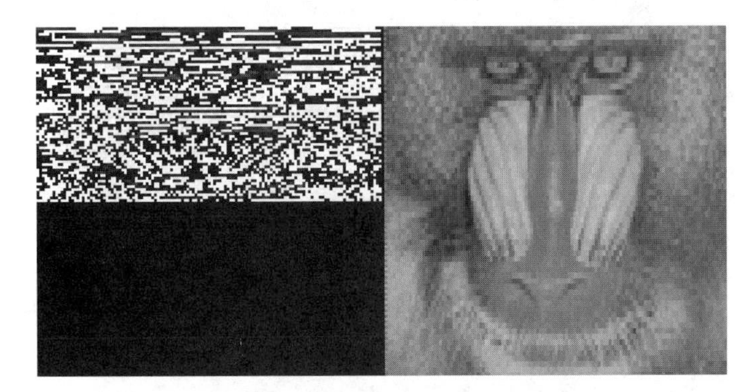

Fig. 15-8. Backward Lossy Integer Haar Transform

Fig. 15-9 shows the effect of eliminating 75% of the detail from the transformed image.

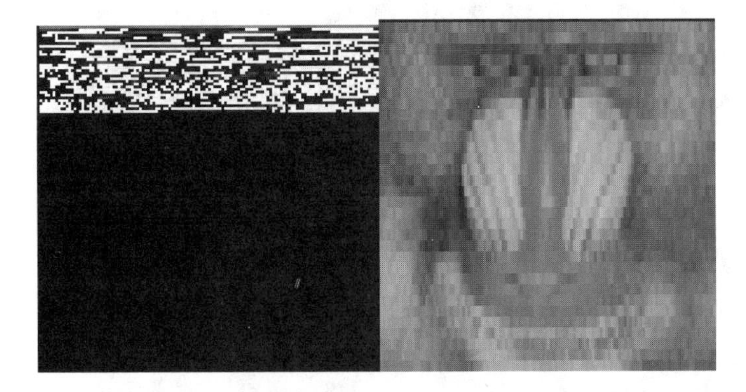

Fig. 15-9. Using 25% of the Detail for a Reconstruction

There is also a non-standard decomposition/reconstruction technique available. This is based on the sub-band layout given in Fig. 15-5. Using this technique we can clear out sub-bands as they are arranged, recursively, into quadrants.

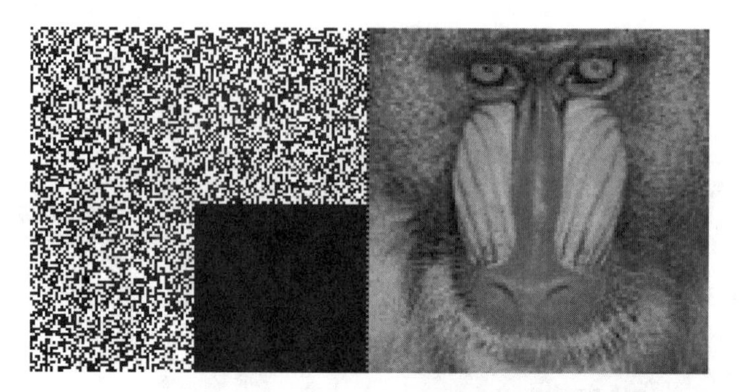

Fig. 15-10. Using 75% of the Detail for a Non-Standard Reconstruction

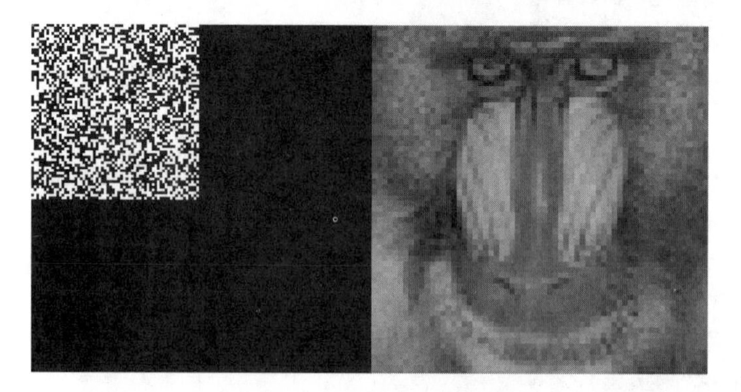

Fig. 15-11. Using 25% of the Detail for a Non-Standard Reconstruction

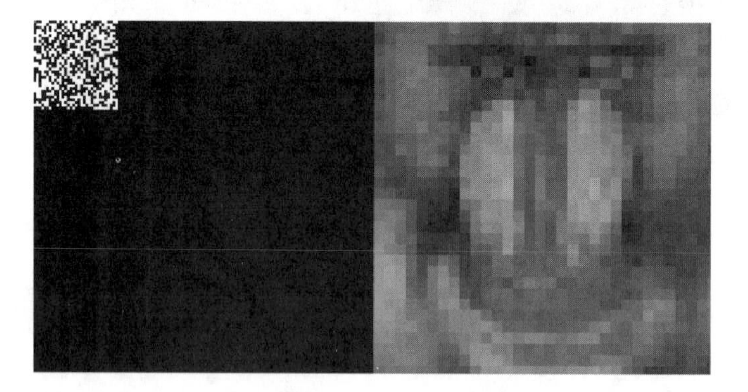

Fig. 15-12. Using 12% of the Detail for a Non-Standard Reconstruction

Fig. 15-10, Fig. 15-11, and Fig. 15-12 show 75%, 25% and 12% of the detail being preserved for a non-standard reconstruction of the image. Fig. 15-12 is saved with a

16:1 compression ratio. The compression was measured by comparing file sizes of the gzipped vs. the non-gzipped, full-resolution image. Better compression can be had with a better quality image using the search algorithm described in question 8 in the Problems section.

15.6. Writing a Wavelet Encoded File

One of the problems with the details obtained from an integer Haar transform is that the details have a larger dynamic range than the original image. This makes sense, since the integer Haar transform is computing the sum and difference between 8-bit numbers. Both the sum and the difference require 9-bit results. This means that we cannot store the wavelet transform results in a standard image file format. Instead, we write the three two-dimensional arrays of shorts to a file, after encoding through a standard compressor. One way of writing out the image is to use *object serialization* . With object serialization, in Java, an instance of an object (such as the array of *short*) may be written directly to a stream. An example of this appears in the *saveAsShortgz* method which appears in the *SaveFrame* in the *gui* package:

```
public void saveAsShortgz(String fn) {
    Timer t = new Timer();
    t.start();
    try {
        FileOutputStream fos = new FileOutputStream(fn);
        ZIPOutputStream gos = new GZIPOutputStream(fos);
        ObjectOutputStream oos = new ObjectOutputStream(gos);
        oos.writeObject(r);
        oos.writeObject(g);
        oos.writeObject(b);
        oos.close();
        gos.finish();

       } catch(Exception e) {
        System.out.println("Save saveAsShortgz:"+e);
       }
     t.stop();
     t.print(" saveAsShortgz in ");
}
```

A compelling reason to use Java is the compact nature of code such as this. The *writeObject* method will work with any type of primitive array (not just arrays of *short*). Further, this greatly simplifies the image reader. The *OpenFrame* resides in the *gui* package and has a method for getting a g-zipped compressed file of short array instances:

```
public void getAsShortgz() {
   String fn = getReadFileName();
   if (fn == null) return;
    Timer t = new Timer();
    t.start();
    try {
         FileInputStream fis = new FileInputStream(fn);
         GZIPInputStream gis = new GZIPInputStream(fis);
               ObjectInputStream ois = new
ObjectInputStream(gis);
         r = (short[][])ois.readObject();
         g = (short[][])ois.readObject();
         b = (short[][])ois.readObject();
         ois.close();
        } catch(Exception e) {
         System.out.println("Open getAsShortgz Exception:"+e);
        }
     t.stop();
     t.print(" getAsShortgz in ");
     width = r.length;
     height =r[0].length;
     setSize(width,height);
     short2Image();
}
```

The writing of the arrays of short does cause an expansion of the image when saved to disk.

15.7. Summary

This chapter showed that the fast Fourier transform is useful for filtering images. We have also seen that for smaller images with small kernels, direct convolution can be a faster approach.

Two algorithms implemented in the Kahindu program include the Radix-2 Cooley-Tukey FFT and the prime factor algorithm (PFA). We saw that the prime factor algorithm was very good at taking an FFT when the image dimensions are non-integral powers of two (a basic constraint of the Radix-2 FFT). Further, we saw that the Radix-2 FFT is generally faster than the PFA, but is limited in that it can only work on the integral power of two image sizes (e.g., 128x128, 256x256, etc.).

In the second half of this chapter, we were introduced to the simplest of wavelet transforms, the Haar transform. We found that this transform yielded details that were floating point. As a result, we implemented the integer Haar transform. The integer Haar transform was shown to enable a perfect reconstruction from the decomposed image. The drawback was that the dynamic range of the detail coefficients required 9 bits rather than the 8 bits per pixel used in the original image.

The integer Haar transform is just one of an infinite set of integer to integer wavelet transforms. As far as this author knows, there is no byte to byte wavelet transform; perhaps the Haar comes closest. Other transforms may be better at compression, but even a sub-optimal integer Haar transform makes it easy to distribute multi-resolution images on the Web. There are also applications in the areas of pattern matching, image data-base browsing and image sequence indexing.

15.8. Projects

In the speed-up projects, be sure to use the latest version of the code. As the code is updated, it will surely become faster. When comparing run-times, include the display of the result and the intermediate computations. Do not include the reading in of an image. Be sure to use the same hardware and virtual machine when comparing benchmarks.

1. Section 15.1 showed the Radix-2 FFT. Create an optimized version of the Radix-2 FFT. Place your code in your own package. For full credit, you must to make your FFT algorithm 10% faster than the one given in the *gui* package.

2. Select another FFT algorithm and implement it. For example, take the radix-4, decimation-in-time FFT, as outlined in [Jähne]. Implement the new FFT and benchmark it on several image sizes. Now benchmark it against the FFT in Kahindu .

3. A technique for filtering using the FFT and multiplying by an image was disclosed in Section 15.2. Using this technique, develop a series of filters that filter in the frequency domain using mathematical formulas. Be sure to derive the Fourier transform of the formulas in the continuous domain, then, using their discrete versions, perform the filtering in the frequency domain. Be sure to display your filters as images and to print them as part of your report. Use the PFA to compute your filters.

4. Section 15.2 disclosed a Maple procedure for generating optimized DFT's of any length. The *FFT1d* class in the *fft* package shows a mixed radix FFT implementation for numbers up to 10. Devise your own variation on the mixed radix algorithm, extending the numbers for optimization beyond 10 (e.g., 11, 13, 16). Test your program on an image that has dimensions which are evenly divided by your radix. For example, an image that is 320 pixels wide should benefit from an optimized radix 32 FFT. Test your program to make sure the outputs are correct and time your performance. Is your program faster? Can you account for the difference in execution speed?

5. Section 15.2 showed that the assumption of a real sample sequence on input to the FFT algorithm could result in a speed up. Implement the real-valued FFT, as described in that section. Measure the speed up of your implementation over the complex valued FFT. How much of a difference is there?

6. The problem with the FFT is that it requires a complex storage facility, in addition to complex multiplications. This is quite memory intensive, particularly for large images. One way around the problem is to use the Hartley transform,

$$F(u,v) = \frac{1}{WH} \sum_{x=0}^{W-1} \sum_{y=0}^{H-1} f(x,y) \left[\cos\left(\frac{2\pi ux}{W} + \frac{2\pi vy}{H} \right) + \sin\left(\frac{2\pi ux}{W} + \frac{2\pi vy}{H} \right) \right] \quad (15.17)$$

and its inverse,

$$f(x,y) = \sum_{x=0}^{W-1} \sum_{y=0}^{H-1} F(u,v) \left[\cos\left(\frac{2\pi ux}{W} + \frac{2\pi vy}{H} \right) + \sin\left(\frac{2\pi ux}{W} + \frac{2\pi vy}{H} \right) \right] \quad (15.18)$$

Since the Hartley transform is a real-valued transform, no imaginary storage is required. The transform complicates convolution, but cuts the memory requirement in half. A Radix-4 algorithm for the Hartley transform is given in [Bracewell]. The symmetry of the transform pair means that one method can do both the forward and inverse Hartley transform. Implement the discrete Hartley transform using the Radix-4 algorithm. Compare your results with the PFA algorithm.

7. Section 15.5 showed that the integer Haar transform created 9-bit data and that this was stored in the arrays of short integers. After this, it was found that saving the arrays caused an expansion relative to the 24-bit ppm g-zipped file (experiments showed file expansion of about 25%). According to Said and Pearlman [Said], it is possible to have perfect reconstruction using an integer transform called the *S+P transform*. The S+P (**S**equential + **P**rediction) transform uses a linear prediction on the low pass filter coefficients. The linear prediction creates high pass filter coefficients after the S transform. If *c[i]* is the one-dimensional input sequence, then

$$\begin{aligned} l[i] &= \lfloor (c[2i] + c[2i+1]) / 2 \rfloor \\ h[i] &= c[2i] - c[2i+1] \\ \Delta l[i] &= l[i-1] - l[i] \\ \hat{h}[i] &= \alpha_1 \Delta l[i] + \alpha_2 \Delta l[i+1] - \alpha_3 h[i+1] \end{aligned} \quad (15.19)$$

The values obtained by Said and Pearlman for the coefficients in (15.19) were determined experimentally for natural images to be

$$\alpha_1 = 1/4, \alpha_2 = 5/16, \alpha_3 = 1/8 \quad (15.20).$$

The best values for (15.20) depend on the image. Write a program that implements (15.19) and that searches for the values in (15.20) so that the SNR for an image is maximized.

8. Write a program that eliminates *N* pixels from a wavelet transformed image. During the process of eliminating the pixels, you must maximize the SNR. Thus, for each pixel in the image you will create a floating point number that corresponds to the change in SNR if the pixel were to be deleted. Then you will delete the *N* pixels that correspond to the smallest change in the SNR. Now repeat your experiment using the YIQ color space. Which offers the higher SNR for a given *N*?

Appendix A. Book Resources On-line

The journey begins
when you look for the end.

– DL, 1983, in the song *The Study Rot Waltz*

The Web page for this book is located at:

`<http://www.DocJava.com>`

Updates for the code in this book are available, via anonymous ftp, at:

`<ftp://vinny.bridgeport.edu//pub/ipij/>`

A mailing list has been established for this book. To find out more, visit:

`<http://vinny.bridgeport.edu/cgi-bin/lwgate/java-LIST/>`

A list of lists run by the author for various classes may be found at:

`<http://vinny.bridgeport.edu/MailArchives/index.html>`

Of particular interest are the University of Bridgeport course lists, CS410x-list and CS411x-list, *Java Programming* and *Advanced Java Programming*. To glimpse into the world of the professor-students interaction, see:

`<http://vinny.bridgeport.edu/MailArchives/cs410x-list/index.html>`

and

`<http://vinny.bridgeport.edu/MailArchives/cs411x-list/index.html>`

For lecture notes on the course, see the CD-ROM . For updates, visit:

`<http://www.DocJava.com/java/java.html>`

Please feel free to e-mail the author directly at:

`lyon@DocJava.com`

This book contains a number of projects that have been proposed for further work. Anyone interested in contributing original code to the Kahindu program (working and tested!) is most welcome to do so. The authors retain the copyright but must sign a release to give permission to have the code redistributed. All accepted contributions will be acknowledged in future editions of this book.

Appendix B. The Kahindu Interface

Wantonly hacked by an endless stream of nameless,
faceless undergraduates,
both men and women,
often by more than one at the same time,
Kahindu fell into a hell-hole of depravity.

– DL, 1998

Writing an interface that provides a satisfying experience for the user is an art, at best. In fact, the design of a graphic user interface is a topic of great discussion.

In the design of the Kahindu program, I have experimented with the new Java class API called the *Swing* classes (also known as the Java Foundation Classes). These have shown themselves to be painfully slow on platforms where only a Java implementation (i.e., no native implementation) exists. As a result, and in the search for something better, I have devised my own scheme for creating an interface. The result is an interface that is both easy to use and is self-contained. One of the basic design goals was to provide icon resources that are built into the program. I did not want to have a program that required files to be located in particular places on the disk.

There are several excellent reasons to avoid locating icon resources on the disk. First, they load slowly over the Net, unless they are bundled in a zip file. Second, should the files move relative to the program, the program breaks. Third, by locating them in static storage in a class, the icons become a part of the class file. Thus, they have their storage allocated at compile time and load very quickly.

My scheme gives the programmer the ability to create icons in one of three ways: by grabbing the icon using a screen grabbing program, by drawing the icon using either the Kahindu drawing tools or another paint program and finally, by entering the icon using teletype graphics.

B.1. Icon Design by Grabbing

There are often icons that are available as system resource which can be freely copied. These icons start as bit-maps and may be scaled to size to suite the application. The Kahindu program will accept icons of any size; however, the automatic layout facilities work best with icons that are all of the same size. There are several applications that are able to take snapshots of the computer screen. These application vary from platform to platform. For example, Silicon Graphics workstations have an application called *Snap* which will save a snapshot of the screen. On the Mac there is a keyboard shortcut, <cntrl-^M-4>, which changes the cursor into a cross-hair and allows the user to click and drag across the screen.

Depending on the platform, the screen shot will be saved to a file or to the *clipboard.* The clipboard enables the screen short to be pasted into another application. Unfortunately, clipboard pasting of graphics is not currently possible in Java, as far as I know. Further, Kahindu is rather limited in the number of file types that it currently supports. Thus, a third-party application is required to convert the snapshot into an image that Kahindu supports. Currently, this means GIF, PPM or JPEG (using Kahindu v1.5). Suppose, for example, we wanted to grab the icon image from the system to symbolize magnification. Fig. B-1 shows an image of the magnifier icon. This was grabbed using the screen capture facilities on a Mac.

Fig. B-1. The Magnifier Icon

Using an application called Debabelizer [Debabelizer], we save the icon image to a GIF file and open it with the Kahindu program.

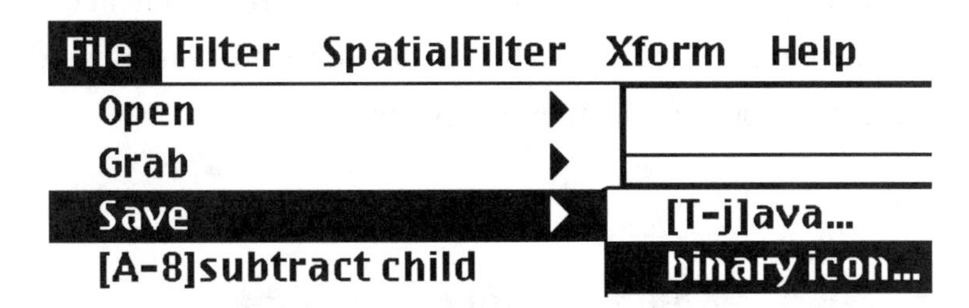

Fig. B-2. Save As Binary Icon

Fig. B-2 shows that the Kahindu program has a menu for saving the binary icon image.

```
public static byte iconName[][] = {

{1,1,1,1,1,1,1,1,1,1,1,1,1,1,1,1},
{1,1,1,1,1,1,1,1,1,1,1,1,1,1,1,1},
{1,1,1,1,1,0,0,0,0,1,1,1,1,1,1,1},
{1,1,1,0,0,1,1,1,1,0,0,1,1,1,1,1},
{1,1,1,0,1,1,1,1,1,1,0,1,1,1,1,1},
{1,1,0,1,1,1,1,1,1,1,1,0,1,1,1,1},
{1,1,0,1,1,1,1,1,1,1,1,0,1,1,1,1},
{1,1,0,1,1,1,1,1,1,1,1,0,1,1,1,1},
{1,1,0,1,1,1,1,1,1,1,1,0,1,1,1,1},
{1,1,1,0,1,1,1,1,1,1,0,1,1,1,1,1},
{1,1,1,0,0,1,1,1,1,0,0,0,1,1,1,1},
{1,1,1,1,1,0,0,0,0,1,0,0,0,1,1,1},
{1,1,1,1,1,1,1,1,1,1,1,0,0,0,1,1},
{1,1,1,1,1,1,1,1,1,1,1,1,0,0,0,1},
{1,1,1,1,1,1,1,1,1,1,1,1,1,0,0,0},
{1,1,1,1,1,1,1,1,1,1,1,1,1,1,0,0},
};
```

Fig. B-3. Binary Icon Output as Java

Fig. B-3 shows the binary icon output at the console as a two-dimensional static byte array. Such data takes very little space in the program (15x15 = 225 bytes) and its space is allocated at compile time.

B.2. Icon Design By Drawing

Another method for obtaining a binary icon is to draw it. Several excellent computer paint programs are available for this task. Kahindu has some built-in drawing tools that also enable the creation (or modification) of an icon.

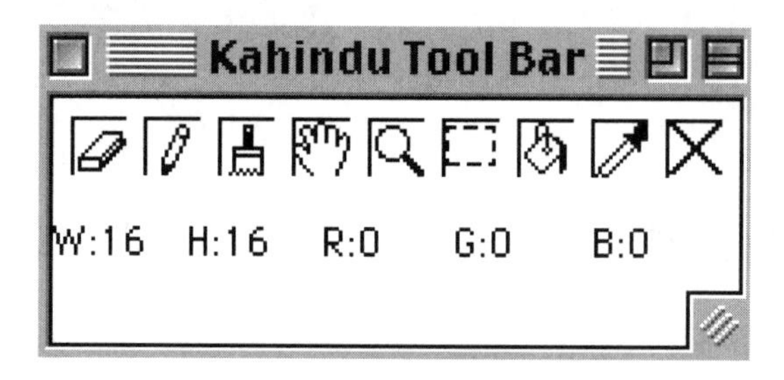

Fig. B-4. The Kahindu Toolbar

Fig. B-4 shows an image of the Kahindu toolbar. From left to right, the icons are identified as the eraser, pencil, paint brush, hand, magnifier, marquee, paint can, eye dropper, and marker. To draw or modify an existing icon, use the eraser, pencil, brush, magnifier and eye dropper. The eraser will clear pixels in the icon. The pencil will set pixels to the value selected with the eye dropper. The brush will set a larger array of pixels than the pencil. Normally, the icons are small. The Kahindu toolbar icons are all 15x15 pixels in size. As a result, the magnifier is used to make the image easier to work on.

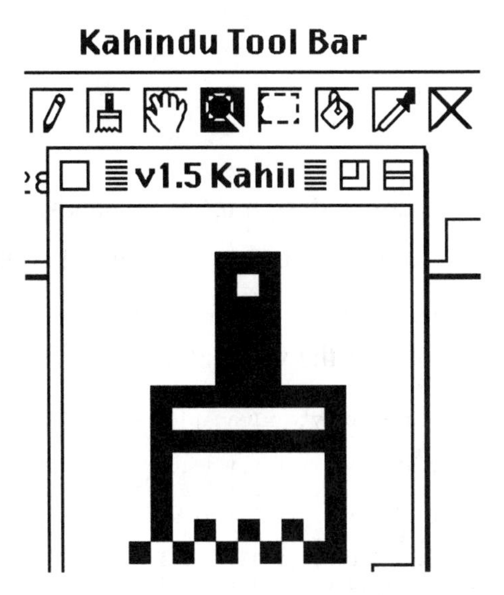

Fig. B-5. Using the Magnifier to Expand the Paint Brush

Fig. B-5 shows an example of the use of the magnifier icon to enlarge the paint brush image. This type of enlargement works by doubling the number of pixels in the icon, and thus growing the icon. To keep the icon the same size, but to enlarge it on the screen, simply resize the frame. Once the frame is enlarged, use the drawing tools to modify the icon.

Fig. B-6. Elements of One Icon Can Be Used in Another

Fig. B-6 shows how the bristles of the brush icon were reused to make a face icon. The design of icons is an art and it takes great care to find icons that have cross-cultural meaning. For example, icons should probably not contain English language characters, as these are not well known in all cultures.

B.3. Icon Design By Typing

As a final approach, the programmer can design icons by hand-keying them into the byte array. The *IconFrame* class resides in the *gui* package. It contains several icons, some of which were typed by hand. Sometimes this is the easiest way to enter an icon, as it gives the greatest control over the value and location of each pixel.

```
private static byte xImage[][] = {
{1,1,1,1,1,1,1,1,1,1,1,1,1,1,1,1},
{0,0,0,1,1,1,1,1,1,1,1,1,1,1,0,0},
{1,1,0,0,1,1,1,1,1,1,1,1,1,0,0,1},
{1,1,1,0,0,1,1,1,1,1,1,1,0,0,1,1},
{1,1,1,1,0,0,1,1,1,1,1,0,0,1,1,1},
{1,1,1,1,1,0,0,1,1,1,0,0,1,1,1,1},
{1,1,1,1,1,1,0,0,1,0,0,1,1,1,1,1},
{1,1,1,1,1,1,1,0,0,0,1,1,1,1,1,1},
{1,1,1,1,1,1,1,0,0,0,1,1,1,1,1,1},
{1,1,1,1,1,1,0,0,1,0,0,1,1,1,1,1},
{1,1,1,1,1,0,0,1,1,1,0,0,1,1,1,1},
{1,1,1,1,0,0,1,1,1,1,1,0,0,1,1,1},
{1,1,1,0,0,1,1,1,1,1,1,1,0,0,1,1},
{1,1,0,0,1,1,1,1,1,1,1,1,1,0,0,1},
{0,0,0,1,1,1,1,1,1,1,1,1,1,1,0,0},
{1,1,1,1,1,1,1,1,1,1,1,1,1,1,1,1},
};
```

Fig. B-7. Hand Typing an Icon

Fig. B-7 shows the marker icon as it was hand-encoded into the *IconFrame*. The shift by exactly one pixel during each entry of the zeros in the array, as well as the precise centering of the mark, are all made much easier by hand entry. In fact, some programs (such as Maple) will output teletype graphics, upon command. The Net is a good source of teletype graphics, which are still e-mailed with wild abandon. Wastrel authors (during lost weekends of drunken excess) have been known to even put them in books. ;)

B.4. Saving the Icon as Java

Now that you have obtained the Java source code needed for an image, you must assign it a name. This is done in the *IconFrame* when the static array is formulated in the Java code. For example:

```
private static byte pencil[][] = {

{1,1,1,1,1,1,1,1,1,1,1,1,1,1,1,1},
{1,1,1,1,1,1,1,1,1,1,1,1,1,1,1,1},
{1,1,1,1,1,1,1,1,1,1,0,0,1,1,1,1},
{1,1,1,1,1,1,1,1,1,0,1,1,0,1,1,1},
{1,1,1,1,1,1,1,1,1,0,1,1,0,1,1,1},
{1,1,1,1,1,1,1,1,0,1,0,0,0,1,1,1},
{1,1,1,1,1,1,1,1,0,1,1,0,1,1,1,1},
{1,1,1,1,1,1,1,0,1,1,1,0,1,1,1,1},
{1,1,1,1,1,1,1,0,1,1,0,1,1,1,1,1},
{1,1,1,1,1,1,0,1,1,1,0,1,1,1,1,1},
{1,1,1,1,1,1,0,1,1,0,1,1,1,1,1,1},
{1,1,1,1,1,0,1,1,1,0,1,1,1,1,1,1},
{1,1,1,1,1,0,1,1,0,1,1,1,1,1,1,1},
{1,1,1,1,1,0,0,0,0,1,1,1,1,1,1,1},
{1,1,1,1,1,0,0,0,1,1,1,1,1,1,1,1},
{1,1,1,1,1,0,0,1,1,1,1,1,1,1,1,1},
};
```

Once the byte array is formulated, an *IconComponent* is created, using the *getIconComponent* method:

```
package gui;
import java.awt.*;
import java.awt.image.*;
import java.awt.event.*;
import java.util.*;

public class IconFrame
   extends ClosableFrame implements ActionListener {

   private Panel iconPanel = new Panel(new FlowLayout());


   IconComponent eraserIcon =
               getIconComponent(eraser);
   IconComponent pencilIcon =
               getIconComponent(pencil);
```

The *IconComponent*instances are added, using:

```
private void addIcons() {
        addIcon(eraserIcon,iconPanel);
        addIcon(pencilIcon,iconPanel);
        addIcon(brushIcon,iconPanel);
        addIcon(handIcon, iconPanel);
        addIcon(magnifyingGlassIcon,iconPanel);
```

Once the icons are added to the *iconPanel* instance, they will appear whenever the Kahindu program starts up.

Event processing occurs both in the *IconFrame* instance and in the frames that are interested in the icon state. For example, in the *PaintFrame*, there is a method called *mouseDragged* which checks the *iconFrame* instance for the selected icon:

```
public void mouseDragged(MouseEvent e) {
                e.consume();
        IconComponent ic = iconFrame.getSelectedIcon();
        if (ic == iconFrame.eraserIcon) erasePoint();
        if (ic == iconFrame.brushIcon) brushPoint();
        if (ic == iconFrame.pencilIcon) pencilPoint();
        if (ic == iconFrame.eyeDropperIcon) getColor();

              setP1(e);
              repaint();
}
```

The *IconComponent* instances act just like other components in the AWT, except that they know how to invert their own appearance. They can be added to panels and layout managers can arrange them, just like other components. The *IconComponent* instance keeps a private copy of an image of its own appearance. While the present implementation is for a bi-level image, color icons are also possible.

Appendix C. The Structure of Kahindu

Do not confine your children to your own learning,
for they were born in another time.

– Hebrew proverb

The main classes in the Kahindu program reside in the *gui* package. Fig. C-1 shows the relationship between some of the main classes involving the *Frame* instances that perform the computations. The Kahindu *Frame* hierarchy starts with the *ClosableFrame*.

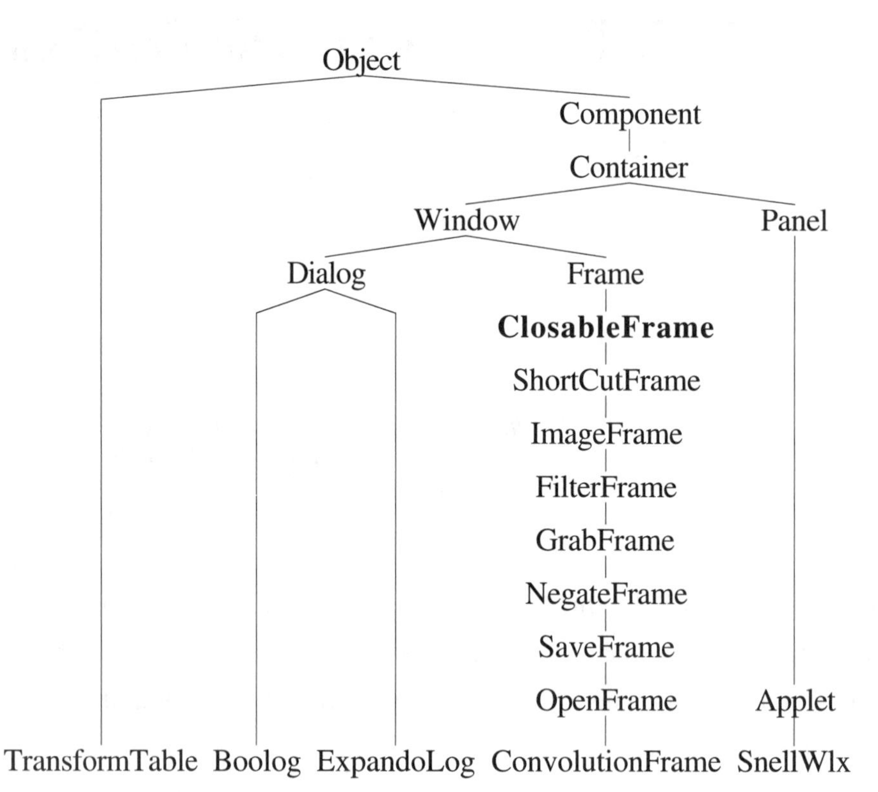

Fig. C-1. Some of the Main Classes in the *gui* package.

Fig. C-2 shows the relationship between the frame subclasses in the *gui* package. When new frames are inserted, they are always extended by the *TopFrame*. The *TopFrame* remains on top because it is always invoked from the *Main* class. The *Main* class resides in the default package and provides the means for invoking the *TopFrame*.

gui.OpenFrame
|
gui.ConvolutionFrame
|
gui.SpatialFilterFrame
|
gui.EdgeFrame
|
gui.MorphFrame
|
gui.BoundaryFrame
|
gui.PaintFrame
|
gui.MartelliFrame
|
gui.ColorFrame
|
gui.XformFrame
|
gui.FFTFrame
|
gui.WaveletFrame
|
gui.TopFrame

Fig. C-2. The Relationship of the Frames in the *gui* Package.

The classes are stacked deeply in order to better organize the code. For example, to add a new save file format, the *SaveFrame* is augmented. To add a new open file format, the *OpenFrame* is expanded. To add a new feature that is not available at all in the program, e.g., pattern recognition, the frame is inserted just below the *TopFrame*. The reason for this is that the new frame inherits all the methods and variables of the super classes.

These super class provide a library of methods that can ease the development of the new feature.

A list of the 564 methods available from the *TopFrame* follows:

```
public class TopFrame
public static void ClosableFrame.main(java.lang.String[])
```

```
public static void ImageFrame.main(java.lang.String[])
public static void FilterFrame.main(java.lang.String[])
public static void GrabFrame.main(java.lang.String[])
public static void NegateFrame.main(java.lang.String[])
public static void SaveFrame.main(java.lang.String[])
public static void OpenFrame.main(java.lang.String[])
public static int ConvolutionFrame.rand(int,int)
public static double
        SpatialFilterFrame.gauss(double,double,double,double,
        double)
public static void SpatialFilterFrame.printGaussKernel(
        int,int,double,double)
public static void
        SpatialFilterFrame.printGaussKernel(int,int,double)
public static float[][]
        SpatialFilterFrame.getGaussKernel(int,int,double)
public static void SpatialFilterFrame.testQuickSort()
public static void SpatialFilterFrame.quickSort(int[])
public static void
        SpatialFilterFrame.printMedian(short[][],java.lang.Strin
        g)
public static double SpatialFilterFrame.mean(int[])
public static double SpatialFilterFrame.variance(int[])
public static double
        SpatialFilterFrame.coefficientOfVariation(int[])
public static short SpatialFilterFrame.median(java.util.Vector)
public static void SpatialFilterFrame.testVariance()
public static void
        SpatialFilterFrame.testCoefficientOfVariation()
public static void SpatialFilterFrame.testOutlier()
public static boolean SpatialFilterFrame.outlierHere(int[])
public static void SpatialFilterFrame.main(java.lang.String[])
public static void SpatialFilterFrame.printMaple(float[][])
public static double EdgeFrame.laplaceOfGaussian(
        double,double,double,double,double)
public static void EdgeFrame.main(java.lang.String[])
public static void
        EdgeFrame.printLaplaceOfGaussianKernel(int,int,double)
public static float[][] EdgeFrame.getLaplaceOfGaussianKernel(
        int,int,double)
public static void MorphFrame.main(java.lang.String[])
public static void
        BoundaryFrame.PrintContainer(java.awt.Container)
public static void BoundaryFrame.main(java.lang.String[])
public static void PaintFrame.main(java.lang.String[])
public static void MartelliFrame.main(java.lang.String[])
public static void XformFrame.main(java.lang.String[])
public static int FFTFrame.gcd(int,int)
public static void WaveletFrame.main(java.lang.String[])
```

```
public static void TopFrame.main(java.lang.String[])
public final native java.lang.Class java.lang.Object.getClass()
public native int java.lang.Object.hashCode()
public boolean java.lang.Object.equals(java.lang.Object)
public java.lang.String java.awt.Component.toString()
public final native void java.lang.Object.notify()
public final native void java.lang.Object.notifyAll()
public final native void java.lang.Object.wait(long) throws
        java.lang.InterruptedException
public final void java.lang.Object.wait(long,int) throws
        java.lang.InterruptedException
public final void java.lang.Object.wait() throws
        java.lang.InterruptedException
public java.lang.String java.awt.Component.getName()
public void java.awt.Component.setName(java.lang.String)
public java.awt.Container java.awt.Component.getParent()
public java.awt.peer.ComponentPeer java.awt.Component.getPeer()
public final java.lang.Object java.awt.Component.getTreeLock()
public java.awt.Toolkit java.awt.Window.getToolkit()
public boolean java.awt.Component.isValid()
public boolean java.awt.Component.isVisible()
public boolean java.awt.Window.isShowing()
public boolean java.awt.Component.isEnabled()
public void java.awt.Component.setEnabled(boolean)
public void java.awt.Component.enable()
public void java.awt.Component.enable(boolean)
public void java.awt.Component.disable()
public void java.awt.Component.setVisible(boolean)
public void java.awt.Window.show()
public void java.awt.Component.show(boolean)
public void java.awt.Component.hide()
public java.awt.Color java.awt.Component.getForeground()
public void java.awt.Component.setForeground(java.awt.Color)
public java.awt.Color java.awt.Component.getBackground()
public void java.awt.Component.setBackground(java.awt.Color)
public java.awt.Font java.awt.Component.getFont()
public synchronized void
        java.awt.Component.setFont(java.awt.Font)
public java.util.Locale java.awt.Window.getLocale()
public void java.awt.Component.setLocale(java.util.Locale)
public java.awt.image.ColorModel
        java.awt.Component.getColorModel()
public java.awt.Point java.awt.Component.getLocation()
public java.awt.Point java.awt.Component.getLocationOnScreen()
public java.awt.Point java.awt.Component.location()
public void java.awt.Component.setLocation(int,int)
public void java.awt.Component.move(int,int)
public void java.awt.Component.setLocation(java.awt.Point)
public java.awt.Dimension java.awt.Component.getSize()
```

```
public java.awt.Dimension java.awt.Component.size()
public void java.awt.Component.setSize(int,int)
public void java.awt.Component.resize(int,int)
public void java.awt.Component.setSize(java.awt.Dimension)
public void java.awt.Component.resize(java.awt.Dimension)
public java.awt.Rectangle java.awt.Component.getBounds()
public java.awt.Rectangle java.awt.Component.bounds()
public void java.awt.Component.setBounds(int,int,int,int)
public void java.awt.Component.reshape(int,int,int,int)
public void java.awt.Component.setBounds(java.awt.Rectangle)
public java.awt.Dimension java.awt.Container.getPreferredSize()
public java.awt.Dimension java.awt.Container.preferredSize()
public java.awt.Dimension java.awt.Container.getMinimumSize()
public java.awt.Dimension java.awt.Container.minimumSize()
public java.awt.Dimension java.awt.Container.getMaximumSize()
public float java.awt.Container.getAlignmentX()
public float java.awt.Container.getAlignmentY()
public void java.awt.Container.doLayout()
public void java.awt.Container.layout()
public void java.awt.Container.validate()
public void java.awt.Container.invalidate()
public java.awt.Graphics java.awt.Component.getGraphics()
public java.awt.FontMetrics
        java.awt.Component.getFontMetrics(java.awt.Font)
public synchronized void
        java.awt.Component.setCursor(java.awt.Cursor)
public java.awt.Cursor java.awt.Component.getCursor()
public void PaintFrame.paint(java.awt.Graphics)
public void java.awt.Container.update(java.awt.Graphics)
public void java.awt.Component.paintAll(java.awt.Graphics)
public void java.awt.Component.repaint()
public void java.awt.Component.repaint(long)
public void java.awt.Component.repaint(int,int,int,int)
public void java.awt.Component.repaint(long,int,int,int,int)
public void java.awt.Container.print(java.awt.Graphics)
public void java.awt.Component.printAll(java.awt.Graphics)
public boolean java.awt.Component.imageUpdate(
        java.awt.Image,int,int,int,int,int)
public java.awt.Image java.awt.Component.createImage(
        java.awt.image.ImageProducer)
public java.awt.Image java.awt.Component.createImage(int,int)
public boolean java.awt.Component.prepareImage(
        java.awt.Image,java.awt.image.ImageObserver)
public boolean java.awt.Component.prepareImage(
        java.awt.Image,int,int,java.awt.image.ImageObserver)
public int java.awt.Component.checkImage(
        java.awt.Image,java.awt.image.ImageObserver)
public int java.awt.Component.checkImage(
        java.awt.Image,int,int,java.awt.image.ImageObserver)
```

```
public boolean java.awt.Component.contains(int,int)
public boolean java.awt.Component.inside(int,int)
public boolean java.awt.Component.contains(java.awt.Point)
public java.awt.Component
        java.awt.Container.getComponentAt(int,int)
public java.awt.Component java.awt.Container.locate(int,int)
public java.awt.Component
        java.awt.Container.getComponentAt(java.awt.Point)
public void java.awt.Container.deliverEvent(java.awt.Event)
public final void
        java.awt.Component.dispatchEvent(java.awt.AWTEvent)
public boolean java.awt.Window.postEvent(java.awt.Event)
public synchronized void
        java.awt.Component.addComponentListener(java.awt.event.Co
        mponentListener)
public synchronized void
        java.awt.Component.removeComponentListener(java.awt.event
        .ComponentListener)
public synchronized void java.awt.Component.addFocusListener(
        java.awt.event.FocusListener)
public synchronized void
        java.awt.Component.removeFocusListener(
        java.awt.event.FocusListener)
public synchronized void java.awt.Component.addKeyListener(
        java.awt.event.KeyListener)
public synchronized void java.awt.Component.removeKeyListener(
        java.awt.event.KeyListener)
public synchronized void java.awt.Component.addMouseListener(
        java.awt.event.MouseListener)
public synchronized void
        java.awt.Component.removeMouseListener(
        java.awt.event.MouseListener)
public synchronized void
        java.awt.Component.addMouseMotionListener(
        java.awt.event.MouseMotionListener)
public synchronized void
        java.awt.Component.removeMouseMotionListener(
        java.awt.event.MouseMotionListener)
public boolean java.awt.Component.handleEvent(java.awt.Event)
public boolean
        java.awt.Component.mouseDown(java.awt.Event,int,int)
public boolean
        java.awt.Component.mouseDrag(java.awt.Event,int,int)
public boolean
        java.awt.Component.mouseUp(java.awt.Event,int,int)
public boolean
        java.awt.Component.mouseMove(java.awt.Event,int,int)
public boolean java.awt.Component.mouseEnter(
        java.awt.Event,int,int)
```

```
public boolean java.awt.Component.mouseExit(
        java.awt.Event,int,int)
public boolean java.awt.Component.keyDown(
        java.awt.Event,int)
public boolean java.awt.Component.keyUp(
        java.awt.Event,int)
public boolean java.awt.Component.action(
        java.awt.Event,java.lang.Object)
public void java.awt.Frame.addNotify()
public void java.awt.Container.removeNotify()
public boolean java.awt.Component.gotFocus(
        java.awt.Event,java.lang.Object)
public boolean java.awt.Component.lostFocus(
        java.awt.Event,java.lang.Object)
public boolean java.awt.Component.isFocusTraversable()
public void java.awt.Component.requestFocus()
public void java.awt.Component.transferFocus()
public void java.awt.Component.nextFocus()
public synchronized void
        java.awt.Component.add(java.awt.PopupMenu)
public synchronized void
        java.awt.Frame.remove(java.awt.MenuComponent)
public void java.awt.Component.list()
public void java.awt.Component.list(java.io.PrintStream)
public void java.awt.Container.list(java.io.PrintStream,int)
public void java.awt.Component.list(java.io.PrintWriter)
public void java.awt.Container.list(java.io.PrintWriter,int)
public int java.awt.Container.getComponentCount()
public int java.awt.Container.countComponents()
public java.awt.Component java.awt.Container.getComponent(int)
public java.awt.Component[] java.awt.Container.getComponents()
public java.awt.Insets java.awt.Container.getInsets()
public java.awt.Insets java.awt.Container.insets()
public java.awt.Component java.awt.Container.add(
        java.awt.Component)
public java.awt.Component java.awt.Container.add(
        java.lang.String,java.awt.Component)
public java.awt.Component java.awt.Container.add(
        java.awt.Component,int)
public void java.awt.Container.add(
        java.awt.Component,java.lang.Object)
public void java.awt.Container.add(
        java.awt.Component,java.lang.Object,int)
public void java.awt.Container.remove(int)
public void java.awt.Container.remove(java.awt.Component)
public void java.awt.Container.removeAll()
public java.awt.LayoutManager java.awt.Container.getLayout()
public void java.awt.Container.setLayout(java.awt.LayoutManager)
public void java.awt.Container.paintComponents(java.awt.Graphics)
```

```
public void java.awt.Container.printComponents(java.awt.Graphics)
public synchronized void
        java.awt.Container.addContainerListener(
        java.awt.event.ContainerListener)
public void java.awt.Container.removeContainerListener(
        java.awt.event.ContainerListener)
public boolean java.awt.Container.isAncestorOf(
        java.awt.Component)
public void java.awt.Window.pack()
public void java.awt.Frame.dispose()
public void java.awt.Window.toFront()
public void java.awt.Window.toBack()
public final java.lang.String java.awt.Window.getWarningString()
public synchronized void java.awt.Window.addWindowListener(
        java.awt.event.WindowListener)
public synchronized void java.awt.Window.removeWindowListener(
        java.awt.event.WindowListener)
public java.awt.Component java.awt.Window.getFocusOwner()
public java.lang.String java.awt.Frame.getTitle()
public synchronized void
        java.awt.Frame.setTitle(java.lang.String)
public java.awt.Image java.awt.Frame.getIconImage()
public synchronized void
        java.awt.Frame.setIconImage(java.awt.Image)
public java.awt.MenuBar java.awt.Frame.getMenuBar()
public synchronized void
        java.awt.Frame.setMenuBar(java.awt.MenuBar)
public boolean java.awt.Frame.isResizable()
public synchronized void java.awt.Frame.setResizable(boolean)
public synchronized void java.awt.Frame.setCursor(int)
public int java.awt.Frame.getCursorType()
public void ClosableFrame.windowClosing(
        java.awt.event.WindowEvent)
public void ClosableFrame.windowClosed(
        java.awt.event.WindowEvent)
public void ClosableFrame.windowDeiconified(
        java.awt.event.WindowEvent)
public void ClosableFrame.windowIconified(
        java.awt.event.WindowEvent)
public void ClosableFrame.windowActivated(
        java.awt.event.WindowEvent)
public void ClosableFrame.windowDeactivated(
        java.awt.event.WindowEvent)
public void ClosableFrame.windowOpened(
        java.awt.event.WindowEvent)
public java.lang.String ShortCutFrame.getPS()
public boolean
        ShortCutFrame.match(java.awt.AWTEvent,java.awt.MenuItem)
public void ShortCutFrame.keyPressed(java.awt.event.KeyEvent)
```

```
public void ShortCutFrame.keyReleased(java.awt.event.KeyEvent)
public void TopFrame.actionPerformed(java.awt.event.ActionEvent)
public void ShortCutFrame.keyTyped(java.awt.event.KeyEvent)
public java.awt.MenuItem
        ShortCutFrame.addMenuItem(java.awt.Menu,java.lang.String)
public void ImageFrame.grabNumImage()
public void ImageFrame.setImageResize(java.awt.Image)
public void ImageFrame.setImage(java.awt.Image)
public void ImageFrame.setImageNoShort(java.awt.Image)
public java.awt.Image ImageFrame.getImage()
public java.lang.String ImageFrame.getFileName()
public void WaveletFrame.clip()
public void ImageFrame.short2Image()
public void ImageFrame.pels2Image(int[])
public void ImageFrame.image2Short()
public void ImageFrame.openGif()
public void ImageFrame.setFileName(java.lang.String)
public void ImageFrame.openGif(java.lang.String)
public void ImageFrame.revert()
public void ImageFrame.int2Short(int[])
public void FilterFrame.gray()
public void GrabFrame.testPattern()
public void GrabFrame.netImageSelector()
public void GrabFrame.grab(java.awt.Container)
public void GrabFrame.grabTestPattern()
public void NegateFrame.auhe()
public void NegateFrame.drawMosaic()
public void NegateFrame.auhe(int,int)
public void NegateFrame.assembleMosaic(NegateFrame,int,int)
public void NegateFrame.drawMosaic(int,int)
public NegateFrame NegateFrame.subFrame(int,int,int,int)
public void NegateFrame.printTT()
public void NegateFrame.add10()
public void NegateFrame.histogram()
public void NegateFrame.negate()
public void NegateFrame.powImage(double)
public double[]
        NegateFrame.average(double[],double[],double[])
public void NegateFrame.unahe()
public void NegateFrame.rnahe(double)
public void NegateFrame.enahe(double)
public double[] NegateFrame.getAverageCMF()
public void NegateFrame.applyLut(short[])
public void NegateFrame.wellConditioned()
public short NegateFrame.inRange(short,int,int)
public short NegateFrame.linearMap(short,double,double)
public void NegateFrame.linearTransform()
public void NegateFrame.linearTransform(double,double)
public void NegateFrame.linearTransform2(double,double)
```

```
public void NegateFrame.computeStats()
public void NegateFrame.printPMFr()
public void NegateFrame.printCMFs()
public void NegateFrame.printPMFg()
public void NegateFrame.printPMFb()
public void NegateFrame.printPMFs()
public void NegateFrame.printStats()
public double NegateFrame.getRBar()
public double NegateFrame.getGBar()
public double NegateFrame.getBBar()
public int NegateFrame.getMin()
public int NegateFrame.getMax()
public void NegateFrame.eponentialLog()
public void NegateFrame.rayleighLog()
public void NegateFrame.linearLog()
public java.lang.String
          SaveFrame.getSaveFileName(java.lang.String)
public void SaveFrame.saveAsPPM()
public void SaveFrame.saveAsPPM(java.lang.String)
public void SaveFrame.saveAsPPMgz(java.lang.String)
public void SaveFrame.saveAsPPMgz()
public void SaveFrame.saveAsShortgz()
public void SaveFrame.saveAsShortZip()
public void SaveFrame.saveAsShortZip(java.lang.String)
public void
          SaveFrame.writeArray(short[][],java.io.DataOutputStream)
          throws java.io.IOException
public void SaveFrame.writeHeader(java.io.DataOutputStream)
          throws java.io.IOException
public void SaveFrame.saveAsShortgz(java.lang.String)
public void SaveFrame.saveAsGif()
public void SaveFrame.saveAsGif(java.lang.String)
public void SaveFrame.saveAsJava()
public void SaveFrame.saveAsJava(java.lang.String)
public void SaveFrame.printIcon()
public void SaveFrame.saveAsJava(java.io.PrintWriter)
public void OpenFrame.getAsShortgz()
public void OpenFrame.openImage()
public void OpenFrame.getShortImageZip(java.lang.String)
public void OpenFrame.readArray(
          short[][],java.io.DataInputStream)
          throws java.io.IOException
public void OpenFrame.readHeader(java.io.DataInputStream)
          throws java.io.IOException
public void OpenFrame.openPPMgz(java.lang.String)
public void OpenFrame.openPPM(java.lang.String)
public void OpenFrame.openPPM()
public StreamSniffer OpenFrame.openAndSniffFile()
public int ConvolutionFrame.cx(int)
```

```
public int ConvolutionFrame.cy(int)
public short[][] ConvolutionFrame.convolveBrute(
        short[][],float[][])
public short[][] ConvolutionFrame.convolve(short[][],float[][])
public short[][] ConvolutionFrame.convolveNoEdge(
        short[][],float[][])
public void SpatialFilterFrame.convolve(float[][])
public void SpatialFilterFrame.makeChild()
public short[][] SpatialFilterFrame.copyArray(short[][])
public void SpatialFilterFrame.subtractChild()
public void SpatialFilterFrame.subtract(SpatialFilterFrame)
public void SpatialFilterFrame.outlierEstimate()
public void SpatialFilterFrame.saltAndPepper(int)
public void SpatialFilterFrame.average()
public void SpatialFilterFrame.hp1()
public void SpatialFilterFrame.hp2()
public void SpatialFilterFrame.hp3()
public void SpatialFilterFrame.hp4()
public void SpatialFilterFrame.hp5()
public void SpatialFilterFrame.usp1()
public void SpatialFilterFrame.lp1()
public void SpatialFilterFrame.lp2()
public void SpatialFilterFrame.lp3()
public void SpatialFilterFrame.mean9()
public void SpatialFilterFrame.mean3()
public void SpatialFilterFrame.gauss3()
public void SpatialFilterFrame.gauss7()
public void SpatialFilterFrame.gauss15()
public void SpatialFilterFrame.gauss31()
public int SpatialFilterFrame.numberOfNonZeros(short[][])
public short SpatialFilterFrame.getMax(short[])
public short SpatialFilterFrame.getMin(short[])
public void SpatialFilterFrame.copyRedToGreenAndBlue()
public void SpatialFilterFrame.medianSquare3x3()
public void SpatialFilterFrame.medianSquare5x5()
public void SpatialFilterFrame.medianOctagon5x5()
public void SpatialFilterFrame.medianDiamond7x7()
public void SpatialFilterFrame.medianCross7x7()
public void SpatialFilterFrame.medianSquare7x7()
public void SpatialFilterFrame.medianCross3x3()
public void SpatialFilterFrame.median(short[][])
public void SpatialFilterFrame.medianBottom(
        short[][],short[][],short[][])
public void SpatialFilterFrame.medianLeft(
        short[][],short[][],short[][])
public void SpatialFilterFrame.medianRightAndTop(
        short[][],short[][],short[][])
public short[][] SpatialFilterFrame.median(
        short[][],short[][])
```

```
public short[][] SpatialFilterFrame.medianNoEdge(
          short[][],short[][])
public void SpatialFilterFrame.testMedian()
public int SpatialFilterFrame.median(int[])
public short[][]
          SpatialFilterFrame.medianSlow(short[][],short[][])
public void EdgeFrame.colorToRed()
public void EdgeFrame.medianSquare2x2()
public void EdgeFrame.median2x1()
public void EdgeFrame.median1x2()
public void EdgeFrame.roberts2()
public void EdgeFrame.shadowMask()
public void EdgeFrame.sizeDetector()
public short[][] EdgeFrame.sizeDetector(short[][])
public void EdgeFrame.sobel3()
public void EdgeFrame.separatedPixelDifference()
public void EdgeFrame.prewitt()
public void EdgeFrame.freiChen()
public void EdgeFrame.pixelDifference()
public void EdgeFrame.templateEdge(float[][],float[][])
public void EdgeFrame.printMaple(float[][],java.lang.String)
public void EdgeFrame.laplacian5()
public void EdgeFrame.laplacian3()
public void EdgeFrame.laplacian3Prewitt()
public void EdgeFrame.laplacian3_4()
public void EdgeFrame.laplacian3Minus()
public void EdgeFrame.tGenerator(int,int)
public void EdgeFrame.thresh()
public void EdgeFrame.convolveZeroCross(float[][])
public void EdgeFrame.zeroCross()
public short[][]
          EdgeFrame.convolveZeroCross(short[][],float[][])
public short[][] EdgeFrame.zeroCross(short[][])
public void EdgeFrame.laplacian9()
public void EdgeFrame.hat13v2()
public void EdgeFrame.hat13()
public void EdgeFrame.horizontalSegment()
public void EdgeFrame.verticalSegment()
public void EdgeFrame.threshLog()
public void EdgeFrame.doit(double[])
public void EdgeFrame.kgreyThresh(double)
public void EdgeFrame.thresh4(double[])
public void MorphFrame.colorPyramid(float[][])
public void MorphFrame.resample2(int)
public void MorphFrame.resample(int)
public short[][] MorphFrame.resampleArray(short[][],int)
public void MorphFrame.colorOpen(float[][])
public void MorphFrame.colorClose(float[][])
public void MorphFrame.open(float[][])
```

```
public void MorphFrame.close(float[][])
public void MorphFrame.serra(float[][])
public short[][] MorphFrame.intersect(short[][],short[][])
public short[][] MorphFrame.complement(short[][])
public void MorphFrame.dilate(float[][])
public void MorphFrame.erode(float[][])
public void MorphFrame.colorDilateErode(float[][])
public void MorphFrame.colorDilate(float[][])
public void MorphFrame.colorErode(float[][])
public void MorphFrame.insideContour(float[][])
public void MorphFrame.outsideContour(float[][])
public void MorphFrame.middleContour(float[][])
public void MorphFrame.clip(short[][],short,short)
public short[][] MorphFrame.subtract(short[][],short[][])
public void MorphFrame.thin()
public void MorphFrame.skeleton()
public boolean MorphFrame.skeletonRedPassSuen(boolean)
public void MorphFrame.deleteFlagedPoints()
public short[][] MorphFrame.erode(short[][],float[][])
public short[][] MorphFrame.erodegs(short[][],float[][])
public short[][] MorphFrame.dilategs(short[][],float[][])
public short[][] MorphFrame.dilate(short[][],float[][])
public int MorphFrame.numberOfNeighbors(boolean[])
public java.util.Vector BoundaryFrame.getPolyList()
public void BoundaryFrame.setPolyList(java.util.Vector)
public void BoundaryFrame.grayPyramid(float[][])
public void XformFrame.copyToChildFrame()
public void BoundaryFrame.displayHoughOfRed()
public java.awt.Point[] BoundaryFrame.getTheLargestPoints(int)
public void BoundaryFrame.drawSomeBigPoints()
public void BoundaryFrame.computeHoughAndDraw()
public void BoundaryFrame.andWithChild()
public void BoundaryFrame.drawThePoints(java.awt.Point[])
public void BoundaryFrame.drawHoughLines(java.awt.Point[])
public short[][] BoundaryFrame.hough()
public short[][] BoundaryFrame.houghGray2()
public void BoundaryFrame.drawHoughLine(int,int,short[][])
public void BoundaryFrame.drawHoughLineGray(int,int,short[][])
public void BoundaryFrame.houghEdge()
public void BoundaryFrame.houghDetect()
public void BoundaryFrame.andHough(BoundaryFrame)
public void BoundaryFrame.inverseHoughToRed()
public void BoundaryFrame.inverseHough()
public void BoundaryFrame.drawLineRed2(int,int,int,int)
public void BoundaryFrame.grabChild()
public void BoundaryFrame.drawLineRed(int,int,int,int)
public void BoundaryFrame.testDrawLineRed()
public short[][] BoundaryFrame.trim(int,int,short[][])
public void BoundaryFrame.bugWalk()
```

```
public void BoundaryFrame.printPolys()
public void BoundaryFrame.listPolys(java.util.Vector)
public void BoundaryFrame.filterPolys()
public java.awt.Polygon BoundaryFrame.thinPoly(java.awt.Polygon)
public boolean BoundaryFrame.onLine(int,int,int,int,int,int)
public int BoundaryFrame.nextClosestPoly(java.awt.Polygon)
public boolean BoundaryFrame.combinePolys(
        java.awt.Polygon,java.awt.Polygon)
public double BoundaryFrame.distance(
        java.awt.Polygon,java.awt.Polygon)
public void BoundaryFrame.polyStats()
public void BoundaryFrame.drawPoly(java.awt.Polygon)
public void BoundaryFrame.drawPolys()
public void PaintFrame.resizeFrame()
public void PaintFrame.showIconFrame()
public void PaintFrame.eraseShapes()
public void PaintFrame.paintShapes(java.awt.Graphics)
public void PaintFrame.mousePressed(java.awt.event.MouseEvent)
public void PaintFrame.mouseExited(java.awt.event.MouseEvent)
public void PaintFrame.mouseEntered(java.awt.event.MouseEvent)
public void PaintFrame.mouseClicked(java.awt.event.MouseEvent)
public void PaintFrame.mouseReleased(java.awt.event.MouseEvent)
public void PaintFrame.magnify()
public void PaintFrame.erasePoint()
public void PaintFrame.pencilPoint()
public void PaintFrame.brushPoint()
public void PaintFrame.mouseDragged(java.awt.event.MouseEvent)
public void PaintFrame.mouseMoved(java.awt.event.MouseEvent)
public void MartelliFrame.averageWithChild()
public Edgel MartelliFrame.minOpenNode()
public int MartelliFrame.countPath(Edgel)
public java.awt.Polygon MartelliFrame.getPath(Edgel)
public void MartelliFrame.drawPath(Edgel)
public Edgel MartelliFrame.searchFromPoint(java.awt.Point)
public void ColorFrame.printSNR()
public double ColorFrame.getSNRinDb()
public float ColorFrame.getSNR()
public float ColorFrame.getTotalSignalPower()
public float ColorFrame.getTotalNoisePower()
public void ColorFrame.copyToFloatPlane()
public void ColorFrame.subSampleChroma4To1()
public void ColorFrame.subSampleChroma2To1()
public void ColorFrame.rgb2Ccir601_2cbcr()
public void ColorFrame.ccir601_2cbcr2rgb()
public void ColorFrame.rgb2xyzd65()
public void ColorFrame.xyzd652rgb()
public void ColorFrame.rgb2iyq()
public void ColorFrame.iyq2rgb()
public void ColorFrame.rgb2hsb()
```

```
public void ColorFrame.hsb2rgb()
public void ColorFrame.rgb2yuv()
public void ColorFrame.yuv2rgb()
public void ColorFrame.rgb2hls()
public void ColorFrame.hls2rgb()
public void ColorFrame.medianCut(int)
public void ColorFrame.linearCut(int,int,int)
public void ColorFrame.linearCut(short[][],int)
public int[] ColorFrame.getPels()
public void ColorFrame.printNumberOfColors()
public int ColorFrame.computeNumberOfColors()
public void ColorFrame.printColors()
public void XformFrame.setPose(double,double,double)
public void XformFrame.turn(double)
public void XformFrame.mirror()
public void XformFrame.turn90()
public void XformFrame.turn180()
public void XformFrame.showAffineFrame()
public java.awt.Point XformFrame.invertMap(
        double[][],double,double)
public java.awt.Point XformFrame.interpolateCoordinates(double[])
public void XformFrame.applyAffineFrame2()
public Mat3 XformFrame.infer3PointA(
        java.awt.Polygon,java.awt.Polygon)
public double[][] XformFrame.infer4PointA(
        java.awt.Polygon,java.awt.Polygon)
public double XformFrame.quadraticRoot(double,double,double)
public double[] XformFrame.inverseMap4(
        double[][],double,double)
public void XformFrame.applyBilinear4Points(
        java.awt.Polygon,java.awt.Polygon)
public void XformFrame.applyBilinear4PointsFeedback(
        java.awt.Polygon,java.awt.Polygon)
public void XformFrame.applyBilinear4Points()
public void XformFrame.applyBilinear4PointsFeedback()
public void XformFrame.inverseBilinearXform(double[][])
public void XformFrame.colorize()
public void XformFrame.zedSquare()
public void XformFrame.inverseBilinearXformfeedback(double[][])
public void XformFrame.applyAffineFrameThreePoints()
public void XformFrame.rotate()
public void XformFrame.scale(int)
public void XformFrame.fishEye()
public void XformFrame.fishEye(double)
public void XformFrame.fishEye(int,int,double)
public void XformFrame.polarTransform()
public void XformFrame.sqrt()
public void XformFrame.xform(Mat3)
public void XformFrame.xformFeedback(Mat3)
```

```
public void FFTFrame.test1DFFTvs()
public void FFTFrame.fftpfa()
public void FFTFrame.fftipfa()
public void FFTFrame.fftR2()
public void FFTFrame.complexMult()
public void FFTFrame.ifftR2()
public void FFTFrame.rgb2Complex()
public void WaveletFrame.demo2d()
public void WaveletFrame.demo1d()
public void WaveletFrame.print(short[][])
public void WaveletFrame.forwardHaar()
public void WaveletFrame.liftingForwardHaar()
public void WaveletFrame.liftingBackwardHaar()
public void WaveletFrame.fh(short[][])
public void WaveletFrame.backwardHaar()
public int[][] WaveletFrame.short2Int(short[][])
public short[][] WaveletFrame.int2Short(int[][])
public void WaveletFrame.stats()
public void WaveletFrame.ulawEncode()
public void WaveletFrame.ulawDecode()
public void WaveletFrame.haarCompress()
public void WaveletFrame.stripimage()
public void WaveletFrame.clearQuad1()
public void WaveletFrame.clearQuad2()
public void WaveletFrame.clearQuad3()
public void WaveletFrame.clearLowerHalf()
public void WaveletFrame.clearLower34()
public void WaveletFrame.clearQuad(int,int,int,int)
public short WaveletFrame.strip(short,short)
public void TopFrame.printMethods(java.lang.reflect.Method[])
public void TopFrame.printClasses()
```

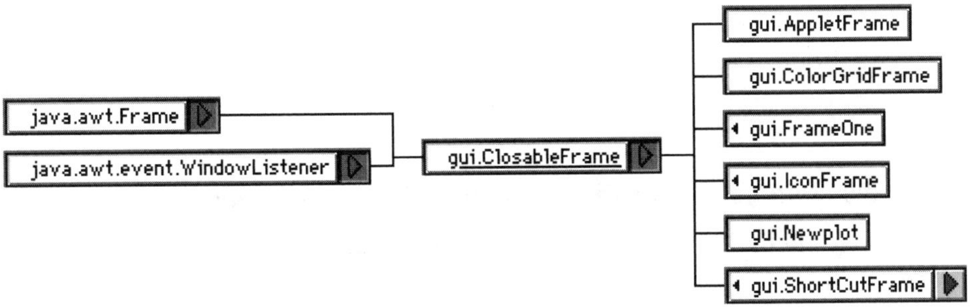

Fig. C-3. The *WindowListener* Hierarchy

Fig. C-3 shows the *WindowListener* hierarchy. The *WindowListener* is an interface that requires the implementation of the following window handling events:

```
public void ClosableFrame.windowClosing(
        java.awt.event.WindowEvent)
public void ClosableFrame.windowClosed(
        java.awt.event.WindowEvent)
public void ClosableFrame.windowDeiconified(
        java.awt.event.WindowEvent)
public void ClosableFrame.windowIconified(
        java.awt.event.WindowEvent)
public void ClosableFrame.windowActivated(
        java.awt.event.WindowEvent)
public void ClosableFrame.windowDeactivated(
        java.awt.event.WindowEvent)
public void ClosableFrame.windowOpened(
        java.awt.event.WindowEvent)
```

In the case of sub-classes of the *ClosableFrame* overriding the window handling events, it is suggestion that the super class version of the method be invoked, before the method returns.

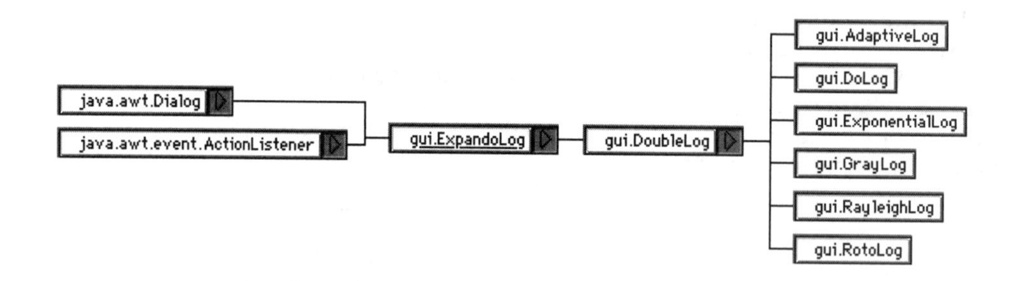

Fig. C-4. The *ActionListener* Hierarchy

Fig. C-4 shows the *Dialog* methods that implement the *ActionListener*. As a result, the *actionPerformed* method is implemented in the *ExpandoLog* class.

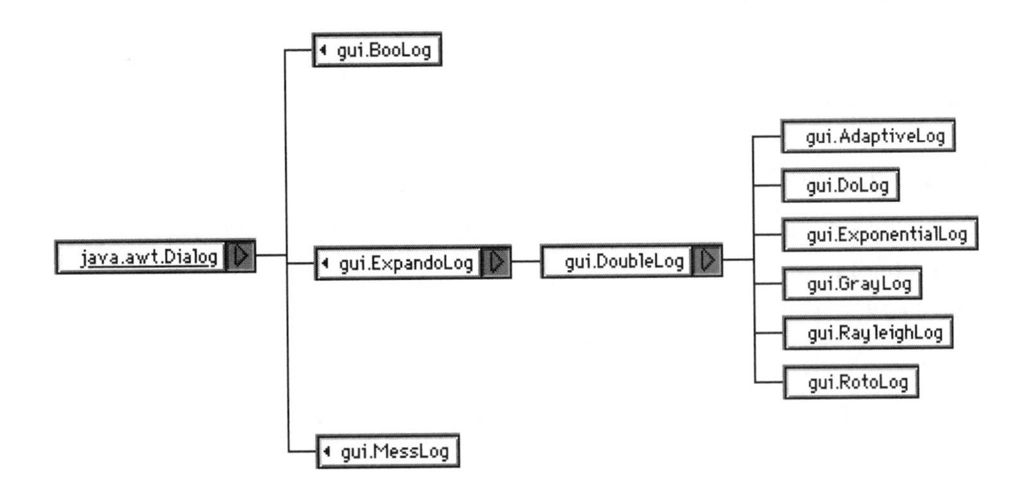

Fig. C-5. The *Dialog* Hierarchy

Fig. C-5 shows a diagram of the *Dialog* hierarchy.

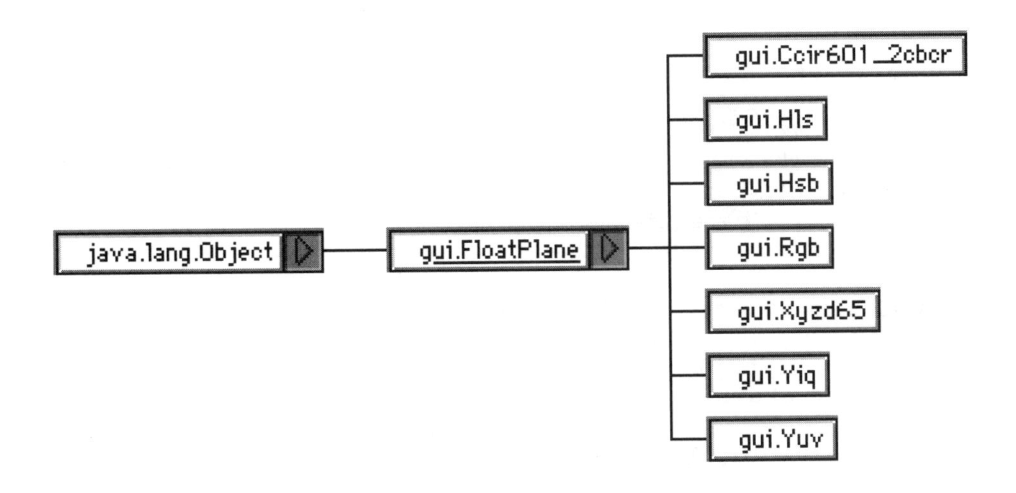

Fig. C-6. The *FloatPlane* Hierarchy

Fig. C-6 shows a diagram of the *FloatPlane* hierarchy. The *FloatPlane* is used where floating point pixel values are needed for image processing. This is very useful when performing color conversions.

To obtain the class diagrams and method lists, we add the *TopFrame* class. The trick in the *TopFrame* class is that it uses the *reflect* package to find out which classes and methods are contained in the system.

```
1.    package gui;
2.
3.    // here is the frame that should always be on
4.    // top.
5.    // To add frames, insert them below TopFrame.
6.    // 10/4/98 - DL
7.    import java.util.*;
8.    import java.awt.*;
9.    import java.awt.event.*;
10.    import java.lang.reflect.*;
11.    public class TopFrame extends WaveletFrame {
12.     MenuItem printInfo_mi =
13.         addMenuItem(fileMenu,"printInfo");
14.     TopFrame(String title) {
15.         super(title);
16.     }
17.     public static void main(String args[]) {
18.         String title = "Kahindu  by D. Lyon";
19.         if (args.length == 1)
20.             title = args[0];
21.     TopFrame tf =
22.         new TopFrame(title);
23.     tf.setVisible(true);
24.     }
25.     public void actionPerformed(ActionEvent e) {
26.         if (match(e,printInfo_mi)) {
27.             printClasses();
28.              return;
29.         }
30.         super.actionPerformed(e);
31.     }
32.     public void printMethods(Method methods[]) {
33.         for (int i=0; i < methods.length; i++)
34.             System.out.println(methods[i]);
35.     }
36.     public void printClasses() {
37.         Vector v = new Vector();
```

```
38.             for (Class c = this.getClass(); c !=
        null;c=c.getSuperclass())
39.                 v.addElement(c);
40.             for (int i=v.size()-1;i >=0 ; i--)
41.                 System.out.println(
                         "("+((Class)v.elementAt(i)).getName());
42.             for (int i=v.size();i >=0; i--)
43.                 System.out.print(")");
44.             System.out.println();
45.             for (int i=0;i >=0 ; i--) {
46.                 Class c = (Class)v.elementAt(i);
47.                 System.out.println(c);
48.                 printMethods(c.getMethods());
49.             }
50.         }
51.     }
```

The class hierarchy, as automatically generated by the JavaDoc program, follows:

- ¥ class java.lang.Object
    - o class gui.Box
    - o class gui.Byte
    - o interface htmlconverter.CText
    - o class vs.ColorUtils
    - o interface gui.Comparable
    - o class java.awt.Component (implements java.awt.image.ImageObserver, java.awt.MenuContainer, java.io.Serializable)
        - class java.awt.Container
            - class java.awt.Panel
                - class java.applet.Applet
                    - class gui.NetImageSelector
                    - class gui.SnellWlx
                - class htmlconverter.ButtonControlPanel (implements java.awt.event.ActionListener)
                - class htmlconverter.ColorControlField
                - class htmlconverter.ColorControlPanel
                - class htmlconverter.FileControlField
                - class htmlconverter.FileControlPanel
                - class htmlconverter.TargetControlPanel
            - class java.awt.Window
                - class java.awt.Dialog
                    - class gui.BooLog (implements java.awt.event.ActionListener)

- class gui.ExpandoLog (implements java.awt.event.ActionListener)
    - class gui.DoubleLog
        - class gui.AdaptiveLog
        - class gui.DoLog
        - class gui.ExponentialLog
        - class gui.GrayLog
        - class gui.RayleighLog
        - class gui.RotoLog
- class gui.MessLog (implements java.awt.event.ActionListener)
- class java.awt.Frame (implements java.awt.MenuContainer)
    - class gui.ClosableFrame (implements java.awt.event.WindowListener)
        - class gui.AppletFrame
        - class gui.ColorGridFrame
        - class gui.IconFrame (implements java.awt.event.ActionListener)
        - class gui.MatLog
        - class gui.ShortCutFrame (implements java.awt.event.KeyListener, java.awt.event.ActionListener)
            - class gui.AffineFrame (implements java.awt.event.MouseListener, java.awt.event.MouseMotionListener)
            - class gui.BeanFrame (implements java.beans.BeanInfo, java.awt.event.ActionListener)
            - class gui.BeanTester
        - class gui.EventTester
        - class gui.FileMenu (implements java.awt.event.ActionListener)
        - class gui.Histogram (implements java.awt.event.ActionListener)
        - class gui.ImageFrame

- class gui.FilterFrame
- class gui.GrabFrame
- class gui.NegateFrame
- class gui.SaveFrame
- class gui.OpenFrame
- class gui.ConvolutionFrame
- class gui.SpatialFilterFrame
- class gui.EdgeFrame (implements gui.Doable)
- class gui.MorphFrame
- class gui.BoundaryFrame
- class gui.PaintFrame (implements java.awt.event.MouseListener, java.awt.event.MouseMotionListener)
- class gui.MartelliFrame
- class gui.ColorFrame
- class gui.XformFrame
- class gui.FFTFrame
- class gui.WaveletFrame
- class gui.TopFrame

class gui.DrawTest
class gui.Graph
class htmlconverter.HtmlGenerator (implements java.awt.event.ActionListener)
class gui.ProgressFrame
class gui.IconComponent
interface htmlconverter.CplusplusText
class java.util.Dictionary

- ¥ class java.util.Hashtable (implements java.lang.Cloneable, java.io.Serializable)
    - o class gui.ColorHash

class gui.Edgel
class fft.FFT2d
class gui.FFTImage
class gui.FFTRadix2
class gui.FloatPlane

- ¥ class gui.Ccir601_2cbcr
- ¥ class gui.Hls
- ¥ class gui.Hsb
- ¥ class gui.Rgb

- ¥ class gui.Xyzd65
- ¥ class gui.Yiq
- ¥ class gui.Yuv

class gui.Haar
class fft.IFFT1d
class fft.IFFT2d
class fft.ImageUtils
class vs.ImageUtils
class htmlconverter.JavaHtmlString
interface htmlconverter.JavaText
class gui.Lifting
class gui.LoopInterchange
class Main
class gui.Mat
class gui.Mat3
class gui.Mat4
class gui.MedianCut
class gui.Multiplot
class gui.NumImage
class gui.Pixel (implements gui.Comparable)
class gui.Point2d
class java.io.Reader

- ¥ class java.io.BufferedReader
  - o class htmlconverter.JavaStream (implements htmlconverter.JavaText, htmlconverter.CText, htmlconverter.CplusplusText)

class gui.Search
class gui.Sort
class gui.StreamSniffer
class fft.TempVars

- ¥ class fft.FFT1d

class ThreadTest
class gui.Timer
class gui.TransformTable
class gui.UByte
class gui.Wavelet
class vs.WriteGIF
class vs.WritePPM
class gui.Wu1
class vs.vsFFT
class vs.vsFFT1D
class vs.vsTimer

Appendix D. Index of Fields and Methods

The best and the wisest is to focus on the good and the beautiful, as represented by working code that is properly indented.

– Unknown

This appendix contains a list of all public methods and fields, sorted by name, in the Kahindu v1.5 program. This is likely to change as the Kahindu program matures, but is correct, as of this writing.

A

actionPerformed(ActionEvent). Method in class gui.AdaptiveLog
actionPerformed(ActionEvent). Method in class gui.AffineFrame
actionPerformed(ActionEvent). Method in class gui.BeanFrame
actionPerformed(ActionEvent). Method in class gui.BooLog
actionPerformed(ActionEvent). Method in class gui.BoundaryFrame
actionPerformed(ActionEvent). Method in class htmlconverter.ButtonControlPanel
actionPerformed(ActionEvent). Method in class gui.ColorFrame
actionPerformed(ActionEvent). Method in class gui.DoLog
actionPerformed(ActionEvent). Method in class gui.EdgeFrame
actionPerformed(ActionEvent). Method in class gui.EventTester
actionPerformed(ActionEvent). Method in class gui.ExpandoLog
actionPerformed(ActionEvent). Method in class gui.ExponentialLog
actionPerformed(ActionEvent). Method in class gui.FFTFrame
actionPerformed(ActionEvent). Method in class gui.FileMenu
actionPerformed(ActionEvent). Method in class gui.FilterFrame
actionPerformed(ActionEvent). Method in class gui.GrabFrame
actionPerformed(ActionEvent). Method in class gui.Graph

magnitudeSpectrum(float[], float[]). Method in class vs.vsFFT
Main(). Constructor for class Main
main(String[]). Static method in class gui.AffineFrame
main(String[]). Static method in class gui.BeanTester
main(String[]). Static method in class gui.BooLog
main(String[]). Static method in class gui.BoundaryFrame
main(String[]). Static method in class gui.ClosableFrame
main(String[]). Static method in class gui.ColorHash
main(String[]). Static method in class gui.DoubleLog
main(String[]). Static method in class gui.DrawTest
main(String[]). Static method in class gui.EdgeFrame
main(String[]). Static method in class gui.EventTester
main(String[]). Static method in class gui.ExpandoLog
main(String[]). Static method in class gui.FilterFrame
main(String[]). Static method in class gui.GrabFrame
main(String[]). Static method in class gui.Graph
main(String[]). Static method in class gui.Haar
main(String[]). Static method in class
htmlconverter.HtmlGenerator
main(String[]). Static method in class gui.IconFrame
main(String[]). Static method in class gui.ImageFrame
main(String[]). Static method in class gui.Lifting
main(String[]). Static method in class gui.LoopInterchange
main(String[]). Static method in class Main
main(String[]). Static method in class gui.MartelliFrame
main(String[]). Static method in class gui.Mat
main(String[]). Static method in class gui.Mat3
main(String[]). Static method in class gui.MessLog
main(String[]). Static method in class gui.MorphFrame
main(String[]). Static method in class gui.Multiplot
main(String[]). Static method in class gui.NegateFrame
main(String[]). Static method in class gui.OpenFrame
main(String[]). Static method in class gui.PaintFrame
main(String[]). Static method in class gui.ProgressFrame
main(String[]). Static method in class gui.SaveFrame
main(String[]). Static method in class gui.Search
main(String[]). Static method in class gui.Sort
main(String[]). Static method in class gui.SpatialFilterFrame
main(String[]). Static method in class ThreadTest
main(String[]). Static method in class gui.TopFrame
main(String[]). Static method in class gui.UByte
main(String[]). Static method in class vs.vsFFT1D
main(String[]). Static method in class gui.Wavelet
main(String[]). Static method in class gui.WaveletFrame
main(String[]). Static method in class gui.XformFrame

T

toString(). Method in class gui.StreamSniffer
toString(byte). Static method in class gui.Byte
toString(byte, int). Static method in class gui.Byte
transform(Polygon). Method in class gui.Mat3
TransformTable(int). Constructor for class gui.TransformTable
transpose(). Method in class gui.FloatPlane
transpose(). Method in class gui.Mat3
transpose(). Method in class gui.Mat4
transpose(int[][]). Static method in class gui.Wavelet
trim(int, int, short[][]). Method in class gui.BoundaryFrame
turn(double). Method in class gui.XformFrame
turn180(). Method in class gui.XformFrame
turn90(). Method in class gui.XformFrame
TYPENOTFOUND. Static variable in class gui.StreamSniffer

U
ui(byte). Static method in class gui.UByte
ulawDecode(). Method in class gui.WaveletFrame
ulawEncode(). Method in class gui.WaveletFrame
unahe(). Method in class gui.NegateFrame
UNIX_COMPRESS. Static variable in class gui.StreamSniffer
update(Graphics). Method in class gui.DrawTest
update(Graphics). Method in class gui.Graph
update(Graphics). Method in class gui.Histogram
update(Graphics). Method in class gui.NetImageSelector
updateParent(). Method in class gui.FloatPlane
updateParent(float). Method in class gui.FloatPlane
us(byte). Static method in class gui.UByte
usp1(). Method in class gui.SpatialFilterFrame
UUENCODED. Static variable in class gui.StreamSniffer

V
variance(int[]). Static method in class gui.SpatialFilterFrame
verticalSegment(). Method in class gui.EdgeFrame
vsFFT(). Constructor for class vs.vsFFT
vsFFT1D(). Constructor for class vs.vsFFT1D
vsTimer(). Constructor for class vs.vsTimer

W
warp. Static variable in class gui.IconFrame
Wavelet(). Constructor for class gui.Wavelet
wellConditioned(). Method in class gui.NegateFrame
width. Variable in class gui.IconFrame
width. Variable in class gui.ImageFrame
windowActivated(WindowEvent). Method in class gui.ClosableFrame

windowClosed(WindowEvent). Method in class gui.ClosableFrame
windowClosing(WindowEvent). Method in class gui.ClosableFrame
windowDeactivated(WindowEvent). Method in class gui.ClosableFrame
windowDeiconified(WindowEvent). Method in class gui.ClosableFrame
windowIconified(WindowEvent). Method in class gui.ClosableFrame
windowOpened(WindowEvent). Method in class gui.ClosableFrame
Write(OutputStream). Method in class vs.WriteGIF
writeArray(short[][], DataOutputStream). Method in class gui.SaveFrame
WriteGIF(byte[][], byte[][], byte[][]). Constructor for class vs.WriteGIF
WriteGIF(Image). Constructor for class vs.WriteGIF
writeHeader(DataOutputStream). Method in class gui.SaveFrame
WritePPM(int, int). Constructor for class vs.WritePPM

X
x. Variable in class gui.Point2d
xform(Mat3). Method in class gui.XformFrame
xformFeedback(). Method in class gui.AffineFrame
xformFeedback(Mat3). Method in class gui.XformFrame
XformFrame(String). Constructor for class gui.XformFrame
Xyzd65(ColorFrame). Constructor for class gui.Xyzd65
xyzd652rgb(). Method in class gui.ColorFrame

Y
y. Variable in class gui.Point2d
yesButton. Variable in class gui.BooLog
yesButton. Variable in class gui.MessLog
Yiq(ColorFrame). Constructor for class gui.Yiq
Yuv(ColorFrame). Constructor for class gui.Yuv
yuv2rgb(). Method in class gui.ColorFrame

Literature Cited

Many men go fishing all of their lives
without knowing that it is not fish they are after.

– Henry David Thoreau

Some of the references cited are available at

<http://www.DocJava.com/book/ipij.html>.

[@Home] At Home Corporation, a distributer of a hybrid fibre coax network service, http://www.home.net/corp/network.html

[Aghajan] 1993. "Sensor Array Processing Techniques for Super Resolution Multi-Line-Fitting and Straight Edge Detection", by Hamid K. Aghajan and Thomas Kailath, IEEE TIP, vol. 2, No. 4, October, pp. 454-465.

[Allen] 1990. 740 pps., *Probability, Statistics, and Queueing Theory with Computer Science Applications,* Second Edition, by Arnold Allen. AP.

[Amazigo] 1980. 407 pps., *Advanced Calculus And Its Applications to the Engineering and Physical Sciences,* by John C. Amazigo and Lester A. Rubenfeld. JW.

[Anton] 1977. 315 pps., *Elementary Linear Algebra,* by Howard Anton. JW.

[Baker] 1991. 308 pps., *More C Tools for Scientists and Engineers,* by Louis Baker. MH.

[Ballard] 1981. "Generalizing the Hough Transform to Detect Arbitrary Shapes", by D. H. Ballard, *Pattern Recognition* vol. 13, No. 2 pp. 111-122.

[Banerjee] 1991. 347 pps., *Principles of Applied Optics*, by Partha P. Banerjee and Ting-Chung Poon, Aksen Associates Inc., Boston, MA.

[Barr] 1981. 409 pps., *The Handbook of Artificial Intelligence*, by Avron Barr and Edward A. Feigenbaum. AW.

[Bentley] 1979. "Data Structures for Range Searching", by Jon Louis Bentley and Jerome H. Friedman, *Computing Surveys*, vol. 11, No. 4, December, pp. 397-409.

[Bishop] 1997. 391 pps., *Java Gently*, by Judy Bishop. AW.

[Boomgaard] 1992. "Methods for Fast Morphological Image Transforms Using Bitmapped Binary Images", by Rein Van Den Boomgaard and Richard Van Balen, CVGIP, vol. 54, No. 3, May, pp. 252-258.

[Boyce] 1977. *Elementary Differential Equations and Boundary Value Problems*, by William E. Boyce and Richard C. DiPrima. JW.

[Bracewell] 1995. 689 pps., *Two-Dimensional Imaging*, by Ronald N. Bracewell. PH.

[Calderbank] "Lossless Image Compression Using Integer to Integer Wavelet Transforms", by A. R. Calderbank, Ingrid Daubechies, Wim Sweldens and Boon-Lock Yeo, pps. 4. Available at: <http://cm.bell-labs.com/who/wim/papers/papers.html>.

[Campione and Walwalrath] 1996. 831pps., *The Java Tutorial*, by Campione and Walwalrath. AW. CD.

[Canny] 1983. 145 pps., *Finding Edges and Lines in Images*, by John Francis Canny. Technical Report No. 720, AI-TR-720, MIT Artificial Intelligence Laboratory, 545 Technology Square, Cambridge, MA 02139.

[Canny2] 1986. "A Computational Approach to Edge Detection", by John Canny, *Transactions on Pattern Analysis and Machine Intelligence*, vol. PAMI-8, No. 6, November, pps. 679-698, IEEE.

[Carlson] 1986. 686 pps., *Communication Systems*, by A. Bruce Carlson. MH.

[Castleman] 1996. 667 pps., *Digital Image Processing,* by Kenneth R. Castleman. PH.

[CCIR-601-2] 1990. “Encoding Parameters of Digital Television for Studios”, Recommendation 601-2, available as scanned Tiff images from <http://www.igd.fhg.de/icib/tv/ccir/rec_601-2/scan.html>.

[Chan and Lee] 1996. 1660 pps., *The Java Class Libraries*, by Chan and Lee. AW. On-line materials.

[Chan and Lee2] 1998. 1682 pps., *The Java Class Libraries,* Second Edition, vol. 2, by Chan and Lee. AW. On-line material.

[Chan et al.] 1998. 2050 pps., *The Java Class Libraries,* Second Edition, vol. 1, by Chan, Lee and Kramer. AW. On-line material.

[Char et al.] 1991. *Maple V Language Reference Manual*, by Bruce W. Char, Kieth O. Geddes, Gaston H. Gonnet, Benton L. Leong, Michael B. Monagan and Stephen M. Watt. SV.

[Churchill] 1976. 332 pps., *Complex Variables and Applications,* by Ruel V. Churchill, James W. Brown and Roger F. Verhey. MH.

[Clocksin] 1981. *Programming in Prolog*, by W. F. Clocksin and C. S. Mellish, SV.

[Cohen] 1993. *Radiosity and Realistic Image Synthesis*, by Michael F. Cohen and John R. Wallace, AP.

[Cornell and Horstmann] 1997. 776 pps., *Core Java*, Second Edition, by Gary Cornell and Cay S. Horstmann, PH. CD.

[Cowan et al.] 1985. “Colour Perception Tutorial Notes”, by William Cowan and Colin Ware, ACM SIGGRAPH, July 22-26, San Francisco, CA.

[Crane] 1997. 317 pps., *A Simplified Approach to Image Processing Classical and Modern Techniques in C*, by Randy Crane, PH. Floppy.

[DataViz] *MacLinkPlus.* DataViz Inc., 55 Corporate Dr., Trumbull CT 06611. Phone: (203)268-0030

[Debabelizer] Debabelizer is a useful program for batch image conversion and processing, available from: Equilibrium, 475 Gate Five Road, #225, Sausalito, CA 94965. Phone: (415)332-4343.

[Déforges] 1997. “Recursive Morphological Operators for Gray Image Processing. Application in Ganulometry Analysis”, by O. Déforges and N. Normand, ICIP.

[Deitel and Deitel] 1997. 1050 pps., *Java: How to Program,* by Deitel and Deitel. PH. On-line materials.

[DeWitt and Lyon] 1995. "Three-Dimensional Microscope Using Diffraction Grating", Optcon, SPIE - International Society for Optical Engineering, Co-authored with Thomas D. DeWitt, Philadelphia, PA, October 24, 2599B-35. Also available from <http://www.DocJava.com>.

[Dougherty] 1992. 161 pps., "An Introduction to Morphological Image Processing", SPIE.

[Dyer] 1983. "Gauge Inspection Using Hough Transform", Charles R. Dyer, IEEE PAMI vol. 5, No. 6, November. pp. 621-623.

[Embree] 1991.456 pps., *C Language Algorithms for Digital Signal Processing,* by Paul M. Embree and Bruce Kimble. Floppy. PH.

[Espeset] 1996. 480 pps., *Kick Ass Java*, by Tonny Espeset. CD. Coriolis Group Books.

[Feitelson] 1989. 393 pps., *Optical Computing*, by Dror G. Feitelson. MIT.

[Feller] 1968. 509 pps., *An Introduction to Probability Theory and its Applications*, by William Feller, JW.

[Forney] 1989. "Introduction to Modem Technology: Theory and Practice of Bandwidth Efficient Modulation from Shannon and Nyquist to Date", by G. David Forney, 49 Minute video, Distinguished Lecture Series, vol. II: University Video Communications. PO Box 5129, Stanford, CA 94309.

[Galbiati] 1990. 164 pps., *Machine Vision and Digital Image Processing Fundamentals,* by Louis J. Galbiati, Jr., PH.

[Gamma] 1995. 395 pps., *Design Patterns*, by Erich Gamma, Richard Helm, Ralph Johnson and John Vlissides. AW.

[Geary] 1997. 877 pps., *Graphic Java Mastering the AWT*, Second Edition by David M. Geary. PH. CD.

[Geary and McClellan] 1997. 600 pps., *Graphic Java Mastering the AWT*, by David M. Geary and Alan L. McClellan. PH. CD.

[Gersho] 1978. "Principles of Quantization", IEEE Transactions on Circuits and Systems, vol. CAS-25, No. 7, July, pp. 427-436.

[Glassner] 1989. 329 pps., *An Introduction to Ray Tracing*, edited, by Andrew S. Glassner. AP.

[Gonzalez et al.] 1977. 431 pps., *Digital Image Processing*, by Rafael C. Gonzalez and Paul Wintz. AW.

[Gonzalez and Woods] 1992. 716 pps., *Digital Image Processing*, by Rafael C. Gonzalez and Richard Woods. AW.

[Gosling et al.] 1996. 825 pps., *The Java Language Specification*, by James Gosling, Bill Joy, and Guy Steele. AW.

[Graf] 1987. 246 pps.,*Video Scrambling & Descrambling for Satellite & Cable TV,* by Rudolf F. Graf and William Sheets. SAMS.

[Guil] 1995. "A Fast Hough Transform for Segment Detection", by Nicolas Guil, Julio Villalba and Emilio L. Zapata,

[Gunter] 1995. *The File Formats Handbook,* by Gunter Born, International Thompson, Computer Press, Boston, MA. Floppy.

[Hall 1974] 1974. "Almost Uniform Distributions for Computer Image Enhancement", by Ernest L. Hall, *IEEE Transactions on Computers*, February, pp. 207-208.

[Halliday] 1978. 1186 pps., *Physics*, by David Halliday and Robert Resnick. JW.

[Heckbert] 1990. "Digital Line Drawing", by Paul S. Heckbert, pps. 99-100 and pp. 685, in *Graphics Gems*, edited by Andrew S. Glassner. AP.

[Heckbert 80] 1980. *Color Image Quantization for Frame Buffer Display*, by Paul S. Heckbert, B.S. Thesis, Architecture Machine Group, MIT, Cambridge, MA. Available at <http://www.cs.cmu.edu/~ph>.

[Heckbert 82] 1982. "Color Image Quantization for Frame Buffer Display", by Paul Heckbert, *Computer Graphics*, vol. 16, No. 3, July. pps. 297-307. Available at <http://www.cs.cmu.edu/~ph>.

[Heckbert 86] 1986. "Survey of Texture Mapping", by Paul S. Heckbert, Nov., pp. 56-67. IEEE CGA. Available at <http://www.cs.cmu.edu/~ph>.

[Heckbert 89] 1989. "Fundamentals of Texture Mapping and Image Warping", by Paul Heckbert, Master's Thesis, Dept. of Electrical Engineering and Computer

Science, University of California, Berkeley, CA 94720. UCB/CSD 89/516. Available at <http://www.cs.cmu.edu/~ph>.

[Hennessy et al.] 1996. *Computer Architecture: A Quantitative Approach*, Second Edition, by John Hennessy and David Patterson. MK.

[Hockney] 1996. 129 pps., *The Science of Computer Benchmarking*, by Roger W. Hockney, SIAM.

[Holzmann] 1988. 120 pps., *Beyond Photography,* by Gerard J. Holzmann. PH.

[Huffman] 1952. "A Method for the Construction of Minimum Redundancy Codes", *Proceedings of the Institute of Radio Engineers*, September, vol. 40, No. 9, pp. 1098-1101.

[Hunt] 1991. 313 pps., *Measuring Color*, by R. W. G. Hunt, Second Edition. EH.

[Hussain] 1991. 406 pps., *Digital Image Processing,* by Zahid Hussain. EH.

[Inglis] 1993. *Video Engineering*, by Andrew F. Inglis. MH.

[Jain]. 1989. 569 pps., *Fundamentals of Digital Image Processing*, by Anil K. Jain. PH.

[Jähne] 1993. 383 pps., *Digital Image Processing, Concepts, Algorithms, and Scientific Applications*, by Bernd Jähne, Second Edition, S.V.

[Kasson] 1992. "An Analysis of Selected Computer Interchange Color Spaces", by Kasson, vol. 11, No. 4, October, 1992. pp. 373-405.

[Kay] 1995. 278 pps., *Graphics File Formats,* by David C. Kay and John R. Levine, Windcrest, an imprint of M.H.

[Kientzle] 1995. *Internet File Formats,* by Tim Kientzle. C.G.

[Kruger] "Median-Cut Color Quantization: Fitting True-Color Images into VGA Displays", by Anton Kruger, *Dr. Dobb's Journal of Software Tools*. MT. vol. 19, No. 10, pp. 46. Code available from <http://www.ddj.com/>

[Kwok] 1997. "A fast recursive shortest spanning tree for image segmentation and edge detection", by S. H. Kwok and A. G. Constantinides, *IEEE Transactions on Image Processing*, vol. 6, No. 2, February, pps. 328-332.

[Lai] 1994. 92 pps., *Deformable Contours: Modeling, Extraction, Detection and Classification*, by Kok Fung Lai. Ph.D. Thesis, Electrical Engineering Department, University of Wisconsin-Madison.

[Sweldens] 1995 "The Lifting Scheme: A new philosophy in biorthogonal wavelet constructions", by Wim Sweldens. Edited by A. F. Laine and M. Unser, Wavelet Applications in Signal and Image Processing III, pps. 68-79, Proc. SPIE 2569. Available at <http://cm.bell-labs.com/who/wim/papers/papers.html>

[Laurel] 1990. 523 pps., *The Art of Human-Computer Interface Design*, edited by Brenda Laurel. AW.

[Lea] 1997. 339 pps., *Concurrent Programming in Java*, by Doug Lea. AW.

[Lemay and Perkins] 1996. 527 pps., *Teach Yourself Java in 21 days*, by Laura Lemay and Charles L. Perkins. Sams Net. CD.

[Levine] 1985. *Vision in Man and Machine,* by Martin Levine. MH.

[Lyon 85] 1985. *Raster-To-Vector Conversion with A Vector Ordering Post-process*, by Douglas Lyon, Image Processing Laboratory User Bulletin U-170, June 25, Computer and Systems Engineering Department, RPI, Troy, NY, 12181.

[Lyon 90] 1990. "Ad-Hoc and Derived Parking Curves", by Douglas Lyon, SPIE - International Society for Optical Engineering, Boston MA, November 8.

[Lyon] 1991. *Parallel Parking with Nonholonomic Constraints,* by Douglas Lyon, Ph.D. thesis, RPI, Computer and Systems Engineering Department, RPI, Troy, NY, 12181. Available at <http://www.DocJava.com>.

[Lyon 1995] 1995. "Using Stochastic Petri Nets for Real-time Nth-order Stochastic Composition ", by Douglas Lyon, *Computer Music Journal,* Winter, vol. 19, No. 4, pp. 13-22.

[Lyon 1997] 1997. US Patent Pending "Apparatus and method for the Generation of Nth order Markov Events with improved management of memory and CPU usage," Number 60/034,303, December 23. United States Patent and Trademark Office, Patent Pending.

[Lyon and Rao] 1997. 428 pps., *Java Digital Signal Processing, M&T*. More information available on-line at <http://www.DocJava.com>. CDROM

[LZ77] Ziv J., Lempel A., "A Universal Algorithm for Sequential Data Compression," IEEE Transactions on Information Theory, vol. 23, No. 3, pp. 337-343.

[Maple 98] 1998. Personal Communications with MapleSoft technical support, Scott Rabuka <srabuka@maplesoft.com> and <support@maplesoft.com>.

[Marr] 1980. "Theory of Edge Detection", by D. Marr and E. Hildreth, *Proc. R. Soc. Lond. B* , vol. 207, pps., 187-217.

[Martelli 72] 1972. "Edge Detection Using Heuristic Search Methods", by Alberto Martelli, CGIP, vol. 1, No. 2, August. pps. 169-182.

[Martelli 76] 1976. "An Application of Heuristic Search Methods to Edge and Contour Detection", by Alberto Martelli, CACM, vol. 19, No. 2, February. pps. 73-83.

[Martindale] 1991. "Television Color Encoding and 'Hot' Broadcast Colors", by David Martindale and Alan W. Paeth, in *Graphics Gems vol. II,* edited by James Arvo. AP. pp. 147-158.

[Mattison] 1994. *Practical Digital Video with Programming Examples in C*, by Phillip E. Mattison. JW.

[McGee] 1995. "A Heuristic Approach to Edge Detection in on-line Portal Imaging", by Kiaran P. McGee, Timothy E. Schultheiss, and Eric E. Martin, *Int. J. Radiation Oncology Biol. Phys.*, vol. 32, No. 4., pp. 1185-1192.

[Mehrotra] 1996. "A Computational Approach to Zero-Crossing-Based Two-Dimensional Edge Detection", by Rajiv Mehrotra and Shiming Zhan, *Graphical Models and Image Processing,* vol. 58, No. 1, January. pps. 1-17. AP.

[Mehtre et al.] 1995. "Color Matching for Image Retrieval", by Babu M. Mehtre, Mohan S. Kankanhalli, A. Desai Narasimhalu, Guo Chang Man, *Pattern Recognition Letters* vol. 16, March, pps. 325-331.

[Merlin] 1975. "A Parallel Mechanism for Detecting Curves in Pictures", by Philip M. Merlin and David J. Farber, IEEE Transactions on Computers, January. pps. 96-98

[Meyer] 1997. *Java Virtual Machine, by* Meyer, J. and Downing, T., OR.

[Mitra] 1993. 1268 pps., *Handbook for Digital Signal Processing*, by Sanjit K. Mitra and James F. Kaiser. JW.

[Moore] 1990. 560 pps., *Elements of Computer Music*, by Moore, F.R., PH.

[Moore 64] 1964. 174 pps., *Elementary General Topology*, by Theral O. Moore. PH.

[Mullet] 1995. 273 pps., *Designing Visual Interfaces*, by Kevin Mullet and Darrell Sano. PH.

[Murray] 1996. 894 pps., *Graphics File Formats*, by James D. Murray and William Vanryper. O'Reilly & Associates. CD.

[Myler] 1993. 284 pps., *Computer Imaging Recipes in C*, by Harley R. Myler and Arthur R. Weeks. PH. Floppy.

[Nadler] 1993. 588 pps., *Pattern Recognition Engineering*, by Morton Nadler and Eric P. Smith. JW.

[Nanzetta] 1971. 117 pps., *Set Theory and Topology*, by Nanzetta and Strecker. BQ.

[Netravali] 1988.*Digital Pictures, by* Arun Netravali and Barry Haskell. Plenum Press, NY.

[NetPBM] 1993. A public domain image processing package. You'll find the latest release of Netpbm at <ftp://wuarchive.wustl.edu/graphics/graphics/packages/NetPBM>

[Newman] 1979. 541 pps., *Principles of Interactive Computer Graphics*, by William M. Newman and Robert F. Sproull. MH.

[Nilsson] 1980. 476 pps., *Principles of Artificial Intelligence*, by Nils J. Nilsson. Tioga Publishing Company, Palo Alto, CA.

[Oaks] 1997. 241 pps., *Java Threads*, by Scott Oaks and Henry Wong. O'Reilly.

[Pal] 1993. "A Review on Image Segmentation Techniques", by Nikhil R. Pal and Sankar K. Pal. *Pattern Recognition*, vol. 26, No. 9. PP. pps. 1277-1294.

[Peli] 1982. "A Study of Edge Detection Algorithms", by Tamar Peli and David Malah. *Computer Graphics and Image Processing* , vol. 20., pps. 1-21. AP.

[Pettofrezzo] 1996. 133 pps., *Matrices and Transformations*, by Anthony J. Pettofrezzo. Dover.

[Poole] 1998. 558 pps., *Computational Intelligence, a Logical Approach*, by David Poole, Alan Mackworth and Randy Goebel. Oxford.

[Poynton] 1998 *Frequently Asked Questions about Gamma*, by Charles Poynton, Jan., <http://www.inforamp.net/~poynton>.

[Poynton 96a] 1996. *Frequently Asked Questions about Gamma*, by Charles Poynton, Feb., <http://www.inforamp.net/~poynton/GammaFAQ.html>.

[Poynton 96b] 1996. A Technical Introduction to Digital Video, by Charles Poynton, JW. Chapter 6 available from <http://www.inforamp.net/~poynton/GammaFAQ.html>.

[Pratt]. 1991. 698 pps., *Digital Image Processing,* by William K. Pratt. JW.

[Preparata]. 1985. 390 pps., *Computational Geometry*, by Franco P. Preparata and Michael Ian Shamos. SV.

[Ralston] 1983. 1664 pps., *Encyclopedia of Computer Scicnce and Engineering*, by Anthony Ralston, editor. Second Edition. VNR.

[Resnick]. 1978. 1131 pps., *Physics*, by David Halliday and Robert Resnick, Third Edition. JW.

[RFC1951] 1996. "DEFLATE Compressed Data Format Specification", by L.P. Deutsch, available from <ftp://ftp.uu.net/pub/archiving/zip/doc/>.

[RFC1952]. 1996. "GZIP File format specification version 4.3", by Peter Deutsch. Available from <http://ds.internic.net/rfc/rcf1952.txt> and as HTML format at <ftp://ftp.uu.net/graphics/png/documents/zlib/zdoc-index.html>.

[Roberts] 1984. 606 pps., *Applied Combinatorics*, by Fred S. Roberts. PH.

[Roberts 93] 1993. Roberts, Eric, "Using C in CS1 Evaluating the Stanford Experience", SIGCSE bulletin. vol. 24, no. 2, pp. 117-121.

[RP 37-1969] 1969. SMPTE Recommended Practice, RP 37-1969.

[Said] 1996. "An image multiresolution representation for lossless and lossy image compression", by A. Said and W. A. Pearlman, IEEE TIP, vol. 5, Sept. pp. 1303-1310.

[Schalkoff] 1989.489 pps., *Digital Image Processing and Computer Vision*, by Robert J. Schalkoff, JW.

[Shamma] 1989 "Spatial and Temporal Processing in Central Auditory Networks", by Shamma, in *Methods in Neuronal Modeling* Eds. Koch and Segev, MIT. pps. 247-289.

[Shirai] 1987. 297 pps. *Three-Dimensional Computer Vision*, by Yoshiaki Shirai. SV.

[Shneidermand] 1987. *Designing the User Interface: Strategies for Effective Human-Computer Interaction, by* Shneidermand. *AW.*

[Standish] 1998. 555 pps., *Data Structures in Java*, by Thomas A. Standish. AW.

[Stevens] 1997. 348 pps., *Graphics Programming with Java*, by Roger T. Stevens. CRM.

[Stockham] 1972. "Image Processing in the Context of a Visual Model", by Thomas G. Stockham, Jr., *Proc. IEEE,* vol. 60, July, pp. 828-842.

[Stollnitz] 1996. 245 pps., "Wavelets for Computer Graphics", by Eric Stollnitz, Tony DeRose and David Salesin. MK.

[Sturrock] 1956. 104 pps., *Fundamentals of Light and Lighting*, by Walter Sturrock and K. A. Staley. Bulletin LD-2, General Electric, Large Lamp Department, Albany, NY.

[Subramanian] 1997. "Converting Discrete Images to Partitioning Trees", by Kalpathi R. Subramanian and Bruce F. Naylor, IEEE TVCG, vol. 3, No. 3. July-September pp. 273-288.

[Teevan et al.] 1961. 214 pps., *Color Vision*, Edited, by Richard C. Teevan and Rovert C. Birney. VN.

[Thomas] 1991. "Efficient inverse Color Map Computation", by Spencer W. Thomas, in *Graphics Gems, vol. II,* edited, by James Arvo. pp. 116-125. AP.

[Thompson] 1996. 500 pps., *PowerPC™ Programmers Toolkit*, by Tom Thompson. Hayden. CD.

[Tog] 1992. 331 pps., *Tog on Interface*, by Bruce Tognazzini. AW.

[Torre] 1986. "On Edge Detection", , by Vincent Torre and Tomaso A. Poggio, *IEEE-PAMI,* vol. PAMI-8, No. 2, March. pps. 147-163.

[Travis] 1991. 301 pps., *Effective Color Displays*, by David Travis. AP.

[Umbaugh] 1998. 504 pps., *Computer Vision and Image Processing*, by Scotte E. Umbaugh. AW.

[Uytterhoeven] 1997. "Waili: Wavelets with Integer Lifting", by Geert Uytterhoeven, Filip Van Wulpen, Maarten Jansen, Dirk Roose and Adhemar Bultheel, Report TW262, July, Katholieke Universiteit Lueven, Department of Computer Science, Celestijnenlaan 200 A - B-30001 Heverlee, Belgium. Also available from <http://www.cs.kuleuven.ac.be/publicaties/rapporten/tw/TW262.abs.html>

[Vliet] 1989. "A nonlinear Laplace operator as edge detector in Noisy Images", by Lucas J. van Vliet, Ian T. Young and Guus L. Beckers, *Computer Vision, Graphics and Image Processing*, vol. 45. pp. 167-195.

[Watkins] 1993. 234 pps., *Modern Image Processing: Warping, Morphing and Classical Techniques*, by Christopher Watkins, Alberto Sadun and Stephen Marenka. AP.

[Watson] 1993. 317 pps., *Portable GUI Development with C++*, by Mark Watson. MH.

[Weeks] 1996. 570 pps., *Fundamentals of Electronic Image Processing,* by Arthur R. Weeks, Jr.. IEEE Press.

[Wehmeier] 1989. 524 pps., "Modeling the Mammalian Visual System", by Udo Wehmeier, D. Dong, C. Koch and D. Essen, in *Methods in Neuronal Modeling* Eds. Koch and Segev. MIT. pps. 335-359.

[Widder] 1947. 432 pps., *Advanced Calculus*, by David V. Widder. PH.

[Williams] 1966. "The Effect of Target Specification on Objects Fixated During Visual Search", by L. G. Williams, *Perception and Psychophysics*, vol. 1,pp. 315-318.

[Winston] 1996. 328 pps., *On to Java*, by Patrick Henry Winston and Sundar Narasimhan. AW. On-line material.

[Wolberg] 1990. 318 pps., *Digital Image Warping,* by George Wolberg. IEEE Press.

[Woods and Gonzalez] 1981. "Real-Time Digital Image Enhancement", by Woods and Gonzalez, in *Proceedings of the IEEE*, vol. 69, No. 5, May. pps. 643-657.

[Wright] 1997. "Watershed Pyramids for Edge Detection", ICIP, by Anthony S. Wright and Scott T. Acton.

[Wu] 1991. "Efficient Statistical Computations For Optimal Color Quantization", by Xiaolin Wu, in *Graphics Gems,* vol. II, edited, by James Arvo. AP. pp. 126-133.

[Wu 97] 1997. "Lossless Compression of Continuous-Tone Images via Context Selection, Quantization and Modeling", by Wu. *IEEE TIP*, vol. 6, No. 5. May. pps. 656-664.

[Wyszecki] 1967. 628 pps., *Color Science*, by Günter Wyszecki and W. S. Stiles. JW.

[Yu et al.] 1993. "A New Adaptive Contrast Enhancement Method", by Tian-Hu Yu and S. K. Mitra, *SPIE, vol.* 1903 Image and Video Processing. pps. 103-110.

[Zimuda] 1996. "Efficient Algorithms for the soft Morphological Operators", by Michael A. Zmuda and Louis A. Tamburino. vol. 18, No. 11, November. IEEE PAMI. pps. 1142-1147.

[Ziv et al.] 1977. "A Universal Algorithm for Sequential Data Compression", by Ziv, J. and Lempel, Z., *IEEE Transactions on Information Theory,* vol. IT-23, May, pp. 337-343.

Abbreviations

ACM = Association for Computing Machinery, 1515 Broadway, New York, NY 10036, (212)869-7440.

AP = Academic Press, Inc., Cambridge, MA.

AW = Addison Wesley, New York, NY.

BQ = Bogden and Quigley, Inc., Publishers, Tarrytown-on-Hudson, NY.

CG = Corilois Group Books, The Coriolis Goup, Inc., Scottsdale, AZ.

CRC = CRC Press, New York, NY.

CRM = Charles River Media, Inc., Rockland, MA.

CVGIP = Computer Vision Graphics and Image Processing. AP.

CGIP = Computer Graphics and Image Processing. Now listed under CVGIP. AP.

Dover = Dover Publications, Inc., 180 Varick St., New York, NY 10014.

EH = Ellis Horwood, West Sussex, England.

Hayden = Hayden Books, 201 West 103rd St., Indianapolis, IN 46290, a division of Macmillan Computer Publishing.

HP = Harmony Books, New York, NY.

IEEE CGA = IEEE Computer Graphics and Applications.

IEEE Press = IEEE Computer Society Press, Los Alamitos, CA.

IEEE PAMI = IEEE Transactions on Pattern Analysis and Machine Intelligence.

IEEE TIP = IEEE Transactions on Image Processing.

IEEE TVCG = IEEE Transactions on Vision and Computer Graphics.

ICIP97 = IEEE International Conference on Image Processing, 10/26-10/29, 1997, Santa Barbara, CA.

JW = John Wiley & Sons, New York, NY.

MH = McGraw-Hill, New York, NY.

MIT = The MIT Press, Cambridge, MA.

MT=M&T Books, an imprint of IDG Books Worldwide, Inc., Foster City, CA

OR = O'Reilly and Associates, Inc., 101 Morris St., Sebastopol, CA 95472.

Oxford = Oxford University Press, New York, NY.

PH = Prentice Hall, Upper Saddle River, NJ 07458

PP = Pergamon Press, Ltd., Great Britain.

Que = Que Corp., Indianapolis, IN.

SAMS = A division of Prentice Hall Computer Publishing, 11711 North College, Carmel, IN 46032, USA.

SIAM = Society of Industrial and Applied Mathematics, Philadelphia, PA.

SMPTE = Society of Motion Picture and Television Engineers, 9 East 41st St., New York, NY 10017, Phone: (914)761-1100. E-mail: mktg@smpte.org

SPIE = Society of Photonic and Industrial Engineers, Optical Engineering Press, Bellingham, WA.

SV = Springer-Verlag, New York, NY.

TAB= TAB Books, Inc., Blue Ridge Summit, PA 17214.

TOG=ACM Transactions on Graphics.

VN = Van Nostrand, Princeton, NJ.

VNR = Van Nostrand Reinhold Company, NY.

Index

Colophon

"Them sons of bitches",
Mrs. Hait said in a grim prescient voice without rancor or heat."

– William Faulkner's story, *Mule in the Yard.*

The manuscript for this book was prepared using Microsoft Word 5.1a on a PowerMac 8100/100AV. The astute reader will notice that benchmarks are reported for the PowerPC 601 and the G3. During the course of the writing of the book, the Mac was upgraded with a Crescendo G3 233 Mhz Sonnet card. This resulted in an instability that was more than made up for by the increase in speed.

The Java source was transformed into HTML using the HTML generator (a Java program, included with this book). The HTML was transformed into Word using MacLink Plus 10.0 from DataViz, Inc. The code was typeset with the Courier 10 typeface. Equations were set with the Design Science Equation editor using 12 point MT Extra, Symbol and Times fonts. The body of the text was set with 12 point Times.

Book layout was done in Word (and not by choice). There is an obvious market for a layout program that can import Word (with equations) correctly. Such a program would make an excellent Java project! Swing classes can already read a subset of RTF (a Word file format).

The class tree layout was done with the MetroWerks class browser and with the *TopFrame* class. The *TopFrame* class generated an output that was used as the input to Theo Vosse's TreeParse program, available only for MacOS, as far as we know.

Some of the Graphs were prepared using the Graphing Calculator, built into MacOS 8.1. The plots were performed using data output from Java and pasted into Microsoft Excel 5.0a. The 3-D functions were plotted using Maple.

All development was done using CodeWarrior Pro by MetroWerks, Inc. All programs were tested using the compiler produced by Sun Microsystems, Inc. The programs were tested on MacOS, Sun Solaris, Windows 95 and Windows NT. The CD-ROM was made by the author using Toast-Pro.

Portions of the book were written with QuickTime for Java. Sorry to say, the Apple lawyers would not give permission to publish anything about it in time for press. As a result, all mention of QuickTime for Java was stripped from the book.

This book was written while drinking vast amounts of Kahindu Kenya AA. I French roasted all the Kahindu in a hot-air popper. See [Lyon and Rao] for detailed instructions. For a source of small quantities of Kahindu (and other flavors) of coffee (green or roasted) contact Doug Nicolaisen at the Kind Coffee Company, Syracuse, NY, Phone 315-425-0035, Email: dougnic@earthlink.net.

Now that the book is done, I can sleep. Good Night!

LICENSE AGREEMENT AND LIMITED WARRANTY

READ THE FOLLOWING TERMS AND CONDITIONS CAREFULLY BEFORE OPENING THIS DISK PACKAGE. THIS LEGAL DOCUMENT IS AN AGREEMENT BETWEEN YOU AND PRENTICE-HALL, INC. (THE "COMPANY"). BY OPENING THIS SEALED DISK PACKAGE, YOU ARE AGREEING TO BE BOUND BY THESE TERMS AND CONDITIONS. IF YOU DO NOT AGREE WITH THESE TERMS AND CONDITIONS, DO NOT OPEN THE DISK PACKAGE. PROMPTLY RETURN THE UNOPENED DISK PACKAGE AND ALL ACCOMPANYING ITEMS TO THE PLACE YOU OBTAINED THEM FOR A FULL REFUND OF ANY SUMS YOU HAVE PAID.

1. **GRANT OF LICENSE:** In consideration of your payment of the license fee, which is part of the price you paid for this product, and your agreement to abide by the terms and conditions of this Agreement, the Company grants to you a nonexclusive right to use and display the copy of the enclosed software program (hereinafter the "SOFTWARE") on a single computer (i.e., with a single CPU) at a single location so long as you comply with the terms of this Agreement. The Company reserves all rights not expressly granted to you under this Agreement.

2. **OWNERSHIP OF SOFTWARE:** You own only the magnetic or physical media (the enclosed disks) on which the SOFTWARE is recorded or fixed, but the Company retains all the rights, title, and ownership to the SOFTWARE recorded on the original disk copy(ies) and all subsequent copies of the SOFTWARE, regardless of the form or media on which the original or other copies may exist. This license is not a sale of the original SOFTWARE or any copy to you.

3. **COPY RESTRICTIONS:** This SOFTWARE and the accompanying printed materials and user manual (the "Documentation") are the subject of copyright. You may not copy the Documentation or the SOFTWARE, except that you may make a single copy of the SOFTWARE for backup or archival purposes only. You may be held legally responsible for any copying or copyright infringement which is caused or encouraged by your failure to abide by the terms of this restriction.

4. **USE RESTRICTIONS:** You may not network the SOFTWARE or otherwise use it on more than one computer or computer terminal at the same time. You may physically transfer the SOFTWARE from one computer to another provided that the SOFTWARE is used on only one computer at a time. You may not distribute copies of the SOFTWARE or Documentation to others. You may not reverse engineer, disassemble, decompile, modify, adapt, translate, or create derivative works based on the SOFTWARE or the Documentation without the prior written consent of the Company.

5. **TRANSFER RESTRICTIONS:** The enclosed SOFTWARE is licensed only to you and may not be transferred to any one else without the prior written consent of the Company. Any unauthorized transfer of the SOFTWARE shall result in the immediate termination of this Agreement.

6. **TERMINATION:** This license is effective until terminated. This license will terminate automatically without notice from the Company and become null and void if you fail to comply with any provisions or limitations of this license. Upon termination, you shall destroy the Documentation and all copies of the SOFTWARE. All provisions of this Agreement as to warranties, limitation of liability, remedies or damages, and our ownership rights shall survive termination.

7. **MISCELLANEOUS:** This Agreement shall be construed in accordance with the laws of the United States of America and the State of New York and shall benefit the Company, its affiliates, and assignees.

8. **LIMITED WARRANTY AND DISCLAIMER OF WARRANTY:** The Company warrants that the SOFTWARE, when properly used in accordance with the Documentation, will operate in substantial conformity with the description of the SOFTWARE set forth in the Documentation. The Company does not warrant that the SOFTWARE will meet your requirements or that the operation of the SOFTWARE will be uninterrupted or error-free. The Company warrants that the media on which the SOFTWARE is delivered shall be free from defects in materials and workmanship under normal use for a period of thirty (30) days from the date of your purchase. Your only remedy and the Company's only obligation under these limited warranties is, at the Company's option, return of the warranted item for a refund of any amounts paid by you or replacement of the item. Any replacement of SOFTWARE or media under the warranties shall not extend the original warranty period. The limited warranty set forth above shall not apply to any SOFTWARE which the Company determines in good faith has been subject to misuse, neglect, improper installation, repair, alteration, or damage by you. EXCEPT FOR THE EXPRESSED WARRANTIES SET FORTH ABOVE, THE COMPANY DISCLAIMS ALL WARRANTIES, EXPRESS OR IMPLIED, INCLUDING WITHOUT LIMITATION, THE IMPLIED WARRANTIES OF MERCHANTABILITY AND FITNESS FOR A PARTICULAR PURPOSE. EXCEPT FOR THE EXPRESS WARRANTY SET FORTH ABOVE, THE COMPANY DOES NOT WARRANT, GUARANTEE, OR MAKE ANY REPRESENTATION REGARDING THE USE OR THE RESULTS OF THE USE OF THE SOFTWARE IN TERMS OF ITS CORRECTNESS, ACCURACY, RELIABILITY, CURRENTNESS, OR OTHERWISE.

IN NO EVENT, SHALL THE COMPANY OR ITS EMPLOYEES, AGENTS, SUPPLIERS, OR CONTRACTORS BE LIABLE FOR ANY INCIDENTAL, INDIRECT, SPECIAL, OR CONSEQUENTIAL DAMAGES ARISING OUT OF OR IN CONNECTION WITH THE LICENSE GRANTED UNDER THIS AGREEMENT, OR FOR LOSS OF USE, LOSS OF DATA, LOSS OF INCOME OR PROFIT, OR OTHER LOSSES, SUSTAINED AS A RESULT OF INJURY TO ANY PERSON, OR LOSS OF OR DAMAGE TO PROPERTY, OR CLAIMS OF THIRD PARTIES, EVEN IF THE COMPANY OR AN AUTHORIZED REPRESENTATIVE OF THE COMPANY HAS BEEN ADVISED OF THE POSSIBILITY OF SUCH DAMAGES. IN NO EVENT SHALL LIABILITY OF THE COMPANY FOR DAMAGES WITH RESPECT TO THE SOFTWARE EXCEED THE AMOUNTS ACTUALLY PAID BY YOU, IF ANY, FOR THE SOFTWARE.

SOME JURISDICTIONS DO NOT ALLOW THE LIMITATION OF IMPLIED WARRANTIES OR LIABILITY FOR INCIDENTAL, INDIRECT, SPECIAL, OR CONSEQUENTIAL DAMAGES, SO THE ABOVE LIMITATIONS MAY NOT ALWAYS APPLY. THE WARRANTIES IN THIS AGREEMENT GIVE YOU SPECIFIC LEGAL RIGHTS AND YOU MAY ALSO HAVE OTHER RIGHTS WHICH VARY IN ACCORDANCE WITH LOCAL LAW.

ACKNOWLEDGMENT

YOU ACKNOWLEDGE THAT YOU HAVE READ THIS AGREEMENT, UNDERSTAND IT, AND AGREE TO BE BOUND BY ITS TERMS AND CONDITIONS. YOU ALSO AGREE THAT THIS AGREEMENT IS THE COMPLETE AND EXCLUSIVE STATEMENT OF THE AGREEMENT BETWEEN YOU AND THE COMPANY AND SUPERSEDES ALL PROPOSALS OR PRIOR AGREEMENTS, ORAL, OR WRITTEN, AND ANY OTHER COMMUNICATIONS BETWEEN YOU AND THE COMPANY OR ANY REPRESENTATIVE OF THE COMPANY RELATING TO THE SUBJECT MATTER OF THIS AGREEMENT.

Should you have any questions concerning this Agreement or if you wish to contact the Company for any reason, please contact in writing at the address below.

Robin Short
Prentice Hall PTR
One Lake Street, Upper Saddle River, New Jersey 07458

About the CD-ROM

The CD-ROM which accompanies this book contains Java source code for a full-blown image processing application, called Kahindu.

System Requirements

The CD will run on Windows 9x/NT, MacOS and SunOS.

The Java source code requires JDK 1.1 or better. This code is tested by graduate students on a variety of platforms and development software systems. It is production-level image processing code, which will be immediately useful to programmers, researchers and students.

How to Use the CD

Programmers will be able to layer new applications directly on top of the CD-ROM code base. This eases the pain of writing applications because the code core is stable and well-tested.

Also included on the CD-ROM is a complete World Wide Web mirror of 3 graduate-level Distance Learning courses taught by the author, a large image data-base and QuickTime videos.

The QuickTime videos, and their sound tracks, were composed and recorded by the author. The videos were synthesized using the software that comes with this book. The QuickTime videos require Quicktime 3.0, available for Windows 9x/NT or MacOS.

Documentation for the classes is on the CD-ROM and takes the form of a complete HTML API listing in JavaDOC format.

Technical Support

Prentice Hall does not offer technical support for the software on the CD-ROM. However, if there is a problem with the CD, you may obtain a replacement copy by sending an email describing your problem to:

`disc_exchange@phptr.com`

Send bug reports, contributions and questions to: `lyon@DocJava.com`

Updates are available at `http://www.DocJava.com`